# Urban Geography
## in America, 1950–2000

# Urban Geography
## in America, 1950–2000

Paradigms and Personalities

Edited by

# Brian J. L. Berry
# and James O. Wheeler

Routledge
Taylor & Francis Group

NEW YORK AND LONDON

Published in 2005 by
Routledge
Taylor & Francis Group
711 Third Avenue
New York, NY 10017

Published in Great Britain by
Routledge
Taylor & Francis Group
2 Park Square
Milton Park, Abingdon
Oxon OX14 4RN

International Standard Book Number-10: 0-415-95190-9 (Hardcover) 0-415-95191-7 (Softcover)
International Standard Book Number-13: 978-0-415-95190-6 (Hardcover) 978-0-415-95191-3 (Softcover)
Library of Congress Card Number 2004029780

### Library of Congress Cataloging-in-Publication Data

Urban geography in America 1950-2000 : paradigms and personalities / editors, Brian J.L. Berry, James O. Wheeler.
    p. cm.
    Collections of papers presented at various conferences pertaining to the field of urban geography and published in various issues of Urban geography.
    Includes bibliographical references and index.
    ISBN 0-415-95190-9 (hbk. : alk. paper) -- ISBN 0-415-95191-7 (pbk. : alk. paper)
    1. Urban geography--United States--History. 2. Human geography--United States--History. I. Berry, Brian Joe Lobley, 1934- II. Wheeler, James O. III. Urban geography.

GF503.U73 2005
307.76'0973--dc22                                                                                    2004029780

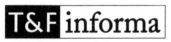

Taylor & Francis Group
is the Academic Division of T&F Informa plc.

Visit the Taylor & Francis Web site at
http://www.taylorandfrancis.com

**and the Routledge Web site at**
**http://www.routledge-ny.com**

# Contents

## Section III
## Urban Geography in the 1970s

## Section IV
## Urban Geography in the 1980s

# Preface

We are proud to present this 50-year history of urban geography in America as told by its participants. The project began as an outgrowth of a paper session organized by James O. Wheeler at the 2001 annual meeting of the Association of American Geographers (AAG) in New York entitled "Urban Geography in the 1960s." The paper presenters in this session (John S. Adams, William A. V. Clark, James O. Wheeler, and Maurice Yeates) all obtained Ph.D.s in the 1960s. Their formative years, first as graduate students and then as assistant professors, coincided with the development and maturation of urban spatial analysis. For them, this was the urban decade, when human geography first became recognized as a legitimate part of the social sciences. This recognition was guided and advanced by Brian J. L. Berry and his many doctoral students at the University of Chicago, who together established the "Chicago School" of urban geography. Berry's students latched onto his innovative ways of thinking about and doing geography—new theories, concepts, methodologies, and policy insights. As a result, this was a decade of paradigm gained and molded. The paper presentations were published, along with Berry's commentary on "The Chicago School in Retrospect and Prospect," in *Urban Geography*, 2001, Vol. 22, No. 6.

The Los Angeles meeting of the AAG in 2002 provided the opportunity for Wheeler and Berry to organize a second paper session, "Urban Geography in the 1970s." Presenters included Martin Cadwallader, Larry R. Ford, Patricia Gober, Peter G. Goheen, and Risa I. Palm, whose Ph.D.s were awarded in the 1970s. These papers, with Berry's commentary, "Paradigm Lost," were published in *Urban Geography* Vol. 23, 2002, No. 5. They revealed that the 1970s was a decade of experimentation and challenge to the spatial analysis paradigm.

The New Orleans meeting of the AAG in 2003 was the forum for two additional sessions organized by Wheeler and Berry: "Urban Geography in the 1980s" and "Urban Geography in the 1990s." The former session comprised presentations by Larry S. Bourne, Sallie A. Marston and Geraldine Pratt, David R. Meyer, and Michael Pacione, with insightful commentary by Robert W. Lake. Paul L. Knox also contributed a paper, although it was not presented in the session. The second session included papers by Trevor J. Barnes, Michael Dear, Susan Hanson, Robert W. Lake, and Helga Leitner and Eric Sheppard. Published versions of the papers from these two sessions were published in *Urban Geography*, Vol. 24, 2003, Nos. 4 and 6, respectively.

The Chicago School of the 1960s did not arise in a vacuum. Papers by Chauncy D. Harris, Elisabeth Lichtenberger and Edward J. Taaffe previously published in *Urban Geography* describe the foundations on which the superstructure of the 1960s was constructed. Taken together, these papers and the four sets of AAG presentations constitute an important historical resource that should be of value to members of the geographic profession, now and in the future, and it has been our pleasure to draw them together in one place. Here are the voices of those who created and transformed modern urban geography in North America.

In gathering the papers we did not act alone. We give an especially hearty thanks to Jodie Traylor Guy for her important editorial role in preparing the manuscripts for the four special issues of *Urban Geography*. As assistant editor for these issues, she corresponded with the authors, checked and copyedited their work, and prepared digital-ready manuscripts for the publisher of *Urban Geography*, Bellwether Publishing, Ltd., Columbia, Maryland. Bellwether, in turn, not only approved of the preparation of this collection but also provided the manuscripts in their final form. We also thank Kim Hawkins for word processing the index.

**Brian J. L. Berry**
*Richardson, TX*

**James O. Wheeler**
*Athens, GA*

# Introduction

During the last half of the 20th century, urban geography came of age and became a recognized part of the social sciences, significantly expanding its scholarly horizons, analytical and narrative approaches, and interdisciplinary scope. Before 1950 the field was a different creature. Most urban geographic studies focused on a single city, using historical approaches to explain how the city grew and evolved, treating both the physical environment and the human geography, relying heavily on field observation and measurement because of the paucity of available published data, and focusing on such concepts as "site" in the morphology of settlement and "situation" in role and function. Many of these studies were completed as doctoral dissertations at the University of Chicago under the tutelage of Charles C. Colby.

The seeds of change appeared in the 1930s, when three students found themselves together at the University of Chicago. In chapter 1, Chauncy Harris recalls:

> At the university I met Harold M. Mayer and Edward L. Ullman, both intensely interested in urban geography. In retrospect I do not believe that Colby was responsible for Ullman, Mayer, or Harris going into urban geography. We developed interest in this field on our own. But Colby did arouse our curiosity, shape our interests, and improve our writing. He became my mentor ... The simultaneous presence of three youngsters all deeply curious about urban phenomena heralded a new period in which an initial critical mass was reached. Thereafter dissertations in urban geography were not isolated individual efforts. Lively student discussions

ranged over the state of the field, appropriate approaches to the subject, the factors in the rise and distribution of cities, functions performed by cities, and many other topics....

Of their three dissertations—Harris, 1940, *Salt Lake City: A Regional Geography*; Mayer, 1943, *The Railway Pattern of Metropolitan Chicago*; and Ullman, 1943, *Mobile: Industrial Seaport and Trade Center*—the third was the most original. Ullman measured Mobile's tributary region by using newspaper circulation, banking services, and retail and wholesale trade, forming "principles [that were] later refined as complementarity, transferability and interviewing opportunity" (Fournier, 2004, p. 321) and "spatial interaction" (Ullman, 1954a, 1954b, 1956). Ullman's interest in principle had already been sharpened in a meeting with the German location economist August Lösch when he visited Harvard University on the eve of World War II. Lösch referred the young geographer to Walter Christaller's doctoral dissertation, *Die zentralen Orte in Süddeutschland* (1933), and Ullman (1941) responded by introducing Christaller's central-place theory to American readers in an article "A Theory of Location for Cities," interestingly published in the *American Journal of Sociology* rather than in a geographical publication.

Harris reentered the scene in 1943, when he published "A Functional Classification of Cities in the United States" in the *Geographical Review*, a paper that set in motion a cottage industry of data-intensive work on city classification, soon followed by his joint effort with Ullman, "The Nature of Cities," published in 1945. Both he and Ullman, newly appointed lieutenants, had been seconded to the Office of Strategic Services when he was asked by Robert B. Mitchell, executive director of the Philadelphia City Planning Commission, to contribute a 5,000-word paper to a projected special issue of the *Annals of the American Academy of Political and Social Science*. Mitchell noted that "when I discussed this with Louis Wirth, he thought you would do it particularly well." Harris responded that

> I should like to participate in the writing of "The Nature of Cities"; it was very kind of you to think of me in connection with this topic. The thought has occurred to me that a joint paper with Dr. Edward L. Ullman might be a good deal stronger than one I might write alone. Dr. Ullman has been a keen student of American cities. He is at present a lieutenant in the Navy stationed in Washington with the Office of Strategic Services; hence collaboration would be easy.

In turn, Mitchell wrote

> I have received your letter of June 21 accepting assignment for the *Annals* volume. I know you will do a good job. If you wish to collaborate with Dr. Ullman I shall be delighted, but expect that you will take personal responsibility for the article. Harold Mayer, who is very busily engaged directing research for this Commission, sends you his greetings. Harold will guide one or two of the articles for this issue.

Harris recalled in his oral history that

> We sat down on a number of weekends in Washington, D.C. ... and discussed how cities are distributed across the country and how internal patterns are arranged within them. We were interested in generalizations not description of individual cities. We worked out a typology. Perhaps, I spent less time on that article than any other I've ever written.

Later, he added

> By myself I probably would have written a more traditional article, but Ullman suggested the idea of presenting a more theoretical generalization. With respect to the patterns of distribution of cities, Ullman certainly contributed the central-place concept and the alignment of cities on transport lines, while I brought in the clustered cities of mining and industrial districts. Ullman was familiar with the work of Ernest Burgess in sociology on concentric zones in cities and of Homer Hoyt, the real-estate economist, in describing sectors of residential development. The multiple-nuclei pattern was suggested by me on the basis of observation that not all activities of a city cluster around a single center, but that different activities have different points of attachment.

It was the elegance and simplicity of the summary diagrams designed to illustrate the ideas that mattered. They put together in one place the concepts that became central to American urban geography in the next half-century, either as research foci during the emergence of a new "Chicago School" of urban geography during the 1960s or as items of opprobrium as a later generation challenged that School's precepts. Their diffusion is recounted in chapters 2 and 3.

The next steps came after World War II. Harris was hired by the University of Chicago and quickly moved into administration. When Colby retired, Mayer was brought back from the Philadelphia City Planning Department to teach urban and transportation geography. Ullman joined the faculty at Harvard, but left for the University of Washington in Seattle when the Harvard geography program was terminated. Another Chicago graduate, Malcolm Proudfoot, who completed his Ph.D. in 1936, had joined the department at Northwestern University, and one of his students was William L. Garrison, who joined the Washington department in 1950. Clyde F. Kohn, a Michigan Ph.D., taught at Northwestern from 1945 to 1958, when he moved to Iowa. Upon Proudfoot's premature death in 1955, Northwestern hired Edward J. Taaffe, another Chicago graduate, in 1956.

Mayer and Kohn were to prepare the chapters on urban and settlement geography for *American Geography Inventory and Prospect* in 1954, and they combined to prepare the influential *Readings in Urban Geography* in 1959, soon after Garrison's student Berry had been added to the Chicago faculty, but already radical change had occurred at the University of Washington. Drawn by Ullman's presence, a group of bright young students subsequently gravitated to Garrison and sparked geography's ill-named "quantitative revolution." Taaffe reminds us that this so-called revolution represented, in fact, a definitional and theoretical change, as well as the embracing of statistical and mathematical methods. This story and its ramifications have been told many times (e.g., Berry in chapter 5; Berry 2001; Holt-Jensen, 1988; Johnston, 1983) and need not be repeated here. After Berry's relocation to Chicago in 1958, his fellow students moved to Northwestern, Michigan and the Department of Regional Science at Pennsylvania. Another cluster arose around Harold McCarty at Iowa. Later, Garrison, Marble, Dacey and Thomas concentrated at Northwestern, and Taaffe took an Iowa graduate, King, to reshape the department at Ohio State.

The full implications of the quantitative revolution for urban geography were to be revealed at the University of Chicago. Yeates (chapter 6) notes that "the urban-oriented research themes ... came to the fore in the Department of Geography at The University of Chicago during the 1960s" (referred to as the Chicago School). Wheeler (chapter 9) adds, "The introduction of spatial analysis in geography initially began slowly but gained remarkable momentum throughout the 1960s, ... nowhere was this energy more charged than in the Chicago School of urban geography." He continued that it was predominantly "Berry's steadfast, unrivaled, and highly influential scholarship" that created and sustained the School, along with a long list of doctoral students who took Berry's ideas and approach to research to other universities in North America and elsewhere. The implications were, however, even broader. King (1993, p. 544) notes that "the case for geography as a social science gained momentum throughout the 1960s" ... [and] "by the mid-1970s, the professionalization of geography as a social science in the United States was well advanced" (p. 545). Geographers contributed to a variety of interdisciplinary journals, such as *Papers in Regional Science, The Journal of Regional Science, Land Economics, Environment and Planning A, The Journal of Urban Affairs, The Journal of Urban and Regional Planning, The International Regional Science Review, The American Journal of Sociology,* and *Urban Studies,* to note a few. Urban research also became prominent in general geography journals such as the *Annals of the Association of American Geographers, The Geographical Review, The Canadian Geographer, Economic Geography, The Professional Geographer,* and *The Transactions of the Institute of British Geographers,* again to name only a few. Culminating the surge of urban research, *Urban Geography* began publication in 1980, joining such new journals as *Urban Economics* (where Berry was a founding member of the editorial board), and urban geography grew to be the second largest of the more than 40 specialty groups of the Association of American Geographers (AAG), second only to Geographic Information Science. The annual meetings of the AAG quickly came to be filled with presentations on myriad topics in urban geography, as were the regional meetings and their division-sponsored publications, such as *The Southeastern Geographer, Yearbook of the Association of Pacific Coast Geographers,* and *The Southwestern Geographer.* As chapters 7 through 13 reveal, the 1960s and 1970s were a time of both exploration and consolidation, of personal growth, and of reinforcement of the social scientific paradigm.

The first signs of reaction came in the 1970s, following publication of *Social Justice and the City* by David Harvey (1973), amidst the turmoil of the Vietnam War. Reflecting the ethos of the antiwar radicals and his own Marxist orientation, Harvey challenged the social scientific paradigm on several grounds. First, he opposed the underlying presupposition that urban dynamics are grounded in the logic of decentralized market-based democratic processes; for him, there were different versions of space, each created by and for the benefit of the few. Second, he rejected the notion of a science of urbanism in its own right; rather, he argued, examination of the processes that concentrate wealth and power in cities are of interest because of the light that is shed on broader issues of societal and political concern, but the structure and growth of cities themselves was of little interest. Thirdly, he argued that social practice directed toward the transformation of inequities is the only legitimate source of theoretical ratification, not some remote brand of methodology and philosophy. Only revolutionary theories, dialectically formulated, offer the prospect of creating truth. Thus, he attempted to refute the theory, empirical research methods, and the policy recommendations of social scientific urban geography.

The challenge led to uncertainty and to an unraveling of the Chicago paradigm: "If the 1960s had been a decade when progress was marked by a paradigm gained, the 1970s was one in which the central paradigm was lost and urban geography began to diverge along a variety of incommurate paths" (Berry 2002, p. 443). There was increasing "fragmentation" (chapter 14), "fragmentation, specialization, and insularity" (Gober, 2002), "competing visions" (chapter 12), and "methodological and philosophic exploration" (chapter 1). Berry (2003, 558–559) writes

> We began the 1960s with an Enlightenment epistemology: a close attention to data sources and to empirical regularities; the search for theory to explain these regularities; the translation of theory into models that illuminate the working of causal variables and provide guidelines to planning and action; and the use of practice to provide use-based information on the limitations of the models and to promote new rounds of empirical investigation and theoretical development as new questions are brought to the surface. It is this epistemology that was alleged to be sycophantic to capitalism by Harvey and his followers, dialecticians committed to Marxism as the proper foundation for a more just society to be achieved by political means. Harvey's dialectics evolved into critical social theory and provided a ready opening for deconstructionist rejection of "positivism," which in turn created an opportunity for the postmoderns, who argue that there are multiple worlds, each subject-centered and contingent, with every characterization merely the perception of those setting it down.

Lake (chapter 20) succinctly captured the essence of urban geography as it moved into the 1980s: "The gravitational pull of sweeping national and global changes, both within and outside the discipline, imposed a strong imprint on urban geography ... Contextuality [became] a common theme in assessments [as did] the closely related theme of agency and structure." He continued

> No more and no less than a mirror of its time, urban geography in the 1980s evinced an antiurban angst that mimicked the broader culture: that angst was revealed in two interrelated observations. First [it] abandoned the material city in a move expressive

of the period's prevailing antiurbanism. Second, ... abandonment of the city reflected, at least in part, the racial dynamics of the times. [While] urban geographers remained busy ... attention had turned elsewhere: outward to the metropolitan scale, and global. We followed "growth" to the suburbs and ignored the city that remained. Our fascinations were with high tech, silicon landscapes, edge cities, financial capital, telecommunications networks, metropolitics, festival marketplaces, the new urbanism, and global cities. Not coincidently, these fascinations addressed the few remaining vestiges of the White city or positioned the city at the metropolitan or global scales where White life now was lived. "Ghetto" became not an urban place but a teenage clothing style.

Knox (chapter 15) similarly recognizes that "the 1980s forced many urban geographers to consider afresh both the nature of their subject matter and their theoretical and methodological approaches to it."

Berry remained the most-cited geographer through the mid-1980s, at which point he was overtaken by Harvey. As the 1990s arrived new forces began to unfold; some, though, were still centered in Harvey's call for dialectics while others reflected the spread of postmodernism.

Hanson (chapter 22) recognizes the "weight of tradition" but also describes "three strands of paths of divergence that ... emerged ... from these traditions: research on globalization, research on gender, and a greater diversity of methodological approaches and data sources used." Barnes (chapter 23) notes how "the new cultural geography" and postmodernism, "finally cracked" urban geography into a "messier, mixed up, and contaminated" research enterprise. Dear (chapter 24), champion of the self-proclaimed "Los Angeles School" of urban geography, led the call for "a comparative urban analysis that utilizes Los Angeles not as a new urban 'paradigm,' but as one of many exemplars of contemporary urban process" (chapter 24). And Leitner and Sheppard (chapter 25) assert that it was "critical urban geography [that] came to dominate knowledge production in urban geography during the 1990s," leading to an "unbounding of the field." Meanwhile, the Chicago School paradigm has been reinvigorated by new ideas from new sources. A creature of its times, the theory of the 1960s was due for change because the models were static, not dynamic, the equilibria were partial, not general, and the processes were first-order, not second-order. New theorists are seeking to address these limitations. Forging new spatially integrated links with other social sciences, new rounds of dynamic modeling are bringing new understanding of urban growth and change.

We thus have entered the 21st century with an urban geography that is a cockpit of competing schools of thought (see Aitken, Mitchell, and Staeheli, 2003; for a British perspective, see Hall, 2003; Peach, 2003). Such diversity can be invigorating, so long as presuppositions are clearly stated, their consequences are understood, and debate is robust and healthy, centering on the clash of ideas, and the pluses and minuses on the balance sheet of alternatives. But healthy debate can be tough and confrontational. Only if we are able to avoid the all-too-frequent tendency to interpret intellectual challenge as personal attack will we be able to stay in communication. With communication, evolution will take care of the rest. The ideas that are best adapted to the needs of society and the aspirations and career opportunities of the next generations of scholars will survive to shape the urban geography of the future. We hope this volume offers future generations of urban geographers the helpful insights of those who were active in creating this history and of the personalities and forces instrumental in shaping and reshaping the discipline that will be their home.

## Literature Cited

(Articles appearing in this book are omitted.)

Aitken, S., Mitchell, D., and Staeheli, L. 2003, Urban geography. In Gary L. Gaile and Cort J. Willmott, eds. *Geography in America at the Dawn of the 21st Century*. Oxford, UK: Oxford University Press, 237–263.

Berry, B. J. L., 2001, The Chicago School in retrospect and prospect. *Urban Geography*. Vol. 22, 559–561.

Berry, B. J. L., 2002, Paradigm lost. *Urban Geography*. Vol. 23, 441–445.

Berry, B. J. L., 2003, The case for debate. *Urban Geography*, Vol. 24, 557–559.

Colby, C.C., 1933, Centrifugal and centripetal forces in urban geography. *Annals of the Association of American Geographers*. Vol. 23, 1–20.

Christaller, W. 1933. *Die Zentralen Orte in Süddeutschland*. Jena: Gustav Fischer Verlag.

Dickinson, Robert E., 1947, *City, Region and Regionalism*. London, UK: K. Paul, Trench, Trubner & Co.

Fournier, Eric J., 2004, Edward Ullman, the port of Mobile, and the birth of modern economic geography. In James O. Wheeler and Stanley D. Brunn, eds. *The Role of the South in the Making of American Geography: Centennial of the AAG. 2004*. Columbia, MD: Bellwether Publishing, 318–326.

Gober, P. and Brunn, S. D., 2004, Introducing voices in the AAG's second century. *The Professional Geographer*, Vol. 56, 1–3.

Hall, P., 2003, Geographers and the urban century. In Ron Johnston and Michael Williams, eds. *A Century of British Geography*. Oxford, UK: Oxford University Press, 545–562.

Harris, Chauncy D. 1940. *Salt Lake City: A Regional Capital*. Chicago, IL: University of Chicago.

Hartshorne, Richard. 1939. *The Nature of Geography*. Lancaster, PA: Association of American Geographers.

Hartshorne, Richard. 1959. *Perspective on the Nature of Geography*. Chicago, IL: Association of American Geographers.

Harvey, D. 1973, *Social Justice and the City*. London, UK: Edward Arnold.

Holt-Jensen, A., 1988, *Geography: Its History and Concepts*. Totowa, NJ: Barnes & Noble Books.

Johnston, R. J., 1983, *Philosophy and Human Geography: An Introduction to Contemporary Approaches*. London, U.K.: Edward Arnold.

King, L. J., 1993, Spatial analysis and the institutionalization of geography as a social science. *Urban Geography*, Vol. 14, 538–551.

Martin, G. J. and James, P. E. 1993, *All Possible Worlds: A History of Geographical Ideas*. New York, NY: John Wiley & Sons.

Mayer, H. M., 1943, *The Railroad Pattern of Metropolitan Chicago*. Chicago, IL: University of Chicago.

Peach, C., 2003, Geographers and the fragmented city. In Ron Johnston and Michael Williams, eds. *A Century of British Geography*. Oxford, UK: Oxford University Press, 563–582.

Ullman, E. L., 1941, A theory of location for cities. *American Journal by Sociology*, Vol. 46, 853–864.

Ullman, E. L., 1943, *Mobile: Industrial Seaport and Trade Center*. Chicago, IL: University of Chicago.

Ullman, E. L., 1954a, Geography as spatial interaction. *Annals of the Association of American Geographers*, Vol. 44, 283–284.

Ullman, E. L., 1954b, Geography as spatial interaction. *Proceeding, Western Committee on Regional Economic Analysis*, Social Science Research Council, 63–71.

Ullman, E. L., 1956, *American Commodity Flow*. Seattle, WA: University of Washington Press.

Van Cleef, E., 1937, *Trade Centers and Trade Routes*. New York, NY: D. Appleton-Century.

Wheeler, J. O., 2001, Special issue: Urban Geography in the 1960s. *Urban Geography*, Vol. 22, 511–513.

# Contributors

**John S. Adams**
Department of Geography
University of Minnesota
Minneapolis, Minnesota

**Trevor J. Barnes**
Department of Geography
University of British Columbia
Vancouver, British Columbia,
  Canada

**Brian J.L. Berry**
School of Social Sciences
The University of Texas at Dallas
Richardson, Texas

**Larry S. Bourne**
Department of Geography and
  Program in Planning
University of Toronto
Toronto, Ontario, Canada

**Martin Cadwallader**
Vice Chancellor for Research and
  Dean of the Graduate School
University of Wisconsin
Madison, Wisconsin

**William A.V. Clark**
Department of Geography
University of California, Los Angeles
Los Angeles, California

**Michael J. Dear**
Department of Geography
University of Southern California
Los Angeles, California

**Larry R. Ford**
Department of Geography
San Diego State University
San Diego, California

**Patricia Gober**
Department of Geography
Arizona State University
Tempe, Arizona

**Peter G. Goheen**
Department of Geography
Queen's University
Kingston, Ontario, Canada

**Susan Hanson**
School of Geography
Clark University
Worcester, Massachusetts

**Chauncy D. Harris**
Department of Geography
University of Chicago
Chicago, Illinois

**Paul L. Knox**
Dean, College of Architecture and
  Urban Studies
Virginia Polytechnic Institute and
  State University
Blacksburg, Virginia

**Robert W. Lake**
Center for Urban Policy Research
Rutgers University
New Brunswick, New Jersey

**Helga Leitner**
Department of Geography
University of Minnesota
Minneapolis, Minnesota

**Elisabeth Lichtenberger**
Department of Geography
University of Vienna
Vienna, Austria

**Sallie A. Marston**
Department of Geography and
  Regional Development
University of Arizona
Tucson, Arizona

**David R. Meyer**
Department of Sociology
Brown University
Providence, Rhode Island

**Michael Pacione**
Department of Geography
University of Strathclyde
Glasgow, United Kingdom

**Risa Palm**
Executive Vice Chancellor and
  Provost
Office of Academic Affairs
Louisiana State University
Baton Rouge, Louisiana

**Geraldine Pratt**
Department of Geography
University of British Columbia
Vancouver, British Columbia,
  Canada

**Eric Sheppard**
Department of Geography
University of Minnesota
Minneapolis, Minnesota

**Edward J. Taaffe**
Department of Geography
The Ohio State University
Columbus, Ohio

**James O. Wheeler**
Department of Geography
University of Georgia
Athens, Georgia

**Maurice Yeates**
Centre for the Study of
  Commercial Activity
Ryerson University
Toronto, Ontario, Canada

# I
## Foundations

1

Boundaries

# CHAPTER 1

# Urban Geography in the United States
## *My Experience of the Formative Years*

CHAUNCY D. HARRIS

During the period up to and through World War II, doctoral dissertations in the University of Chicago were written in the form of monographs on individual cities. Among the major themes of this period were situation of a city with respect to a productive hinterland; historical geography of original settlement and sequential development; transportational, industrial, trade, and residential activities in individual cities; land utilization patterns within a city; and tributary areas. These themes illustrate the context of the field at the time members of my generation were students. Among the contributions we in our turn made were empirical quantitative studies classifying cities, theoretical typologies of city distributions and internal patterns, field studies of individual cities, and patterns of urban distribution and growth for the United States or other countries.

   With such a galaxy of stars in this symposium, I shall limit my remarks to my own experience of the formative years of urban geography in the United States up to and including 1945. First I shall discuss the context of

3

*Urban Geography*, 1990, 11, 4, pp. 403–417.

the evolving field into which a number of us entered at that time. The corpus of doctoral dissertations in urban geography at the University of Chicago, with which the graduate students in this university became more or less familiar, will illustrate the types of problems that our predecessors and contemporaries had addressed or were addressing. Next I shall turn to the contributions we ourselves were then in the process of making, using primarily myself as an example.

## Context

The first doctoral dissertation in geography in the University of Chicago was in the field of urban geography: "A Geographical Interpretation of New York City" by Frederick V. Emerson (1908–1909). His work is worth examining as an indication of the sorts of questions that were asked early in the century. Though Emerson developed primarily into a physical geographer, he worked on his dissertation with J. Paul Goode, who over the years concerned himself with the regional location and transportation of Chicago (Goode, 1926). Emerson's study analyzed the position of New York City in the local hinterland with respect to population and commerce, access through the Appalachian barrier to the larger hinterland of the Great Lakes region, first through building of canals and later of railroads, the qualities of the harbor, and the trade and transportation of the city. Situation with respect to productive hinterland was the focus.

Historical geography of cities was a theme of a large number of studies over many years, especially of dissertations supervised by Harlan H. Barrows. An excellent study by A. E. Parkins on Detroit examined its sequential evolution, the development of transportation by water and by land, and the rise of manufacturing (1918). In the same tradition Mary Jean Lanier studied the development of commerce in Boston during the first three decades of that city in the early 17th century (1924). Edward N. Torbert traced the evolution of land utilization in Lebanon, New Hampshire from its earliest settlement (1931, 1935). Hubert L. Minton interpreted the evolution of the small city of Conway, Arkansas (1937, 1939). James Glasgow analyzed the stages in the evolution of the lake port of Muskegon, Michigan (1939). Carl F. Carlson mapped in great detail the historical periods of land utilization in Aurora, Illinois (1940). Edward B. Espenshade, Jr., tracked the stages in the evolution of the urban complex of Davenport, Rock Island, and Moline with regard to the factors involved in this development on the Upper Rapids of the Mississippi River (1944). Building on these studies, many students sought to determine the factors in the original location and subsequent rise of cities, e.g., Salt Lake City (Harris, 1940, pp. 93–125; 1941).

Helen M. Strong in her pioneering study of Cleveland advanced the field by closer attention to the contemporary functions of the city in commerce and manufacturing and to the location of distinct districts within the city, all in relation to transportation and the port (1921, 1925). Transport roles of cities were investigated in detail also by Clarence F. Jones for the port of Montreal (1923, 1924, 1925a, 1925b), Richard Hartshorne for the lake traffic of Chicago (1924), and Harold M. Mayer for the railroads of Chicago (1943). Traffic by all forms of transportation was treated by Harris for Salt Lake City (1940, pp. 47–56, pp. 61–92), by Robert L. Wrigley for Pocatello, Idaho (1942, pp. 95–118; 1943), and by Edward L. Ullman for Mobile, Alabama (1943, pp. 108–161). Both Mayer and Ullman later made many contributions to transportation geography. Ullman extended the study of traffic flow to include interstate movements nationwide (1957).

Specific activities in individual cities were the subject of much doctoral research at Chicago: the iron and steel industry of the Calumet district of Chicago by John B. Appleton (1927); mining and manufacturing in Scranton, Pennsylvania by Clifford M. Zierer (1925, 1927); mining in Negaunee, Michigan, by J. Russell Whitaker (1930, 1931); business and manufacturing in St. Louis by Lewis F. Thomas (1927); manufacturing in Joliet, Illinois, by William T. Chambers (1926); in Hamilton, Ontario, by Harold B. Ward (1934); and in Lowell, Massachusetts, by Margaret T. Parker (1940). Outlying business centers of Chicago were investigated by Malcolm J. Proudfoot (1936, 1937, 1938). Robert C. Klove was the first at Chicago to study in detail the residential land-use patterns in a metropolitan suburb: the Park Ridge-Barrington area in the northwestern part of the Chicago metropolitan area (1942).

Mapping and description of land utilization or of functional districts within a city found expression in dissertations by Willis H. Miller on Pomona, California (1933, 1935), Rayburn W. Johnson on Memphis (1936a, 1936b), William F. Christians on Denver (1938, 1944), and James R. Beck on Dover and New Philadelphia, Ohio (1942). Johnson's study of Memphis in one sense was the most successful; on its basis Johnson concluded that future industrial expansion in Memphis would have to be located on land then vacant and not highly valued. He bought up blocks of such land, which he later sold at sufficient profit for him, on retirement, to donate to Memphis State University a sum of money equal to the total salary he had been paid over his many years as a faculty member at that institution. This contribution was used to construct a building named for him in which the Department of Geography has been housed.

Attention was devoted also the service areas, or tributary areas, or hinterlands, of cities, especially by Leslie M. Davis for the small city of Elwood,

Indiana (1935, 1937), by Harris for the inland regional center of Salt Lake City (1940, pp. 23–30), and by Ullman for the port of Mobile, Alabama (1943, pp. 57–77).

Many, but not all, of these dissertations in urban geography were supervised by Charles C. Colby. In his course in urban geography he stimulated students with new ideas in the study of cities. He also wrote a seminal paper, "Centrifugal and Centripetal Forces in Urban Geography" (1933), which was, however, his only published paper in this field.

I arrived at the University of Chicago as a graduate student in the summer of 1933. At the university I met Harold M. Mayer and Edward L. Ullman, both intensely interested in urban geography. In retrospect I do not believe that Colby was responsible for Ullman, Mayer, or Harris going into urban geography. We developed interest in this field on our own. But Colby did arouse our curiosity, shape our interests, and improve our writing. He became my mentor (Harris, 1966). The simultaneous presence of three youngsters all deeply curious about urban phenomena heralded a new period in which an initial critical mass was reached. Thereafter dissertations in urban geography were not isolated individual efforts. Lively student discussions ranged over the state of the field, appropriate approaches to the subject, the factors in the rise and distribution of cities, functions performed by cities, and many other topics. In classifications of cities we considered the important element to be the commercial, transportation, or industrial functions which cities performed, not the physical feature of location on coastlines or river systems, which had formed the basis of some previous classifications. Our minds should focus on the human, especially economic, factors—not on the physical sites. Cities, we noted, arose where there were urban functions to be fulfilled. Man-made railroads loomed as large a transportation element as natural river systems.

As students at this time we, of course, scoured the literature for ideas, concepts, and models. I was particularly intrigued by some of the early papers of Richard Hartshorne on location factors and industry as they affected cities. That was before he was seduced into the detailed study of the history of geographical thought, which resulted in his monograph on the nature of geography. I was also stimulated by the seminal ideas of Mark Jefferson (Jefferson, 1909, 1931, 1933, and 1939). Jefferson later encouraged me greatly in correspondence on my early papers. A number of ideas came from the work of Robert E. Dickinson on zones of influence of cities (Dickinson, 1930, 1934a, 1934b), later developed into his book (Dickinson, 1947). The writings of William J. Reilly on service areas in retail trade were also provocative (Reilly, 1929, 1931). But what really excited me were the writings of Robert M. Haig and Roswell C. McCrea, two economists, who brought order into patterns of location of activities within the urban area (Haig, 1926; Haig and McCrea, 1927), and of Richard M. Hurd, who examined location of urban functions in relation to real-estate values (Hurd, 1911). It should be remembered that at the time I became a graduate student there were no general treatments in English on the field of urban geography. During the later part of my student days Eugene Van Cleef examined trade centers (Van Cleef, 1937), Lewis F. Thomas joined with the sociologist Stuart A. Queen to produce a joint treatment, *The City* (Queen and Thomas, 1939), and the U.S. National Resources Committee produced a report, *Our Cities: Their Role in National Economy* (1937).

The first Presidential Address to the Association of American Geographers discussing urban geography was "Environment, Village, and City: Genetic Approaches to Urban Geography" by Griffith Taylor (1942), but it was not in the mainstream of development of the field, though Taylor had been a faculty member of the University of Chicago for a few years in the early 1930s. It was not followed until several decades later by another presidential consideration of the field. Taylor's ideas were not based on work by others in the field of urban geography, nor were his concepts widely used in later work in the field. He was *sui generis*, a self-made original thinker.

After World War II, Harold M. Mayer, Brian J. L. Berry, Norton S. Ginsburg, and I all taught at the University of Chicago and engaged in research in various aspects of urban geography, but that is another, later, story, which should be told by others.

## My Contributions 1940–1945

Since the editors have requested specifically that contributors to this symposium emphasize their personal involvement and role in the development of the field, I cast aside normal reticence and describe how I came to write some of my early papers in urban geography.

*Salt Lake City: A Regional Capital* (1940), my doctoral dissertation, arose out of an interest that began during the summer of 1922, when I was eight years old. My father had a copy of the *World Almanac*, which had population figures for cities in the United States. I became curious about why some cities were larger than others and what were the activities that supported the population in larger cities. Specifically, in 1920 Utah County, Utah, consisted of a county seat—my home town of Provo—of about 10,000 people and five other independent cities, three to ten miles apart, each with a population of about 3,000. I wondered what determined their relative size. And why was Salt Lake City in the next county more than ten times as large as Provo, and so on. At that time I decided to become a geographer. Many years later in my doctoral dissertation I returned to this question and used a forerunner of the location quotient to note what proportion of the regional employment in various activities was concentrated in the regional capital and therefore which activities might be presumed to be part of its basic support (Harris, 1940, pp. 3–12). Ullman was interested in similar concepts and, much later, with the help of Michael F. Dacey and Harold Brodsky, developed the minimum requirements approach to the study of basic and nonbasic activities in cities (Ullman, Dacey, and Brodsky, 1971).

When population figures for cities in the United States became available from the 1940 census, I decided to write a short article on the growth of the larger cities during the previous decade. It seemed obvious that the growth would reflect not only regional and other factors but also differences in the functions the cities performed. In the absence of a classification of cities of the United States, I decided to construct one. Soon I changed my focus from growth of cities to their classification. The papers on growth of the larger cities and the metropolitan districts became only very minor publications of ephemeral and limited interest (Harris, 1942a, 1942b). But the paper on city classification became a major effort and had some impact (Harris, 1943). The lesson from this, I suppose, is that one should be prepared to follow one's intellectual interests and the new vistas that open up in the course of research and not be too bound by the initial goal.

It had long been recognized that cities differ in functions. A few examples of various types had often been cited. But no classification existed for all the cities of the United States. The use of comparable quantitative data on major activities as measured by occupational or employment structure seemed to me to offer a possible basis. The data at that time were relatively poor. The 1930 Census of Population did provide occupational data for cities with more than 100,000 inhabitants and also for cities with populations of 25,000 to 100,000, but only separately for males and females. I simply added figures for males and females in my head, wrote down the total, and then calculated on a two-dollar slide rule the percentage of total gainfully occupied persons in each category. But the occupational categories of that census were ill-suited to such a classification, since they were by personal occupation instead of by useful industry groups. Thus the category of manufacturing and mechanical occupations, for example, was blurred for my purpose by the inclusion of carpenters, painters, electricians, mechanics, and other local service groups with workers in industrial establishments that might contribute to the basic support of the city. Employment in retail and wholesale trade was available in the Census of

Business for 1935 in separate volumes for retail and wholesale distribution. Employment in manufacturing was available in the Biennial Census of Manufacturing 1935 in mimeographed press releases for each state. Fortunately these employment figures were handily assembled by cities in B. P. Hayes and G. R. Smith, *Consumer Market Data Handbook* (1939). Thus occupational data were available by place of residence for larger cities for an earlier date, and employment data were available by place of work for retail and wholesale trade and manufacturing for a later date, for cities down to 10,000 population. Some other types of data could be ferreted out, such as student enrollments in colleges and universities; these helped define university towns.

To establish the defining parameters of each type, I simply arrayed the percentages in each occupational or employment group by city from the highest to the lowest, noting the range, the median, and whether the data were continuous over a normal distribution, were skewed, or were clustered in significant ways. Also, in an attempt to distinguish basic from nonbasic, the minimum, maximum, and median figures in this distribution were compared with the average urban figures for the country as a whole, for regions, and for states. Obviously many of those gainfully occupied in manufacturing and mechanical occupations were serving local needs, and the percentage of persons engaged in these occupations had to be relatively high to distinguish manufacturing as the basic support of the city. On the other hand wholesale trade contained relatively few engaged in local service activities, and the percentage needed to distinguish a wholesale center could be much lower.

With the advantage of hindsight I regret not having discussed the methodology more explicitly in the original article, but at the time interest seemed to be more on the results than on the methodology.

This work resulted in my first paper at a meeting of the Association of American Geographers, presented in New York City in December 1941 and published as "A Functional Classification of Cities in the United States" (Harris, 1943). At the meeting this paper was generously received and commented on by J. Russell Whitaker, Richard Hartshorne, and Meredith F. Burrill, all later presidents of the AAG and all remarkably long-lived. Indeed Hartshorne and Burrill were both present 47 years later in the session "Origins and Evolution of Urban Geography in the United States during Its Formative Years" at the annual meeting of the AAG in Baltimore, Maryland, March 21, 1989, at which a draft of this paper was presented. From this I come to the irresistible conclusion that favorable comment on my early papers promoted longevity!

In the paper on classification of cities I attempted to treat whole metropolitan communities as units, as well as individual smaller cities. Since at that time data generally came by political cities or by counties, I was forced to combine data from many central cities and suburbs in each metropolitan area. The data for the noncentral cities formed the basis of another study, "Suburbs," published as the lead article in the *American Journal of Sociology* (Harris, 1944), in which I attempted to distinguish clearly between residential suburbs (dormitory cities) and industrial suburbs; in the former, commuting was then generally to the central city; in the latter, commuting was out from the central city or from other suburban areas to the industrial suburb. Perhaps because this article was not published in a geographical periodical it has received little attention by geographers.

The practical utility of my method of city and suburb classifications was soon recognized by sociologists and political scientists as well as by geographers. Louis Wirth, the sociologist, and Herbert Simon, the political scientists, were involved in advising on its adaptation by Grace M. Kneedler (later with the married name of Grace M. Ohlson) for the *Municipal Year Book* as "Economic Classification of Cities" (Kneedler, 1945), published annually from 1945 through 1950. The political scientist Victor Jones then prepared a later modification, "Economic Classification of Cities and Metropolitan Areas" (Jones, 1953), published annually through 1955. The sociologist Albert J. Reiss, Jr., prepared "A Functional Specialization of Cities (Reiss, 1957). Victor Jones and Andrew Collver made later revisions (1959). Jones with the geographer Richard L. Forstall prepared "Economic and Social Classification of Metropolitan Areas" (1963) and Jones, Forstall, and Collver, "Economic and Social Characteristics of Urban Places" (1963). Forstall presented a "Classification of Places over 10,000: Functional Classification of Cities 1960/63" (1967) and a still later and more advanced "A New Social and Economic Group of Cities" (1970). These reflected the evolving adaptations to the needs of the users of the *Municipal Year Book*, rapidly improving data bases, evolving techniques, and new conceptual tools. To discuss this evolution fully would require a separate article. Since the time of publication of these two original articles, many new possibilities have enriched the study of functional classification of cities and suburbs. In the United States the data base has improved enormously. The 1940 census made available occupational data by useful industry groups. Employment data have been extended to more groups of activities. The availability of data for many variables by cities, counties, and metropolitan districts has been extended appreciably by successive editions of the *County and City Data Book* (1949–1988) and the *State and Metropolitan Area Data Book* (1982–1986), both compiled by the United

States Bureau of the Census. The data are now available in machine-readable form. The quantitative revolution has developed many new statistical tools for analysis. New techniques facilitate combining large numbers of economic and social variables, which are becoming available in increasing numbers. Many social indicators can now be included as well as economic ones in analysis of the human ecology of cities, suburbs, and even neighborhoods. Instead of the old two-dollar slide rule the geographer can now utilize a new $2,000 personal computer, which calculates with lightning speed and makes possible complex analysis of many variables and interrelationships. The basic activities of cities in advanced societies have been transformed, with the decline of secondary activities such as manufacturing and with the remarkable growth of the tertiary sector, the service functions, and even of a quaternary sector.

Among the articles or books that advanced the methodology of city classification in the United States, a few major works may be mentioned. J. Fraser Hart applied the method to cities of the South (1955). Howard Nelson made more explicit statistical criteria for the recognition of types of American cities (1955). He recorded variations of each city from the mean for all cities in nine classes of occupations (industrial classification) from the 1950 United States Census of Population and calculated standard deviations; cities with more than one standard deviation were recognized as specialized cities in his classification. The Swedish geographer, Gunnar Alexandersson (1956), working at the University of Nebraska, and the American sociologists, Otis Dudley Duncan and Albert J. Reiss (1956), both devoted entire monographs to the classification of American cities. Robert H. T. Smith (1965a, 1965b), the Australian geographer, then working at the University of Wisconsin, reviewed the field and applied the methodology. Brian J. L. Berry assembled, edited, and contributed to a volume, *City Classification Handbook: Methods and Applications* (1972). General reviews of city classification in the context of urban geography in general have been provided by Raymond E. Murphy (1966, 1974), Brian J. L. Berry and Frank E. Horton (1970), Maurice H. Yeates and Barry J. Garner (1971, 1980), Dean S. Rugg (1972, 1978), and others (e.g., Wheeler, 1986).

In 1945 The American Academy of Political and Social Science published a special volume on the city in its *Annals*. Harold M. Mayer was involved in the invitation to me to prepare an article on the geography of the city. I asked for permission to have Edward L. Ullman as coauthor. Together we wrote a typology of spatial distribution of cities and of internal structures of cities entitled "The Nature of Cities" (Harris and Ullman, 1945). Edward L. Ullman was fascinated by patterns of regular distribution of central-place settlements and was among those early introducing into American literature the ideas of Walter Christaller (Ullman, 1941), whose contributions were not fully recognized, however, until the rise of a group of model-building geographers a couple of decades later (Berry and Harris, 1968, 1970). I later met Christaller in 1957 in Frankfurt-am-Main, Germany, where he had traveled to hear a presentation in German of my paper on the market as a factor in the location of industry (Harris, 1954), in which he was interested. We also both participated in a seminar in urban geography in Lund, Sweden, in 1960, at which time I snapped a photograph of Christaller and Ullman together (Photo 1). By myself I probably would have written a more traditional article, but Ullman suggested the idea of presenting a more theoretical generalization. With respect to the patterns of distribution of cities, Ullman certainly contributed the central-place concept and the alignment of cities on transport lines, while I brought in the clustered cities of mining and industrial districts. Ullman was familiar with the work of Ernest Burgess in sociology on concentric zones in cities and of Homer Hoyt, the real-estate economist, in describing sectors of residential development. The multiple-nuclei pattern was suggested by me on the basis of observation that not all activities of a city cluster around a single center, but that different activities have different points of attachment. Specifically I had observed in London, England, the focus of commercial activities in the city and of governmental activities in Westminster, the separate centers joined by the Strand, which for centuries ran through open space between them. I had also observed in German cities the separate historical centers of the church, often in the form of a cathedral, of commerce (in the market), and of government in the castle or city hall. In Salt Lake City I had noted that the retail-commercial center was attached to Main Street, focus of automobile, streetcar, and bus transport by road and pedestrian concentration on the sidewalk, whereas the important wholesale and industrial district was attached to the railroad tracks and to truck transport. From these simple observations arose the concept of the multiple-nuclei pattern, later developed much further by others and applied in new ways in social area analysis or factorial ecology (literature in Larkin and Peters, 1983, pp. 168–170). Forty-five years after publication I am still receiving frequent requests for permission to reproduce from this article the diagram of generalizations of the internal structure of cities, particularly for textbooks in geography or in introductions to sociology. It is important to bear in mind that this paper was a typological scheme, not an empirical research finding. Since in different ways Ullman, Mayer, and I were all involved in this paper, I would like to note that in my judgment Edward L. Ullman had the most original mind of his generation of urban geographers (Eyre, 1977; Harris,

1977; Ullman, 1980) and that Harold M. Mayer has had the most comprehensive command of the entire field.

Another early paper of mine was "Ipswich, England" (Harris, 1942c).
Its genesis was quite simple. At Oxford all undergraduates in geography
had to prepare a regional paper. I chose an area in East Anglia, since it contained Ipswich, which I had observed was a city with interesting relationships
between its hinterland and its commercial and industrial activities — relationships which had shifted over time. It represents a study based primarily on field observation and mapping on foot or bicycle.

While in the Department of State, I turned my attention toward the
Soviet Union, which posed such a challenge both in obtaining information
and in presenting an objective treatment. Scanty data from the 1939 census (only population data for cities with more than 50,000 inhabitants)
were combined with extremely crude approximations of economic structure from atlas maps (showing by divisions of a circle major industries of
principal cities) into an analysis of both regional and functional aspects of
urban growth and a crude functional classification of cities of the Soviet
Union (Harris, 1945a). The paper on Soviet cities did, however, attract
favorable notice from the Soviet reviewer O. A. Konstantinov of Leningrad,
a leading early student of Soviet urban geography. Happily, many years
later we met personally and exchanged ideas. This was the first of a series
of papers devoted to the urban geography of the USSR, culminating much
later in a monograph *Cities of the Soviet Union* (Harris, 1970).

I was also fascinated by the ethnic complexity of cities in the western
and southern fringes of the Soviet Union in non-Russian areas into which
the Russian Empire and later the Soviet Union had penetrated over the
centuries. These cities serve as points of administrative and economic
organization, cultural contact, and some degree of assimilation (Harris,
1945b). But the data in the 1926 Census of the USSR have not been super-
seded by later or better comparable data for the country as a whole. In
spite of enormous changes and the current interest in the subject, no more
recent studies have made advances over this paper of 45 years ago.

These early papers were researched and written without any external
financial support. This was long before the creation of such welcome
resources as the National Science Foundation and similar funding agencies.

In 1946 I became Secretary of the Association of American Geographers;
that was in the days before the establishment of a Central Office. This
activity was followed by a series of administrative responsibilities in geography, the social sciences, international studies, Soviet studies, and especially in the University of Chicago, which reduced the portion of my time
which could be devoted to research. Also my attention was diverted to

Edward Ullman (left) and Walter Christaller (right). Photo by Chauncy D. Harris.

other fields of geography, particularly to regional studies in Germany and in the Soviet Union and to the bibliography of geographical sources. But I still look back with nostalgia to this early period of intense activity in the field of urban geography. I am delighted that some of the problems to which I then devoted attention have been greatly advanced by the creative work of later students of the city.

## Literature Cited

### A. Doctoral Dissertations in Urban Geography at the University of Chicago, 1907–1945, and Some Derivative Publications

Appleton, John B., 1927, *The Iron and Steel Industry of the Calumet District: A Study in Economic Geography.* Urbana: The University of Illinois Studies in the Social Sciences, Vol. 13, No. 2.

Beck, James R., 1942, *The Dover-New Philadelphia, Ohio, Area.* Published in full in private edition distributed by the University of Chicago Libraries.

Carlson, Carl F., 1940, *Aurora, Illinois: A Study in Sequent Land Use.* Published in full in private edition distributed by the University of Chicago Libraries.

Chambers, William T., 1926, *A Geographic Study of Joliet, Illinois, an Urban Center Dominated by Manufacturing.*

Christians, William F., 1938, *Land Utilization in Denver.*

___, 1944, *Land Utilization in Denver.* Extended excerpts in private edition distributed by the Department of Geography of the University of Chicago.

Davis, Leslie M., 1935, *Service Areas of Elwood, Indiana.*

___, 1937, *Service Areas of Elwood, Indiana.* Extended excerpts in private edition distributed by the University of Chicago Libraries.

Emerson, Frederick V., 1908–1909, A geographical interpretation of New York City. *Bulletin of the American Geographical Society.* Vol. 40, 587–612, 726–738; Vol. 41, 3–21. Also separately paged.

Espenshade, Edward B., 1944, *Urban Development at the Upper Rapids of the Mississippi.* Published in full in private edition distributed by the Department of Geography of the University of Chicago.

Glasgow, James, 1939, *Muskegon, Michigan: The Evolution of a Lake Port.* Published in full in private edition distributed by the University of Chicago Libraries.

Harris, Chauncy D., 1940, *Salt Lake City: A Regional Capital.* Published in full in private edition distributed by the University of Chicago Libraries.

___, 1941, The location of Salt Lake City. *Economic Geography,* Vol. 17, 204–212.

Hartshorne, Richard, 1924, *The Lake Traffic of Chicago.*

Johnson, Rayburn W., 1936a, *Land Utilization in Memphis.*

___, 1936b, *Land Utilization in Memphis.* Extended excerpts in private edition distributed by the University of Chicago Libraries.

Jones, Clarence F., 1923, *The Port of Montreal.*

___, 1924, Transportation adjustments in the railway entrances and terminal facilities at Montreal. *Bulletin of the Geographical Society of Philadelphia,* Vol. 22, 98–110.

___, 1925a, Geographic influences in the routes and traffic of the Atlantic railway connections at Montreal. *Bulletin of the Geographical Society of Philadelphia,* Vol. 23, 1–12.

___, 1925b, The grain trade of Montreal. *Economic Geography,* Vol. 1, 53–72.

Klove, Robert G., 1942, *The Park Ridge-Barrington Area, A Study of Residential Land Patterns and Problems in Suburban Chicago.* Published in full in private edition distributed by the University of Chicago Libraries.

Lanier, Mary Jean, 1924, *The Earlier Development of Boston as a Commercial Center.*

Mayer, Harold M., 1943, *The Railway Patterns of Metropolitan Chicago.* Published in full in private edition distributed by the Department of Geography of the University of Chicago.

Miller, Willis H., 1933, *The Localization of Functions in the Pomona Area.*

___, 1935, The localization of functions in the Pomona area. California. *Economic Geography,* Vol. 11, 410–425.

Minton, Hubert L., 1937, *The Evolution of Conway, Arkansas.*

___, 1939, *The Evolution of Conway, Arkansas.* Extended excerpts in private edition distributed by the University of Chicago Libraries.

Parker, Margaret Terrell, 1940, *Lowell: A Study of Industrial Development.* New York: Macmillan.

Parkins, Almon. E., 1918, *The Historical Geography of Detroit.* Lansing, MI: The Michigan Historical Commission. University Series, III. Published in full in private edition distributed by the University of Chicago Libraries.

Proudfoot, Malcolm J., 1936, The Major Outlying Business Centers of Chicago.

___, 1937, The outlying business centers of Chicago. *Journal of Land and Public Utility Economics,* Vol. 13, 57–70.

___, 1938, *The Major Outlying Business Centers of Chicago.* Extended excerpts in private edition distributed by the University of Chicago Libraries.

Strong, Helen M., 1921, *The Geography of Cleveland.*

___, 1925, Cleveland: a city of contacts. *Economic Geography,* Vol. 1, 198–205.

Thomas, Lewis F., 1927, *The Localization of Business Activities in Metropolitan St. Louis.* St. Louis: Washington University Studies, New Series, Social and Philosophical Sciences, No. 1.

Torbert, Edward N., 1931, *The Evolution of Land Utilization in Lebanon, New Hampshire.*

___, 1935, The evolution of land utilization in Lebanon, New Hampshire. *Geographical Review,* Vol. 25, 209–230.

Ullman, Edward L., 1943, *Mobile: Industrial Seaport and Trade Center.* Published in full in private edition distributed by the Department of Geography of the University of Chicago.

Ward, Harold B., 1934, *Hamilton, Ontario, as a Manufacturing Center.*

Whitaker, J. Russell, 1930, *Negaunee, Michigan: An Urban Center Dominated by Iron Mining.*

___, 1931, Negaunee, Michigan: an urban center dominated by iron mining. *Bulletin of the Geographical Society of Philadelphia,* Vol. 29,137–174, 215–240, 306–339. Also separately paged.

Wrigley, Robert L., 1942, *The Occupational Structure of Pocatello, Idaho.* Published in full in private edition distributed by the University of Chicago Libraries.

___, 1943, Pocatello, Idaho as a railroad center. *Economic Geography*. Vol. 19, 325–336.
Zierer, Clifford M., 1925, *The Industrial Geography of Scranton*.
___, 1927, Scranton as an urban community. *Geographical Review*, Vol. 17, 415–428.

## B. Other Works Cited

Alexandersson, Gunnar, 1956, *The Industrial Structure of American Cities*. Lincoln, NE: University of Nebraska Press.
Berry, Brian J. L., editor, 1972, *City Classification Handbook*: *Methods and Applications*. New York: Wiley-Interscience.
Berry, Brian J. I. and Harris, Chauncy D., 1968, Central place. *International Encyclopedia of the Social Sciences*, Vol. 2, 365–370.
___, 1970, Walter Christaller: an appreciation. *Geographical Review*, Vol. 60, 116–119.
Berry, Brian J. L. and Horton, Frank E., 1970, *Geographic Perspectives on Urban Systems*. Englewood Cliffs, NJ: Prentice-Hall. Types of cities and the study of urban functions, 106–149.
Colby, Charles C. 1933, Centrifugal and centripetal forces in urban geography. *Annals of the Association of American Geographers*, Vol. 23, 1–20.
Dickinson, Robert E., 1930, The regional functions and zones of influence of Leeds and Bradford. *Geography*, Vol. 15, 548–557.
___, 1934a, Markets and Market Areas of East Anglia. *Economic Geography*, Vol. 10, 172–182.
___, 1934b, The metropolitan regions of the United States. *Geographical Review*, Vol. 24, 278–291.
___, 1947, *City Region and Regionalism*. London: K. Paul, Trench, Trubner & Co.
Duncan, Otis Dudley and Reiss, Albert J., 1956, *Social Characteristics of Urban and Rural Communities*. New York: John Wiley.
Eyre, John D., 1977, *A Man for All Regions*: *The Contributions of Edward L. Ullman to Geography*. Chapel Hill, NC: University of North Carolina at Chapel Hill, Department of Geography, Studies in Geography, No. 11.
Forstall, Richard L., 1967, Classification of places over 10,000: functional classification of cities 1960/63. *The Municipal Year Book 1965*. Chicago: International City Managers' Association, 30–65.
___, 1970, A new social and economic grouping of cities. In *The Municipal Year Book 1970*. Washington, DC: The International City Management Association, 102–170.
Goode, J. Paul, 1926, *The Geographic Background of Chicago*. Chicago: University of Chicago Press.
Haig, Robert M., 1926, Towards an understanding of the metropolis. *Quarterly Journal of Economics*, Vol. 40, 179–208, 402–434.
___ and McCrea, Roswell C, 1927, *Major Economic Factors in Metropolitan Growth and Arrangement*. New York: Regional Plan of New York and Its Environs. Regional Survey of New York and Its Environs, Vol. 1.
Harris, Chauncy D., 1942a, Growth of the larger cities in the United States 1930–1940. *Journal of Geography*, Vol. 41, 313–318.
___, 1942b, The Metropolitan Districts in 1940. *Journal of Geography*, Vol. 41, 340–343.
___, 1942c, Ipswich, England. *Economic Geography*, Vol. 18, 1–12.
___, 1943, A functional classification of cities in the United States. *Geographical Review*, Vol. 33, 86–99.
___, 1944, Suburbs. *American Journal of Sociology*, Vol. 49, 1–13.
___, 1945a, Cities of the Soviet Union. *Geographical Review*, Vol. 35, 107–121.
___, 1945b, Ethnic groups in cities of the Soviet Union. *Geographical Review*, Vol. 35, 466–473.
___, 1954, The market as a factor in the localization of industry in the United States. *Annals of the Association of American Geographers*, Vol. 55, 315–348.
___, 1966, Charles C. Colby 1884–1965. *Annals of the Association of American Geographers*, Vol. 56, 378–382.
___, 1970, *Cities of the Soviet Union*: *Studies in Their Functions, Size, Density, and Growth*. Washington, DC: Association of American Geographers. Monograph Series, No. 5.
___, 1977, Edward Louis Ullman, 1912–1976. *Annals of the Association of American Geographers*, Vol. 67, 595–600.

___ and Ullman, Edward L., 1945, The nature of cities. *Annals of the American Academy of Political and Social Science*, Vol. 242, 7–17.

Hart, John Fraser, 1955, Functions and occupational structure of cities of the American South. *Annals of the Association of American Geographers*, Vol. 45, 269–286.

Hayes, B. P. and Smith, G. R., 1939, *Consumer Market Data Handbook, 1939 edition.* Washington, DC: Government Printing Office. U.S. Bureau of Foreign and Domestic Commerce, Domestic Commerce Series, No. 102.

Hurd, Richard M., 1911, *Principles of City Land Values.* 3rd edition. (1st ed., 1903, 4th ed., 1924).

Jefferson, Mark, 1909, The anthropogeography of some great cities: a study in the distribution of population. *Bulletin of the American Geographical Society*, Vol. 41, 537–566.

___, 1931, Distribution of the world's city folks: a study in comparative civilization. *Geographical Review*, Vol. 21, 446–465.

___, 1933, Great cities of 1930 in the United States with a comparison of New York and London. *Geographical Review*, Vol. 23, 90–100.

___, 1939, The law of the primate city. *Geographical Review*, Vol. 29, 226–232.

Jones, Victor, 1953, Economic classification of cities and metropolitan areas. *The Municipal Year Book 1953.* Chicago: International City Managers' Association, 26–32, 49–57, 70–96. Also in Year Books for 1954 and 1955.

___ and Collver, Andrew, 1959, Economic classification of cities. *The Municipal Year Book 1959.* Chicago: International City Managers' Association, 67–77, 87–115. Also in Year Book for 1960.

___ and Forstall, Richard L., 1963, Economic and social classification of metropolitan areas. *The Municipal Year Book 1963.* Chicago: International City Managers' Association, 31–44.

Kneedler, Grace M. [later Ohlson, Grace M.], 1945, Economic classification of cities. *The Municipal Year Book 1945.* Chicago: International City Managers' Association, 30–38, 48–68. Also in Year Books for 1946, 1947, 1948, 1949, and 1950.___, ___, and Collver, Andrew, 1963, Economic and social characteristics of urban places. *The Municipal Year Book 1963.* Chicago: International City Managers' Association, 85–157.

Larkin, Robert P. and Peters, Gary L., 1983, *Dictionary of Concepts in Human Geography.* Westport, CT: Greenwood Press. Multiple nuclei theory, 168–170.

Murphy, Raymond E., 1966, *The American City: An Urban Geography.* New York: McGraw-Hill. Functional classification of cities, Chapter 8, 113–129.

___, 1974, *The American City: An Urban Geography.* New York: McGraw-Hill. 2nd edition. Classification of cities, Chapter 6, 101–124.

Nelson, Howard J., 1955, A service classification of American cities. *Economic Geography*, Vol. 31, 189–210.

Queen, Stuart A. and Thomas, Lewis F., 1939, *The City: A Study of Urbanism in the United States.* New York: McGraw-Hill Book Co.

Reilly, William J., 1929, Methods for the study of retail relationships. *University of Texas Bulletin*, No. 2944. Bureau of Business Research. Research Monograph, No. 4.

___, 1931, *The Law of Retail Gravitation.* New York: William J. Reilly. Reiss, Albert J. Jr., 1957, Functional specialization of cities. *Municipal Year Book.* Chicago: International City Managers' Association, 54–68.

Rugg, Dean S., 1972, *Spatial Foundations of Urbanism.* Dubuque, IA: Wm. C. Brown. Detailed classification of cities based on the activities of people working in the city, 110–116.

___, 1978, *Spatial Foundations of Urbanism.* 2nd edition. Dubuque, IA: Wm. C. Brown.

Smith, Robert H. T., 1965a, Method and purpose in functional town classification. *Annals of the Association of American Geographers*, Vol. 55, 539–548.

___, 1965b, The functions of Australian towns, *Tijdschrift voor Economische en Sociale Geografie*, Vol. 56, 81–92.

Taylor, Griffith, 1942, Environment, city, and village: a genetic approach to urban geography. *Annals of the Association of American Geographers*, Vol. 32, 1–67.

Ullman, Edward L., 1941, A theory of location for cities. *American Journal of Sociology*, Vol. 46, 853–864.

___, 1957, *American Commodity Flow: A Geographical Interpretation of Rail and Water Traffic Based on Principles of Spatial Interchange.* Seattle: University of Washington Press.

____, 1980, *Geography as Spatial Interaction*. Edited by Ronald R. Boyce. Seattle, WA: University of Washington Press. Urbanization, 169–216.

Ullman, Edward L., Dacey, Michael R, and Brodsky, Harold, 1971, *The Economic Base of American Cities: profiles for the 101 metropolitan areas over 250,000 population based on the minimum requirements for 1960*. Seattle: University of Washington Press. University of Washington, Center for Urban and Regional Research. Monograph No. 1.

U.S. Bureau of the Census, 1949–1988, *County and City Data Book*. A Statistical Abstract Supplement. Washington, DC: U.S. Government Printing Office. Irregular. Issued for years 1949, 1952, 1956, 1962, 1967, 1972, 1977, 1983, and 1988.

____, 1982, 1986, *State and Metropolitan Area Data Book*. A Statistical Abstract Supplement. Washington, DC: U.S. Government Printing Office. Irregular.

U.S. National Resources Committee, 1937, *Our Cities: Their Role in National Economy*. Report of the Urbanism Committee. Washington, DC: U.S. Government Printing Office.

Van Cleef, Eugene, 1937, *Trade Centers and Trade Routes*. New York: D. Appleton–Century.

Wheeler, James O., 1986, Similarities in the corporate structure of American cities. *Growth and Change*, Vol. 17, 13–21.

Yeates, Maurice H. and Garner, Barry J., 1971, *The North American City*. New York: Harper and Row. City types and functions, Chapter 2, 59–87.

____, 1980, *The North American City*. New York: Harper and Row. 3rd edition. San Francisco: Harper & Row. City types and functions, Chapter 5, 97–120.

# CHAPTER 2

# Harris and Ullman's
# "The Nature of Cities"
## *The Paper's Historical Context and Its Impact on Further Research*

ELISABETH LICHTENBERGER

In an effort to model internal patterns within cities and the areal distribution of cities, Edward L. Ullman and I nearly half a century ago wrote a brief paper, "The Nature of Cities" (Harris and Ullman, 1945). The reception of this modest article was far more widespread and more enduring than either of us anticipated at that time (Harris, 1992, p. 41).

This simple statement refers to a story of scientific progress and success. Chauncy Harris, a young geographer then 31 years old, published "The Nature of Cities" together with Edward Ullman in 1945, a time that, from a European perspective, was highly inauspicious for a wide circulation. When World War II ended, the continent was to a large extent devastated, libraries were severely affected by war damage, and the exchange of publications between institutions in the former German Reich and Western

19

*Urban Geography*, 1997, 18, 1, pp. 7–14

states, particularly the United States, had been broken off. Volume 242 (1945) of *The Annals of the American Academy of Political and Social Science* is still missing in many scientific libraries in German-speaking countries. Circulation of the paper was accomplished by means of reprints in readers in two fields, namely urban sociology and urban geography, reflecting its interdisciplinary importance.

The paper examines two basic aspects of the nature of cities—their functional and economic orientation and their internal structure. The first topic was based on research concerned with the distribution and classification of cities in the United States (Harris, 1943). This work is still cited today and the relative or absolute predominance of labor in specific fields has been consistently applied in nomothetic taxonomies of cities. The article's success, however, rests mainly on the second basic aspect, the models of the internal structure of cities on which this paper concentrates.

Harris and Ullman wrote "The Nature of Cities" in, and about, Chicago, the center of urban ecology and of interdisciplinary cooperation between urban sociology and urban geography. It is thus one of the forerunners of a fundamental scientific attitude in which the extensive and fragmented discipline of geography touches upon and cooperates with neighboring disciplines in formulating and studying a research problem, thereby arriving at innovative and original results that are integrated into both disciplines.

The triad of models presented in the Harris and Ullman paper remains a valuable contribution to the history of concepts in urban geography. Theoretical interpretations make use of various mechanisms, including invasion and succession of social groups in Burgess's zonal model, transport-induced ribbon developments along traffic lines in Hoyt's sector model, and collective preferences as to the allocation of workplaces in the Harris-Ullman multiple nuclei model. The third model not only formally recognizes the existence of several nuclei but also, by systematically including workplaces, accords to the division of labor a level of importance similar to the principle of social differentiation of residential areas. Research into internal structures of cities demonstrates that some combination of these models will fit any city studied and can describe its overall structure. The heuristic value of the models was never disputed, although there is wide agreement today that it is difficult to apply them directly in other cultural and political systems.

## The Paper's Position in the History of the Social Sciences

At a time when knowledge tends to become obsolete very quickly because of the accelerated emergence of new ideas, theories, and problems, it is remarkable that the triad of models is still considered basic in all of the large European-language areas and that they are discussed and illustrated in most textbooks and manuals (Schwarz, 1959; Beaujeu-Garnier and Chabot, 1963; Chorley and Haggett, 1967; Claval, 1981; Lichtenberger, 1986, 1993a; Heineberg, 1989). This persistence of knowledge, remaining relevant for half a century, requires explanation, and several reasons might be given. The basic ones seem to be the internal consistency and the simplicity of the system of statements developed, which dispense with categories such as real space, size, density, and hierarchy; the independence of the models' concepts from any subsequent complementary or new approaches; and last but not least, their plausible representation in graphic form. The triad of models of urban structure, moreover, mirrors a specific temporal and spatial context of development so that, as is the case with all discoveries valid at some time in the past, they are securely stored and not prone to the usual process of a rapid disintegration of knowledge. These classical models of social ecology represent ideal types of North American cities of the interwar period.

This leads to the question of the paper's position in the history of the social sciences. When we ponder this question, considerable differences become obvious between the United States and Continental Europe, both with respect to the philosophy of science and in the network formed by urban geography and its neighboring disciplines of urban planning and urban design.

Analysis of the paper's title, "The Nature of Cities," from the point of view of semantics reveals a positivist approach. An apposite translation in German appears to be impossible. *"Die Natur der Städte"* is misleading. The usual translation, *"Das Wesen der Städte,"* implicitly assumes a hermeneutic interpretation, but one must be aware that the hermeneutic tradition that was well established in Continental Europe was lacking in the English-speaking realm in 1945 and was imported more recently. Among the scientific imports to North America at the time, there was Max Weber's theory of society and economy (4th ed., 1956, translation 1978, including the famous chapter on "The Non-Legitimate Domination of the City") that was subsequently re-exported to Continental Europe in the version of neo-Weberianism. A second group of imports concerned locational theories of interurban systems, such as the central place theory of W. Christaller (1933, 1938, translation 1957; cf. Berry and Pred, 1961) and the economic theory of location of A. Lösch (1944, translation 1954). But a quite astonishing discrepancy must be noted between theoretical approaches to interurban and intraurban systems, respectively, in Continental Europe. There were no theories developed on the intraurban system analogous to those on the interurban system.

Although theoretical sociology succeeded in setting international standards, empirical urban sociology was more or less a newcomer in Germany in the 1950s (Mackensen, 1959). In France, Chombart de Lauwe (1952) and his followers—who definitely went far beyond mere "social ethnography" (Caplow, 1982)—made Paris an outstanding center of urban research in Continental Europe, a development from which geographical research generally profited. Caplow (1982, p. 385) wrote, with reference to France, that "human geographers have developed an impressive literature on dwelling types, the distribution of economic functions, and the evolution of street plans," and this statement applies to the German-speaking countries as well. A large number of studies in urban geography based on a cultural-historical tradition adopted the explorative design characteristics of the geosciences and studied the succession of physical structures and urban land use in great detail, focusing on the microlevel of individual structures, blocks of buildings, streets, and quarters. This cultural-historical tradition augmented the priority accorded to regional geography in Germany and France in the interwar period. The idiographic principle was accepted in urban geography as well. Exploration and interpretation of the individuality of cities was the goal, and monographs on cities based on urban-landscape research were legion. In this context, the spillover of social ecology studies into East Central Europe in the interwar period ought to be mentioned. Papers on the socioeconomic structure of Prague (Moscheles, 1937) and Budapest (Beynon, 1943) published in the *Geographical Review* constitute important historical documents for recent studies analyzing ongoing processes (Lichtenberger, 1993b, 1994a, 1994b).

In light of this work, the question arises as to why no theories of urban structure were developed in the theoretically oriented world of the German- and French-speaking countries. In fact, models of socioeconomic structures already had been introduced in the German-speaking realm during the 19th century by disciplines that dealt with urban development, such as geography, political economy, and urban design. As early as 1841, the geographer J. G. Kohl presented a zonal model of the pre-industrial city with its status-oriented society, most probably the very first three-dimensional model. It cannot be proved that Kohl was influenced by von Thünen's model of the solitary city in an isolated state, as were von Wieser (1909) and Naumann (1909) in their theory of urban land rents.

Architects and city planners simultaneously were active in model building, but soon, under the influence of municipal authorities, they became exclusively interested in finding solutions to problems of technical planning and developing public instruments for regulating land use (Böhm, 1986). The development of theoretical approaches to urban modeling was effectively blocked in Continental Europe because of different national and even regional normative principles in urban planning enforced by means of master plans and building regulations. Normative plans for urban development and urban design replaced theoretically oriented urban models. The first master plans were designed in large German cities as early as the last third of the 19th century during the period of urban growth and industrialization, with decreasing building heights from the center toward the periphery and a functional differentiation of residential areas and industrial areas, although the categories of residential areas for specific social groups that could still be found in 1877 in the *"Denkschrift des Österreichischen Ingenieur- und Architektenvereins"* were abandoned very soon thereafter. In the interwar period, the separation of urban functions was made obligatory in the Charter of Athens in 1927. Simultaneously, the discussion of a spatially separate allocation of social groups was tabooed by the municipal authorities, first of all in the cities with a social-democratic majority. This points to another important difference between North America and the European Continent.

With the ideology of social welfare states developing after World War I, urban research in Europe could no longer accept those liberal premises that implicitly still inform both actual urban development and urban research in North America. They rested on two principles: first, private capitalism with *laissez-faire, laissez-aller* politics, considering urban development as a self-governing process in which mechanisms are regulated by economic competition organized on the lines of a division of labor; and second, Social Darwinism, postulating that societal processes, just like biological ones, conform to the law of survival of the fittest and that segregation is to be considered a natural and inevitable process. The paper by Harris and Ullman was informed by this political environment.

### Extension and Reformulation of the Concept

*Extension of the Concept*
Interest in urban modeling continued in the post-war period. Research in the United States progressed along established lines, whereas innovative research structures were developed in Europe as a result of a confrontation of traditional concepts with those of social ecology.

In the United States, the paper by Harris and Ullman formed the keystone within urban geography for the first stage in a school of thought that was labeled the social ecology of Chicago. Two successive research stages followed that extended their original concept. The first was that of social area analysis, based on the concept that a city consists of fairly homogeneous areas in which the standard of living, way of life, ethnic origin, norms, and general behavior all coincide, i.e., are "natural neighborhoods" (Shevky and Bell, 1955). These assumptions mirror the real-world experience of members of the white middle class: the homogeneity of neighborhoods is continuously reinforced by economic institutions such as banks offering mortgages, real estate offices, and building contractors, and homogeneous areas are kept intact by a large number of norms and regulations internalized in the residents' behavior.

The merit of social area analysis derived from its discovery of three dimensions of segregation of the population—the social, the demographic, and the ethnic. This made research into segregation an independent inquiry. Combining the spatial principles of demographic segregation with the zonal model, those of economic segregation with the sector model, and those of ethnic segregation with the multiple nuclei model contributed to the persistence of these earlier models.

In the second stage extending Harris and Ullman's concept, B. J. L. Berry and his colleagues and collaborators in Chicago in the 1960s initiated methodological progress through factorial ecology (Berry and Horton, 1970). This development formed the basis for worldwide research on cities that triggered the internationalization of analytical urban research with worldwide diffusion and innovation even in less-developed countries with reliable statistical data. This also was true of sociology in the German-speaking countries, where Friedrichs (1977) imported theories and methods of social ecology together with other sociological concepts. Social ecology and factorial ecology were introduced in France and Germany in the 1970s as part and parcel of analytical geography but were confronted, first, with a well-established field of urban geography already existing there and, subsequently, with neo-Marxist ideas. The diffusion of ideas and methods to Europe was markedly furthered by Europeans who held positions as guest professors in North America. The University of Ottawa, for example, was a meeting point for several such scholars in 1969/70, among them Gregory from Great Britain, Claval and Rimbert from France, Racine from Switzerland, and Lichtenberger from Austria. The first factorial analysis of a city in the German-speaking countries was made by Sauberer and Cserjan (1972) for Vienna. An extension of the theoretical constructs of factorial ecology (including institutions engaged in building, the segmentation of the housing and labor markets, and the division of the housing function into first and second homes) and of the methodology (by developing a stepwise dynamic factorial analysis) was introduced by Lichtenberger, Fassmann, and Mühlgassner (1987).

*Urban Modeling: European Reformulation*

There is a need for increased emphasis upon cross-cultural comparative studies of cities and metropolitan areas, the better to determine which characteristics and attributes of cities represent cultural variables and which are relatively independent of cultural differences (Mayer, 1965, p. 81).

The present author agrees with Mayer's statement advocating the definition of "cultural variables" within the framework of comparative urban research. The models of social ecology should be referred to once more in this context, as three important assumptions have not been made explicit—the political system of liberalism, the historical "one-dimensionality" of urban development, and the concept of the city as a centered system, with the Central Business District (CBD) as the city center.

Cultural assumptions as such are missing from these assumptions, but a cultural dimension might be detected indirectly. The starting point for this deduction is the paradox of high-rank economic functions in the CBD, the distance decay of land rents from the center toward the periphery and, on the other hand, the rising of the social gradient from the center toward the periphery. The city center is the economic center and simultaneously a place of "social devaluation," its inhabitants having low social status and being socially disintegrated. Such a social inversion can only result from an anti-urban attitude on the part of the society.

This uncovers an important cultural phenomenon in North American society that might implicitly be contained in Mayer's "cultural variable." The more general question that is raised pertains to the standing cities have for their inhabitants, for society, and for the political system, as this provides a basis for normative models that become visible as physical structures and are realized as actually existing social systems. Including the political system as well as the pro- or anti-urban attitudes of society as key variables for explaining city structures and the complex historical development of cities was a decisive step toward progress in the development of models in urban geography.

When studying the importance of cities for society and for political systems, special stress should be placed on analyzing the city centers. Research into European cities and their development clearly showed that there is a quite marked connection between the center's social status and function and the political system (Lichtenberger, 1970, 1976). European urban development can be defined as a succession of four types of political systems and their respective types of cities:

- Feudal territorial state of the Middle Ages, with the burghers' city;
- Absolutist territorial state, with its imperial residence city;
- Liberal state of industrialization, with the industrial city; and
- Social welfare state, with the New Town.

The concept of the city center showed marked differences in each of these periods. The market place was the social center in the burghers' city of the Middle Ages, but the center shifted to the emperor's residence in the imperial city (Lichtenberger, 1988). A social gradient declining from the center toward the periphery is thus the general rule in pre-industrial cities. The traditional role of the "social center" came to an end in the industrial city. This period saw the creation of factories and the development of a centrifugal structure of social patterns. In the British cities, in which there was no urbanization of the nobility during the era of absolutism (a process so very characteristic of Continental cities), a social gradient rising from the center toward the periphery was predominant, and a similar situation was to be found in North American cities at a later date. From the very outset, any social segregation of the inhabitants was barred from the design of New Towns, a fact still influencing today's urban planning ideologies. Their centers were not defined by means of social categories.

North American urban development closely conforms to the industrial-city model, as neither the burghers' city nor the imperial-residence city ever existed there. European city development, in contrast, is very complex: various superpositions and, to a varying degree, persistent historical predecessors caused the diversification of socioeconomic patterns and different social gradients. Wherever there existed a social gradient falling from the center toward the periphery, namely in the former burghers' and imperial-residence cities, it was retained during the growth period of the liberal era as the middle and upper classes continued to consider the city center as the social focus. The growing need for housing therefore led to social upgrading by means of demolition of older physical structures and the rebuilding of new ones. The lower classes were pushed out toward the periphery, so that massive gentrification could take place around the inner city. The much-discussed gentrification of American cities had predecessors in European urban history.

## On the Nature of Cities: The Immanent Question

The paper by Harris and Ullman was written at the end of World War II, and the three models they depicted are stored in a historical dimension. With breathtaking speed, the processes of suburbanization and counterurbanization made the model of the city as a centered system questionable. The city's existence as a "nonplace" is anticipated in some scenarios. New models for the American "urbanlike" system have been developed. During the last 50 years, this new system of suburbia, which was growing extremely rapidly, formed a kind of extensive network, destroying former central place hierarchies, developing the surroundings of metropolitan areas, and halting the process of restructuring of many central cities by means of downtown redevelopment and gentrification. Core cities of metropolises such as Chicago now have a new opportunity for restructuring as nodes in air transport, as centers of finance, business, or the quaternary public sector, or as multicultural centers. The dichotomy between the old ideal types of urban structures and the new megastructures appears to correspond to the abandonment of the old industrial city by postmodern American society.

## Literature Cited

Beaujeu-Garnier, J. and Chabor, B., 1963, *Traite de Geographie Urbaine*. Paris: Armand Colin.
Berry, B.J.L. and Horton, F.E., 1970, *Geographic Perspectives on Urban Systems*. Englewood Cliffs, NJ: Prentice-Hall.
____ and Pred, A., 1961, *Central Place Studies: A Bibliography of Theory and Applications*. Philadelphia: Regional Science Research Institute.
Beynon, E., 1943, Budapest: An ecological study. *Geographical Review*, Vol. 33, 255–275.
Böhm, H., 1986, Soziale und räumliche Organisation der Stadt. Vorstellungen in der geographischen, städtebaulichen und nationalökonomischen Literatur Deutschlands vor 1918. *Colloquium Geographicum, zur empirischen Wirtschaftsgeographie*, Vol. 19, 33–55.
Caplow, T., 1982, Urban structure in France. In G.A. Theodorson, editor, *Studies in Human Ecology*. New York: Harper and Row, 384–389.
Chombart de Lauwe, P., 1952, *Paris et l'agglomeration parisienne*, first edition. Paris: Presses Universitaires de France.
Chorley, R. and Haggett, P., 1967, *Models in Geography*. London: Methuen.
Christaller, W., 1933, *Die zentralen Orte in Süddeutschland*. Jena: Fischer (translated by C.W. Baskin, 1957, Ph.D. thesis, University of Virginia; Englewood Cliffs, NJ: Prentice-Hall, 1966).
____, 1938, *Rapports fonctionnels entre les agglomerations urbaines et les campagnes*. Comptes Rendus, International Geographical Congress, Amsterdam.
Claval, P., 1981, *La logique des villes, Essai d'urbanologie*. Paris: Litec (Librairies Techniques).
Friedrichs, J., 1977, *Stadtanalyse: Soziale und räumliche Organisation der Gesellschaft*. Hamburg: Rowohlt.
Harris, C., 1943, A functional classification of cities in the United States. *Geographical Review*, Vol. 33, 85–99.
____, 1992, Areal patterns of cities through time and space: Technology and culture (The nature of cities further considered). *Colloquium Geographicum, Modelling the City: Cross-Cultural Perspectives*, Vol. 22, 41–53.
____ and Ullman, E., 1945, The nature of cities. *The Annals of the American Academy of Political and Social Science*, Vol. 242, 7–17.
Heineberg, H., 1989, *Stadtgeographie. Grundriss Allgemeine Geographie Teil X*. Paderborn: Schöningh.
Kohl, J., 1841, *Der Verkehr und die Ansiedlungen der Menschen in ihrer Abhängigkeit von der Gestaltung der Erdoberfläche*. Dresden/Leipzig.
Lichtenberger, E., 1970, The nature of European urbanism. *Geoforum*, Vol. 4, 45–62.
____, 1976, The changing nature of European urbanization. In B. J. L. Berry, editor, *Urbanization and Counterurbanization*, Urban Affairs Annual Reviews, Vol. 11. Beverly Hills, CA: Sage, 81–107.
____, 1986, *Stadtgeographie I—Begriffe, Konzepte, Modelle, Prozesse*. Stuttgart: Teubner (second edition 1992).
____, 1988, The socio-ecological structure as spatial differentiation of Vienna in the 16th Century. In C.S. Yadav, editor, *Contemporary City Ecology*. New Delhi: Concepts International Series in Geography, Vol. 6.
____, 1993a, *Geografia dello Spazio Urbano*. Milan: Edizioni Unicopli.
____, 1993b, *Wein-Prag. Metropolenforschung*. Wien: Böhlau-Verlag.
____, 1994a, Vienna and Prague: Political systems and urban development in the postwar period. In M. Barlow, P. Dostal, and M. Hampl, editors, *Development and Administration of Prague*. Amsterdam: Amsterdam University Press, 91–115.
____, 1994b, Urban decay and urban renewal in Budapest and Vienna: A comparison. *Acta Geographia Lovaniensia*, Vol. 34, 469–477.

_____, Fassmann, H., and Mühlgassner, D., 1987, Stadtentwicklung und dynamische Faktorialökologie. In E. Lichtenberger, editor, *Beiträge zur Stadt- und Regionalforschung,* Vol. 8. Wien: Verlag der österreichischen Akademie der Wissenschaften.

_____ and Heinritz, G., 1986, The take-off of suburbia and the crisis of the central city. Proceedings of the International Symposium in Munich and Vienna 1984. *Erdkundliches Wissen,* Vol. 76. Wiesbaden: Steiner.

Lösch, A., 1954, *The Economics of Location* (translated by W.H. Hoglom and W.F. Stolper). New Haven, CT: Yale University Press (German original: *Die räumliche Ordnung der Wirtschaft,* second edition. Jena: Gustav Fischer, 1944.

Mackensen, R., et al., 1959, *Daseinsformen der Grossstadt.* Tübingen: Mohr (Siebeck).

Mayer, H., 1965, A survey of urban geography. In P. Hauser and L. Schnore, editors, *The Study of Urbanization.* New York: John Wiley, 81–113.

Moscheles, J., 1937, The demographic, social and economic regions of Greater Prague. *Geographical Review,* Vol. 27, 414–429.

Naumann, M., 1909, Miete und Grundrente. *Zeitschrift für Volkswirtschaft, Sozialpolitik und Verwaltung,* Vol. 18, 133–146.

Sauberer, M. and Cserjan, K., 1972, Sozialräumliche Gliederung Wien 1961: Ergebnisse einer Factorenanalyse. *Der Aufbau,* Vol. 7/8, 284–306.

Schwarz, G., 1959, *Allgemeine Siedlungsgeographie.* Berlin: De Gruyter.

Shevky, E. and Bell, W., 1955, *Social Area Analysis: Theory, Illustrative Applications, and Computational Procedure.* Palo Alto, CA: Stanford University Press.

von Wieser, F., 1909, Die Theorie der städtischen Grundrente. In W. Mildschuh, editor, *Mietzinse und Bodenwerte in Prag in den Jahren 1869–1902.* Wien-Leipzig: V-XL Wiener Staatsw. Studien, Vol. 9, No. 1.

Weber, M., 1978, *Economy and Society.* Berkeley: University of California Press. (translation of Max Weber, *Wirtschaft und Gesellschaft: Grundriss der verstehenden Soziologie,* fourth ed., 1956. Tübingen: J.C.B. Mohr.)

# CHAPTER 3
# Diffusion of Urban Models
## *A Case Study*

CHAUNCY D. HARRIS

**ABSTRACT**

The analysis of diffusion through the academic landscape of models of internal patterns of cities, as published in "The Nature of Cities" (Harris and Ullman, 1945) in simplified, generalized, and comparative cartodiagrams of concentric zones, sectors, and multiple nuclei, is based on records of 128 citations of the basic article in periodicals and on 309 reproductions of the models in books. The half-life of periodical citations of the article was reached in 1974, 29 years after publication. The half-life of reproduction of the models in books occurred eight years later in 1982, but the peak year was 1995, with 20 new republications.

Major diffusion studies in geography have been made by Hägerstrand (1952, 1953, 1967), Gould (1969), Hudson (1972), Brown (1974–1975,

1981), and Morrill et al. (1988). Diffusion also has been the object of study in other fields, such as anthropology, sociology, economics, urban history, and communications. Specialized studies have been devoted to many topics.

This article is concerned with a very special type of diffusion, that of an idea or concept through the academic landscape. Urban models of the distribution of cities by Christaller (1933, 1966) and of internal patterns of cities by Burgess (1925, 1929), by Hoyt (1936-1937, 1939), and by Harris and Ullman (1945) have played roles in urban geography. It is possible to study their diffusion through intellectual space, over time, and across disciplinary and linguistic barriers. Sometimes a marker helps to identify different streams of diffusion. Thus, different early lines of diffusion of the central-place concepts and models of Christaller into the English-speaking world can be distinguished by the spelling of the author's first name in citations. Ullman (1941), a geographer writing in a sociological journal, and Wehrwein (1942), an agricultural economist writing in a geographical journal, both in the United States, spelled his name correctly but Dickinson (1947), a geographer writing in England, misspelled his first name as Walther. The form of citation by still later authors who used one of these three sources in English may be inferred from how they spelled the first name. The central-place concept has been studied widely and developed further and its diffusion already is well known (Berry and Pred, 1964; Berry and Harris, 1968; Beavon, 1977; King, 1984; and Berry and Parr, 1988, among others).

Specifically, attention here is focused on the Burgess concentric-zone, the Hoyt sector, and the Harris-Ullman multiple-nuclei models of the internal patterns of cities as depicted together in the Harris-Ullman article in 1945 (Figure 3.1). The juxtaposition of the three models in simplified, generalized, comparative cartodiagrams had an appeal far beyond that of the individual models themselves. Incidentally, it may be noted that the principal authors of these different models, after creating them, turned to other interests. Ernest W. Burgess, the sociologist, devoted his later years to predicting success in marriage (even though he himself was a bachelor). Homer Hoyt, the economist, engaged in real estate and housing. Edward L. Ullman pursued the study of amenities as a factor in regional growth of transportation and of spatial interaction. I turned my attention to urbanization in the Soviet Union and other topics. But these three models of internal patterns of cities, particularly in their association, apparently filled an ecological niche in the intellectual landscape of a particular period and place. The urban models, once created, took on a life of their own quite separate from that of their originators.

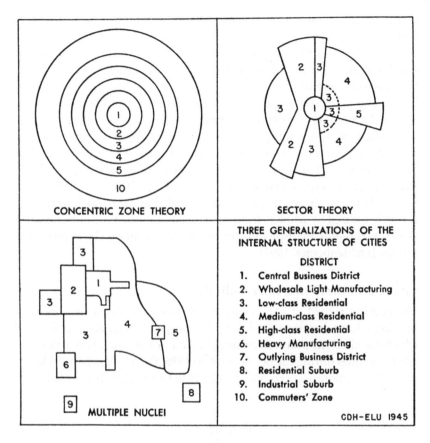

**Figure 3.1** The Harris-Ullman generalized models (cartodiagrams) of internal patterns of cities as originally published (Harris and Ullman, 1945, p. 13, Figure 5. Courtesy of the American Academy of Political and Social Science).

Two different data sets can be utilized to study diffusion of urban models. One data source is the citations in periodical articles as recorded in the *Social Sciences Citation Index (SSCI,* 1956– ), hereafter cited simply as *Index.* This widely available source has been utilized in a number of studies of diffusion of ideas, particularly of authors (Whitehand, 1985; Wrigley and Matthews, 1987; Bodman, 1991), of programs (Turner and Meyer, 1985), of periodicals (Whitehand, 1984, 1990; Turner, 1988), of individual articles and books (Wrigley and Matthews, 1986), and across national and linguistic frontiers (Whitehand and Edmondson, 1977). This paper is a case study confined to citations of a single article. The other data source, available only in a unique copy, records reproductions in books of the urban models in the form of cartodiagrams of concentric zones, sectors, and multiple nuclei.

On the basis of theory, one would expect that citations in periodical articles on the frontiers of research would occur earlier and peak sooner than reproduction of the urban models in synthesizing works or in textbooks. These expectations are confirmed (Figures 3.2 and 3.3).

The *Annals of the American Academy of Political and Social Science,* in which the 1945 article was published, is a general social science periodical with an emphasis on policy questions. Individual numbers are often devoted to special areas, problems, or issues. The number in which the paper appeared, a special issue, "Building the Future City," emphasized the role of planning. The journal is broadly available in American libraries and is widely utilized by scholars from many fields. But the time of publication (1945) was highly inauspicious for wide circulation on the continent of Europe (Lichtenberger, 1997).

### Citations of the Article in Periodicals

The citation record of "The Nature of Cities" is distinguished by large numbers, long duration, and the wide range of journals and fields in which the citations occur. According to data in the *Index,* this article was cited 128 times during the period 1956 to 1996, the 41 years covered by the *Index.* No citations were recorded from 1946 to 1955, the first decade after

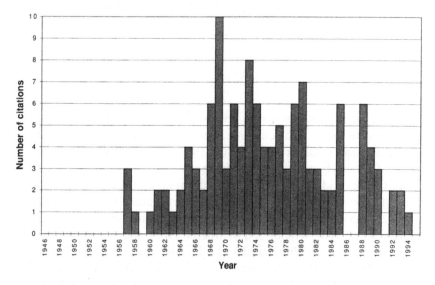

**Figure 3.2** Number of citations in periodicals of Harris-Ullman "Nature of Cities," by year, 1956–1995.

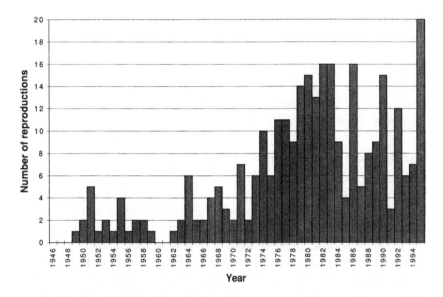

**Figure 3.3** Number of reproductions in books of Harris-Ullman urban models, by year, 1946–1995.

publication of the article, since the *Index* had not been established at that time. The *Index* records 16 citations in the second decade after publication, 52 in the third decade, 41 in the fourth decade, and 18 in the fifth decade (and one thus far in the sixth decade).

Citations reached a peak in the year 1969, 24 years after publication, with 10 citations. The half-life of citations occurred at the beginning of 1974, 29 years after publication (Figure 3.2).

Citations were recorded in articles in 70 diffferent periodicals. Fifteen journals included three or more citations each, led by *The Annals of the Association of American Geographers* (13 citations) and followed in geography by *Economic Geography* (6), *The Professional Geographer* (4), *Transactions of the Institute of British Geographers* (4), *The Geographical Review (3)*, *The Journal of Geography (3)*, *Geografiska Annaler (3)*, *Series B. Human Geography (3)*, and *Tijdschrift voor Economische en Sociale Geografie (3)*. Sociological journals with multiple citations are represented by *The American Journal of Sociology* (4), *Demography* (4), and *The Canadian Review of Sociology and Anthropology* (3). Other specialized periodicals in related fields are *Land Economics (3)*, *Ekistics (3)*, *Urban Studies (3)*, and *Urban History* (3).

Other geographical journals with one or two citations include *The Canadian Geographer, Journal of Historical Geography, Progress in Human Geography, Annales de Géographie, Geographische Zeitschrift, Mitteilungen der Österreichischen Geographischen Gesellschaft,* and *Urban Geography.*

*Urban Geography* was founded only in 1980 and citations from it were included in the *Index* only from 1988, in the late tail of diminished citations of the article.

The wide range of fields represented in journals with citations include ones with many titles, such as sociology (12 different periodicals with citations), general social and behavioral science (8), planning (8), regional science or regional studies (6), and economics (4).

## The Database for Reproduction of the Urban Models in Books

Some of the characteristics of the diffusion of the urban models published together in 1945 can be studied from a unique database. The paper, published in *The Annals of the American Academy of Political and Social Science*, was copyrighted by the Academy in Philadelphia, which required written permission for republication of its text or figures. The Academy in turn would not grant permission without the approval of the authors. Some reproductions were made without such approval, or even knowledge that they were required or appropriate. My file of authors' permissions has been supplemented by records of other reproductions that have come to my attention. This extensive, but still incomplete, file is useful in tracing permissions to reproduce one or more of the figures in book-length publications. This study does not distinguish which of the figures were reproduced. The most common reproduction was of the concentric-zone, sector, and multiple-nuclei patterns alongside one another for comparison, but some books reproduced only one of the patterns, such as that of multiple nuclei or, less often, of the patterns of distribution of cities.

This analysis is not of the concepts involved, for these have been evaluated elsewhere (Lake, 1997). Many good discussions of these urban patterns with related references do not reproduce any of the figures (e.g., Nelson, 1969; Larkin and Peters, 1983); they are not included in this particular database of reproduction of urban models.

These cartodiagrams may be regarded as cultural artifacts with a history of diffusion through intellectual space. They are icons, which in simple graphic form represent complex concepts. They have been found especially useful in didactic works such as textbooks or in generalizing works summarizing broad fields of knowledge. Most, but not all, of the reproductions of the urban models in books also cite the original source article or a secondary source, but these specific citations are not analyzed here.

The number of recorded published reproductions of these urban models in this database is 309. The most intense diffusion streams tended to flow through specialized channels of individual disciplines, particularly sociology and geography, or more specifically urban sociology and urban geography. Disciplinary boundaries and language walls often are barriers to diffusion.

## Role of Collected Works As Secondary Sources of Diffusion

Many of the reproductions of the urban models cite the original 1945 publication, but many obviously have been taken from later intermediate sources, as revealed by various markers such as the form or year of the citation provided or the diagram submitted in the request for permission to reproduce. A secondary source may have suggested to authors the use of the models in some book in preparation, but when permission was sought to reproduce from the secondary source, it was discovered that the copyright was held by the original publisher, to which reference may then have been made in the resulting publication.

Very important as sources of further diffusion of the models were a dozen republications of the basic text and figures of the article in collections of papers in urban geography or urban sociology. These collections brought together papers on closely related subjects. The impact of the assembled papers was substantially greater than that of individual articles scattered through the literature. The collections had advantages of agglomeration (a group of articles on related themes) and of selectivity (in the exclusion of unrelated material). This agglomeration effect was especially important in the early stages when the total corpus of relevant literature was relatively small. These collections appeared at a time before monograph-length syntheses covering the whole field of urban geography or of urban sociology were generally available and at a time of very rapid expansion of these fields when the need for instructional materials was keenly felt.

In urban geography, such collected works gave wide distribution to the models. Of particular importance was the republication of the text and figures in *Readings in Urban Geography* (Mayer and Kohn, 1959), which brought together significant papers of the period by a large number of authors from widely scattered sources and provided a convenient assemblage for study and for consultation. Three other collections in urban geography played key roles in surmounting linguistic and national boundaries. *Geografiya Gorodov* (Pokshishevskiy, 1965), a Russian translation of most of the material in *Readings in Urban Geography,* diffused the models into the Soviet Union and Eastern Europe. *Allgemeine Stadtgeographie,* edited by Schöller (1969), contained a collection of papers in urban geography in German and English; it diffused knowledge of the models in the German-language areas and in Central Europe. *A Geography of Urban Places: Selected Readings* (Putnam et al., 1970) provided published copies of the text and models in Canada, the United Kingdom, Australia, and

New Zealand. On the other hand, the first geographical republication of the article, in *Outside Readings in Geography* (Dohrs et al., 1955), does not appear to have played a major role in diffusion of the models, perhaps because it reproduced the text but not the figures, or because it did not focus on urban geography, or because it appeared before the great expansion in urban geography or in advance of interest in model building, or because the concept "outside readings" was less useful than that of "integrated readings," which characterized some later collections. Other traceable cited secondary sources of diffusion of these urban models within geography include articles by Pred (1964) and Nelson (1969) and books by Haggett (1965) and Rugg (1979).

Collections in urban sociology also played major roles in diffusion of the models. Much earlier than the other collections was the *Reader in Urban Sociology* (Hatt and Reiss, 1951), but it remained for the second edition, *Cities and Society: The Revised Reader in Urban Sociology* (Hatt and Reiss, 1957), which also republished the text and figures, to become the principal point of diffusion of these urban models within the field of sociology. Republication of text and figures in other collections undoubtably played roles as important secondary sources for further waves of diffusion of these urban models (Freedman et al., 1952; Halebsky, 1973; Schwirian, 1974; Cargan and Ballantine, 1979). But it was the Hatt and Reiss (1957) collection of papers in urban sociology that constituted the major traceable source of dissemination within the sociology and urban sociology fields into texts that further reproduced the models, such as Light and Keller (1975) or, more recently, Calhoun et al. (1994), McGahan (1995), and Gelles and Levine (1995). Still other verifiable lines of diffusion came from a library journal in the United States to a public library manual in England or from a sociological source to the Sunday School Board of the Southern Baptist Convention.

## Linguistic and National Boundaries

Linguistic boundaries apparently were bridged more easily in geography than in sociology. Thus, the Mayer and Kohn (1959) collection provided an impetus and a source for the republication of the Harris-Ullman article in Russian (Pokshishevskiy, 1965) and in German (Schöller, 1969). Furthermore, the Mayer and Kohn collection was the source of other reproductions of the urban models in French (Beaujeu-Garnier and Chabot, 1967a; Bailly, 1978; Beaujeu-Garnier, 1980) and in German (Hofmeister; 1969), as well as in English (Beaujeu-Garnier and Chabot, 1967b; Boyce, 1974; Hartshorn, 1980). The French translation of Haggett (1973) provided yet another secondary source of diffusion in French (Derycke, 1995). The urban models in sociological literature in languages other than English apparently occurred somewhat later, as in publication of translations of sociology textbooks into Spanish (Mack and Pease, 1980).

One may speculate that geography is a more worldwide discipline than sociology and that general concepts are readily applied to differing areas and cultures, as when Mabogunje (1968) noted that "the idea of multiple nuclei is fundamental in understanding the nature of Nigerian cities" (p. 179) and "the effect [of modern technology] in Nigeria so far has been to create twin-cities—one traditional and one modern" (p. 183). On the other hand, sociology textbooks pay much more attention than do geography textbooks to the authors of concepts, often including a separate index of persons, which greatly facilitates the tracing of reproductions of these urban models.

### Paths of Diffusion

The paths of diffusion sometimes occurred through several stages, though these may be difficult to trace with certainty. A few cases will illustrate some interesting types of diffusion. In a work by Derycke (1995) in a French encyclopedia of geography, the models are traced to a French translation of Haggett (1973), in turn based on the English original (Haggett, 1965), which in its turn cites an article by Pred (1964). Publication of the models by a Nigerian (Ayeni, 1979) quotes the source as 1965; this presumably refers also to the Haggett (1965) publication, though the author was also familiar with the work of Mabogunje (1968). Particularly intriguing are diffusions across disciplinary lines. A good example is the reproduction of models by Goldfield and Brownell (1979) in an urban studies publication based on McKelvey (1968), an historian, who bases them on Hatt and Reiss (1957), two sociologists who take the models from two geographers.

As the models became well known (Lake, 1997; Agnew, 1997), and to a degree internalized, they were sometimes cited simply as classical or traditional models without attribution or citation (Bourne, 1971; Fielding, 1974; Leser et al., 1995; Getis and Getis, 1995; Fellmann et al., 1997). Also, the cartodiagrams stimulated others to make further modifications, especially of the multiple-nuclei models, as in Jordan and Rowntree (1976) in geography and in Romanos (1976) in sociology.

### Reproductions by Fields

Of the 309 recorded reproductions of the Harris-Ullman urban models (Table 3.1), the largest number have occurred in sociology books, some 165 in all, predominantly in general or introductory textbooks (115) but

**TABLE 3.1** Number of Books Reproducing Harris-Ullman (1945) Urban Models, by Fields

| Total | 309 |
| --- | --- |
| Geography | 128 |
|   Urban geography | 59 |
|   Human geography | 23 |
|   Economic geography | 15 |
|   Cultural geography | 12 |
|   General and regional | 11 |
|   Reference books | 8 |
| Sociology | 165 |
|   General sociology | 115 |
|   Urban sociology | 43 |
|   Applied sociology | 7 |
| Other fields | 16 |

with 43 in the more specialized subfield of urban sociology, and in lesser numbers (7) in related fields such as social welfare, social problems, and criminology. The earliest reproductions of the Harris-Ullman urban models were published in general sociology textbooks, such as Bernard (1949), MacIver and Page (1949), Freedman et al. (1952), and Lundberg et al. (1954), or in sociology of urban communities, such as Hillman (1950), Hallenbeck (1951), Smith and McMahan (1951), Queen and Carpenter (1953), LaGory and Pipkin (1981), and Schwab (1981).

The models are reproduced in 128 books in geography, including 59 in urban geography, 23 human geography, 15 economic geography (including location analysis and land-use studies), 12 cultural and social geography, 11 general or regional textbooks, and 8 geography reference books (2 encyclopedias and 6 dictionaries). Geographers were somewhat later than sociologists in reproducing the cartodiagrams from the 1945 article. My earliest record of a geographical reproduction stems from Lund, Sweden, six years after the original publication, by the Estonian refugee scholar Kant (1951), in a sector analysis of rural–urban interaction.

In the field of urban geography, one may note the following general syntheses of the field, arranged by country: in the United States, Murphy (1966, 1974), Rugg (1971, 1979), Hartshorn (1980, 1992), Palm (1981), Exline et al. (1982), Brunn and Williams (1983, 1993), and Knox (1994); in Canada, Yeates and Garner (1971), Jackson (1973), Nader (1975), and Yeates (1990, 1998); in the United Kingdom, Dickinson (1964), Johnson

(1967), Herbert (1972), Carter (1976, 1995), Gordon and Dick (1980), Herbert and Thomas (1982, 1990), and Davies and Herbert (1993); in France, Beaujeu-Garnier (1980, 1995); in Germany, Schwarz (1966, 1989) and Hofmeister (1969, 1976); in Austria, Lichtenberger (1986, 1991); in Nigeria, Mabogunje (1968); in South Africa, van der Merwe and Nel (1975); and in Australia, Rose (1967).

Among the works in closely related fields are those in economic geography, Boyce (1974), Berry, Conkling, and Ray (1976), Wheeler and Muller (1981, 1986), Hartshorn and Alexander (1988); in cultural geography, Kolars and Nystuen (1974), Jordan and Rowntree (1976), Austin et al. (1987), and Jackson and Hudman (1990); in human geography, Tidswell (1976), Bradford and Kent (1977), de Blij (1977, 1996), Zimolzak and Stanfield (1979, 1983), Larkin et al. (1981), Stoddard et al. (1989), Fellmann et al. (1990, 1997), Bailly (1992), Shelley and Clarke (1994), Bailly and Beguin (1995), Nelson et al. (1995), and Rubenstein (1996); in social geography, Dicken and Lloyd (1981) and Ley (1983); in location analysis, Haggett (1965) and Haggett et al. (1977); in land use, Chapin (1957), Platt (1977), Chapin and Kaiser (1979), and Rhind and Hudson (1980); in regional United States, Holcomb (1988) and Getis and Getis (1995); and in general introductions, Renwick and Rubenstein (1995). Among reference books are geographical encyclopedias (Derycke, 1995) and geographical dictionaries (Johnston et al., 1986, 1994; Leser et al, 1995; Small and Witherick, 1995).

The wide range of replications of these models is indicated by their occurrence also in 16 works in other academic disciplines such as political science, urban history, and biology (ecology), and in applied fields such as town planning, real estate, retailing, and library service. Diffusion of the models typically proceeded from academic fields to applied fields, as from sociology (DeFleur et al., 1971) to social welfare (Federico, 1973).

## Reproductions Over the Years

Diffusion through the academic landscape may take decades, and new reproductions may extend over many years. These urban models published in 1945 were reproduced in 16 books in the first decade following publication, in 17 in the second decade, in 47 in the third decade, in 118 in the fourth decade, in 101 in the fifth decade, and in 10 thus far since 1995. Numbers of books in which the models were reproduced in each individual year of publication are shown in Figure 3.3. The half-life of the reproduction of these urban models in books occurred in 1982, 37 years after publication. The peak year in number of new reproductions (20) was in 1995, just half a century after original publication.

The persistence of the models over the years may be attributed in part to reproduction in successive editions of succeful and widely used basic textbooks, in which each new edition sent additional impulses into the information system. Among the earlier such textbooks in general sociology were those by MacIver and Page (1949, 1962), Lundberg et al. (1954, 1968), and Broom and Selznick (1955, 1968), and in urban sociology, Gist et al. (Gist and Halbert, 1956; Gist and Fava, 1974). Among the more recent such textbooks in general sociology are those by DeFleur et al. (1971, 1984), Mack and Pease (1973), Lloyd et al. (1979), Shepard (1974, 1993), Light and Keller (1975), Calhoun et al. (1994), Rose et al. (1977, 1990), Vander Zanden (1979, 1996), Hagedorn (1980, 1994), Tischler (Tischler et al., 1983; Tischler, 1996), Sullivan (Sullivan and Thompson, 1984; Sullivan, 1995), Brinkerhoff (Brinkerhoff and White, 1985; Brinkerhoff et al., 1997), Thio (1986, 1996), and Alan Johnson (1986, 1996), and in urban sociology, Phillips (Phillips and LeGate, 1981; Phillips, 1996), Schwab (1982, 1992) Teevan (1982; Teevan and Hewitt, 1995), McGahan (1982, 1995), and Kammeyer et al., (1994).

A similar role in geography was played by successive editions of textbooks in general and regional geography, Getis et al. (1981, 1985, 1991; Getis and Getis, 1995); in human geography, Fellmann et al. (1990, 1997) and de Blij (1977, 1995, 1996); and in cultural geography, Jordan (Jordan and Rowntree, 1976; Jordan et al., 1997). In urban geography, one may note particularly the continuing role of Hofmeister (1969, 1980), Yeates (Yeates and Garner, 1971; Yeates, 1990, 1998), Herbert (1972, Herbert and Thomas, 1982, 1990), Carter (1976, 1995), Hartshorn (1980, 1992), Beaujeu-Garnier (1980, 1995), and Lichtenberger (1986, 1991).

## Stages

Seven stages in the diffusion of the Harris-Ullman urban model may be recognized: (1) original publication in a periodical of wide circulation and availability, but not one devoted to an individual discipline; (2) scattered citations of the article or reproduction of the cartodiagrams in the early years; (3) republication of the article in several widely used collections at a time of rapid development in the fields of urban geography and urban sociology and of rising interest in models; (4) reproduction of the cartodiagams, with varying amounts of discussion, in a large number of summarizing works or textbooks in a wide variety of fields; (5) modification of the models by later writers; (6) reproduction of the cartodiagrams as classical models without attribution or citation, in some cases because they have become so well known that this is not considered necessary, or because textbook writers wish to simplify text to reduce the learning burden for students, or in some cases because the authors themselves, taking the cartodiagrams from secondary or tertiary sources, may not even be aware of, or care about, their origin; and (7) elimination of these urban models entirely as interest shifts to other paradigms or approaches. That the stages overlap and have co-existed over long periods of time is reflected in the extended duration of successive reproduction of these urban models in geography and sociology in many parts of the world.

## Literature Cited

Agnew, J., 1997, Commemoration and criticism fifty years after the publication of Harris and Ullman's "The nature of cities." *Urban Geography,* Vol. 18, 4–6.

Austin, C. M., Honey, R., and Eagle, T. C., 1987, *Cultural Geography.* St. Paul, MN: West, 496, Figure 12.9.

Ayeni, B., 1979, *Concepts and Techniques in Urban Analysis.* London, UK: Croom Helm, and New York, NY: St. Martin's Press, 13, Figure 1.2.

Bailly, A. S., 1978, *L'Organisation Urbaine: Théories et Modèles* [Urban Organization: Theories and Models], 2nd ed. Paris, France: Centre de Recherche d'Urbanisme, 111, Figure 4.

_____ , 1992, *Les Concepts de la Géographie Humaine* [Concepts in Human Geography], 2nd ed. Paris, France: Masson, 102, Figure 1.

_____ and Beguin, H., 1995, *Introduction à la Géographie Humaine* [Introduction to Human Geography], 5th ed. Paris, France: Masson, 163, Figure 9.2.

Beaujeu-Garnier, J., 1980, *Géographie Urbaine* [Urban Geography]. Paris, France: Armand Colin, 115, Figure 20.

_____ , 1995, *Géographie Urbaine* [Urban Geography], 4th ed. Paris, France: Armand Colin, 93, Figure 18.

_____ and Chabot, G., 1967a, *Traité de Géographie Urbaine* [Treatise on Urban Geography]. Paris, France: Librairie Armand Colin, 283, Figure 22.

_____ and _____ , 1967b, *Urban Geography.* London, UK: Longman; and New York, NY: John Wiley, 290, Figure 22.

Beavon, K. S. O., 1977, *Central Place Theory: A Reinterpretation.* London, UK, and New York, NY: Longman.

Bernard, J., 1949, *American Community Behavior.* New York, NY: Dryden, 19, Figures 2–5, and 145, Figure 15.

Berry, B. J. L., Conkling, E. C., and Ray, D. M., 1976, *Geography of Economic Systems.* Englewood Cliffs, NJ: Prentice-Hall, 135–139, and 227, Figure 12.2.

_____ and Harris, C. D., 1968, Central place. In D. L. Sills, editor, *International Encyclopedia of the Social Sciences.* New York, NY: Macmillan and Free Press, Vol. 2, 365–370.

_____ and Parr, J. B., 1988, *Geography of Market Centers and Retail Location: Theory and Applications.* Englewood Cliffs, NJ: Prentice-Hall.

_____ and Pred, A., 1964, *Central Place Studies: A Bibliography of Theory and Applications.* Bibliography Series, No. 1. Philadelphia, PA: Regional Science Research Institute (reprint of 1961 edition with additions.)

Bodman, A. R., 1991, Weavers of influence: The structure of contemporary geographic research. *Transactions, Institute of British Geographers,* Vol. 16, 21–37.

Bourne, L. S., editor, 1971, *Internal Structure of the City: Readings on Space and Environment.* New York, NY: Oxford University Press, 71, Figure 1.

Boyce, R. R., 1974, *Bases of Economic Activity: An Essay on the Spatial Characteristics of Man's Economic Activities.* New York, NY: Holt, Rinehart and Winston, 255, Figure 13.1.

Bradford, M. G. and Kent, W. A., 1977, *Human Geography: Theories and their Application.* Oxford, UK: Oxford University Press, 76, Figure 5.7.

Brinkerhoff, D. B. and White, L. K., 1985, *Sociology: An Introduction.* St. Paul, MN: West Publishing, 500–502, Figures 18.3, 18.4, and 18.5.

_____ , _____ , and Riedmann, A. C., 1997, *Sociology,* 4th ed. Belmont, CA: Wadsworth, 634, Figure 22.3.

Broom, L. and Selznick, P., 1955, *Sociology: A Text with Adapted Readings.* Evanston, IL: Row Peterson, 449, Figure XI-9.

_____ and _____ , 1968, *Sociology: A Text with Adapted Readings,* 4th ed. New York, NY: Harper and Row, 445, Figure XIII.2.

Brown, L. A., 1981, *Innovation Diffusion: A New Perspective.* London, UK, and New York, NY: Methuen.

_____ , editor, 1974–1975, Studies in spatial diffusion processes. *Economic Geography,* Vol. 50, 285–374, and Vol. 51, 185–304.

Brunn, S. D. and Williams, J. F., editors, 1983, *Cities of the World: World Regional Urban Development.* New York, NY: Harper and Row, 25, Figure 1.7.

_____ and _____ , editors, 1993, *Cities of the World: World Regional Urban Development,* 2nd ed. New York, NY: HarperCollins College Publishers, 28–29, Figure 1.13.

Burgess, E. W., 1925, The growth of the city. In R. E. Park, E. W. Burgess, and R. D. McKenzie, editors, *The City.* Chicago, IL: University of Chicago Press, 47–62, Chart I, p. 51, and Chart II, p. 55. (Republished 1967 and 1984.)

_____ , 1929, Urban areas. In T. V. Smith and Leonard D. White, editors, *Chicago: An Experiment in Social Science Research.* Chicago, IL: University of Chicago Press, 113–138, fig. on p. 115.

Calhoun, C., Light, D., and Keller, S., 1994, *Sociology,* 6th ed. New York, NY: McGraw-Hill, 529, Figure 20.2.

Cargan, L. and Ballantine, J. H., editors, 1979, *Sociological Footprints: Introductory Readings in Sociology.* Boston, MA: Houghton Mifflin, 342–354, Figures 1–5.

Carter, H., 1976, *The Study of Urban Geography,* 2nd ed. London, UK: Edward Arnold; and New York, NY: John Wiley, 173, Figure 9-1.

_____ , 1995, *The Study of Urban Geography,* 4th ed. London, UK: Edward Arnold; and New York, NY: St. Martin's Press, 126, Figure 7.1, 132.

Chapin, F. S., 1957, *Urban Land Use Planning.* New York, NY: Harper & Brothers, 10-11, Figure 1. (Republished 1963 as 2nd ed. Urbana, IL: University of Illinois Press, 14-21, Figure 2.)

_____ and Kaiser, E. J., 1979, *Urban Land Use Planning,* 3rd ed. Urbana, IL: University of Illinois Press, 32–37, Figure 2.1. (Reprinted 1985.)

Christaller, W., 1933, *Die zentralen Orte in Süddeutschland: Eine ökonomisch-geographische Untersuchung über die Gesetzmässigkeit der Verbreitung und Entwicklung der Siedlungen mit städtischen Funktionen* [Central Places in South Germany: An Economic-Geographic Investigation of the Regularity in the Distribution and Development of Settlements with Urban Functions]. Jena, Germany: Fischer. (Unrevised reprints Darmstadt, Germany: Wissenschaftliche Buchgesellschaft, 1968, 1980.)

_____ , 1966, *Central Places in Southern Germany.* Translated by Carlisle W. Baskin. Englewood Cliffs, NJ: Prentice-Hall.

Davies, W. K. D. and Herbert, D. T., 1993, *Communities within Cities: An Urban Social Geography.* London, UK: Belhaven Press; and New York, NY: Halsted Press of John Wiley & Sons, 41.

de Blij, H. J., 1977, *Human Geography: Culture, Society, and Space.* New York, NY: John Wiley, 272–276, Figures 14-5, 14-6, and 17-7.

_____ , 1995, *The Earth: An Introduction to Its Physical and Human Geography,* 4th ed. New York, NY: John Wiley, 319, Figure 23.4.

_____ , 1996, *Human Geography: Culture, Society, and Space,* 5th ed. New York, NY: John Wiley, 409, Figure 32-4.

DeFleur, M. L., D'Antonio, W. V., and DeFleur, L. B., 1971, *Sociology: Man in Society.* Glenview, IL: Scott Foresman, 291, Figure 9.1.

_____ , _____ , and _____ , 1984, *Sociology: Human Society,* 4th ed. New York, NY: Random House, 586, Figure 15.3.

Derycke, P.-H., 1995, L'organisation de l'espace dans les villes [Organization of space in cities]. In A. Bailly, R. Ferras, and D. Pumain, editors, *Encyclopédie de Géographie* [Encyclopedia of Geography], 2nd ed. Paris, France: Economica, 662, Figure 90.

Dicken, P. and Lloyd, P., 1981, *Modern Western Society: A Geographical Perspective on Work, Home and Well-Being.* London, UK, and New York, NY: Harper & Row, Ltd., 1981, 254–258, Figure 4.19.

Dickinson, R. E., 1947, *City Region and Regionalism.* London, UK: K. Paul, Trench, Trubner, 30.

_____, 1964, *City and Region: A Geographical Interpretation.* London, UK: Routledge and Kegan Paul, 125–131, Figures 17, 18a, and 18b.

Dohrs, F. E., Sommers, L. M., and Petterson, D. R., editors, 1955, *Outside Readings in Geography.* New York, NY: Thomas Y. Crowell, 1955, 659–670.

Exline, C. H., Peters, G. L., and Larkin, R. P., 1982, *The City: Patterns and Processes in the Urban Ecosystem.* Boulder, CO: Westview, 81, Figure 13.5.

Federico, R. C., 1973, *Social Welfare Institution: An Introduction.* Lexington, MA: D. C. Heath, 94–95, Exhibit 6-4.

Fellmann, J. D., Getis, A., and Getis, J., 1990, *Human Geography: Landscapes and Human Activities,* 2nd ed. Dubuque, IA: William C. Brown, 365, Figure 11.26.

_____, _____, and _____, 1997, *Human Geography: Landscapes of Human Activities,* 5th ed. Madison, WI: Brown and Bookmark, 393, Figure 11-26.

Fielding, G., 1974, *Geography as Social Science.* New York, NY: Harper and Row, 166, Figure 5.32.

Freedman, R., Hawley, A. H., Landecker, W. S., and Miner, H. M., 1952, *Principles of Sociology: A Text with Readings.* New York, NY: Henry Holt, 395–400, Figures 35–38.

Gelles, R. J. and Levine, A., 1995, *Sociology: An Introduction,* 5th ed. New York, NY: McGraw-Hill, 573, Figure 16.6.

Getis, A. and Getis, J., 1995, *The United States and Canada: The Land and the People.* Dubuque, IA: William C. Brown, 255–259, Figures 9.14, 9.17, and 9.18.

_____, _____, and Fellmann, J., 1981, *Geography.* New York, NY: Macmillan; and London, UK: Collier Macmillan, 401, Figure 11.17.

_____, _____, and _____, 1985, *Human Geography: Culture and Environment.* New York, NY: Macmillan; and London, UK: Collier Macmillan, 277, Fig 10.21.

_____, _____, and _____, 1991, *Introduction to Geography,* 3rd ed. Dubuque, IA: William C. Brown, 393, Figure 12-24.

Gist, N. P. and Fava, S. F., 1964. *Urban Society,* 5th ed. New York, NY: Thomas Y. Crowell, 110, Figure 10.

_____ and _____, 1974, *Urban Society,* 6th ed. New York, NY: Thomas Y. Crowell, 162, Figure 6.1.

_____ and Halbert, L. A., 1956, *Urban Society,* 4th ed. New York, NY: Thomas Y. Crowell, 85.

Goldfield, D. R. and Brownell, B. A., 1979, *Urban America: From Downtown to No Town.* Boston, MA: Houghton Mifflin, 11, Figure 1.7.

Gordon, G. and Dick, W. J., 1980, *Urban Geography: Models and Concepts.* Edinburgh, Scotland: Holmes McDougall, 30.

Gould, P. R., 1969, *Spatial Diffusion.* Resource Paper, No. 4. Washington, DC: Association of American Geographers, Commission on College Geography.

Hagedorn, R., editor, 1980, *Sociology.* Toronto, Ontario, Canada: Holt, Rinehart and Winston of Canada, 537, Figure 14-7.

_____, editor, 1994, *Sociology,* 5th ed. Toronto, Ontario, Canada: Harcourt Brace, Canada, 552–553, Figure 17.5.

Hägerstrand, T., 1952, The Propagation of Innovation Waves, Lund Studies in Geography. *Series B. Human Geography,* No. 4, 1–20.

_____, 1953, *Innovationsförloppet ur Korologisk Synpunkt* [Innovation Waves from a Chorological Viewpoint]. Lund, Sweden: C. W. K. Gleerup.

_____, 1967, *Innovation Diffusion as a Spatial Process.* Postscript and translation by Allan Pred. Chicago, IL: University of Chicago Press.

Haggett, P., 1965, *Locational Analysis in Human Geography.* London, UK: Edward Arnold; and New York, NY: St. Martin's Press, 178, Figure 6.15.

_____, 1973, *L'Analyse Spatiale en Géographie Humaine* [Spatial Analysis in Human Geography]. Paris, France: Armand Colin.

_____, Cliff, A. D., and Frey, A., 1977, *Locational Analysis in Human Geography,* 2nd ed. New York, NY: Wiley, 226, Figure 6.21.

Halebsky, S., editor, 1973, *The Sociology of the City.* New York, NY: Charles Scribner's Sons, 102–115, Figures 1–5.

Hallenbeck, W. C., 1951, *American Urban Communities*. New York, NY: Harper and Brothers, 543–545, Figures 43 and 44.

Harris, C. D. and Ullman, E. L., 1945, The nature of cities. In R. B. Mitchell, editor, *The Annals of the American Academy of Political and Social Science* (special issue on Building the Future City), Vol. 242, 7–17, Figures 1–5.

Hartshorn, T. A., 1980, *Interpreting the City: An Urban Geography*. New York, NY: John Wiley, 22, Figure 2-3, and 220, Figure 11-12.

_____ , 1992, *Interpreting the City: An Urban Geography,* 2nd ed. New York, NY: John Wiley, 32, Figure 4.1, and 160, Figure 11-2.

_____ and Alexander, J. W, 1988, *Economic Geography,* 3rd ed. Englewood Cliffs, NJ: Prentice-Hall, 328–330, Figures 19.4, 19.5, and 19.6.

Hatt, P. and Reiss, A. J., Jr., editors, 1951, *Reader in Urban Sociology*. Glencoe, IL: The Free Press, 222–232, Figure 1.

_____ and _____ , editors, 1957, *Cities and Society: The Revised Reader in Urban Sociology*. Glencoe, IL: Free Press, 237–247, Figure 1.

Herbert, D. T., 1972, *Urban Geography: A Social Perspective*. Newton Abbot, UK: David and Charles, 71, Figure 1.4.

_____ and Thomas, C. J., 1982, *Urban Geography: A First Approach*. Chichester, UK, and New York, NY: John Wiley & Sons, 20, Figure 1.4.

_____ and _____ , 1990, *Cities in Space; Cities as Place*. London, UK: David Fulton, 134, Figure 5.2.

Hillman, A., 1950, *Community Organization and Planning*. New York, NY: Macmillan, 61.

Hofmeister, B., 1969, *Stadtgeographie* [Urban geography]. Braunschweig, Germany: Georg Westermann Verlag, 78, Figure 6.

_____ , 1976, *Stadtgeographie* [Urban geography], 4th ed. Braunschweig, Germany: Georg Westermann Verlag, 54, Figure 5.

Holcomb, B., 1988, Metropolitan development. In P. L. Knox, E. Bartels, J. Bohland, B. Holcomb, and R. J. Johnston, *The United States: A Contemporary Human Geography*. London, UK: Longman; and New York, NY: John Wiley, 192, Figure 7.1.

Hoyt, H., 1936–1937, City growth and mortgage risk, *Insured Mortgage Portfolio*, Vol. l, Nos. 6–10.

_____ , 1939, *The Structure and Growth of Residential Neighborhoods in American Cities* (U. S. Federal Housing Administration). Washington, DC: Government Printing Office.

Hudson, J. C., 1972, *Geographical Diffusion Theory*. Studies in Geography, No. 19. Evanston, IL: Northwestern University Press.

Jackson, J. N., 1973, *The Canadian City: Space, Form, Quality*. Toronto, Ontario, Canada: McGraw-Hill Ryerson, 124, Figure 8.3.

Jackson, R. H. and Hudman, L. E., 1990, *Cultural Geography: People, Places and Environment*. St. Paul, MN: West, 260, Figure 7-15.

Johnson, A. G., 1986, *Human Arrangements: An Introduction to Sociology*. San Diego, CA: Harcourt Brace Jovanovich, 266, Figure 9.2.

_____ , 1996, *Human Arrangements,* 4th ed. Madison, WI: Brown and Benchmark, 214, Figure 10.2.

Johnson, J. H., 1967, *Urban Geography; An Introductory Analysis*. Oxford, UK: Pergamon, 166, Figure 49.

Johnston, R. J., Gregory, D., and Smith, D. M., editors, 1986, *The Dictionary of Human Geography,* 2nd ed. Oxford, UK: Basil Blackwell, 309.

_____ , _____ , and _____ , 1994, *The Dictionary of Human Geography,* 3rd ed. Oxford, UK: Blackwell Publishers, 402.

Jordan, T. G., Domosh, M., and Rowntree, L., 1997, *The Human Mosaic: A Thematic Introduction to Cultural Geography,* 7th ed. New York, NY: Addison Wesley Longman, 434, Figure 11.17.

_____ and Rowntree, L., 1976, *The Human Mosaic: A Thematic Introduction to Cultural Geography*. San Francisco, CA: Canfield, 390, Figure 9-11.

Kammeyer, K. C. W., Ritzer, G., and Yetman, N. R,, 1994, *Sociology: Experiencing Changing Societies,* 6th ed. Boston, MA: Allyn and Bacon, 606, Figure 17-5.

Kant, E., 1951, Umland Studies and Sector Analysis. Studies in Rural-Urban Interaction. Lund Studies in Geography. *Series B. Human Geography*, No. 3, p. 8, Figure 6.

King, L. J., 1984, *Central Place Theory*. Scientific Geography Series, Vol. l. Beverly Hills, CA: Sage.

Knox, P. L., 1994, *Urbanization: An Introduction to Urban Geography*. Englewood Cliffs, NJ: Prentice-Hall, 117, Figure 5-14.

Kolars, J. F. and Nystuen, J. D., 1974, *Geography: The Study of Location, Culture, and Environment.* New York, NY: McGraw-Hill, 42, Figure 3.8.

LaGory, M. and Pipkin, J., 1981, *Urban Social Space.* Belmont, CA: Wadsworth, 91, Figure 5.5.

Lake, R. W., editor, 1997, Chauncy Harris and Edward Ullman, "The nature of cities": A fiftieth year commemoration. *Urban Geography,* Special Issue, Vol. 18, 1–35.

Larkin, R. P. and Peters, G. L., 1983, *Dictionary of Concepts in Human Geography.* Westport, CT: Greenwood, 168–170.

_____ , _____ , and Exline, C. H., 1981, *People, Environment, and Place: An Introduction to Human Geography.* Columbus, OH: Charles E. Merrill, 328, Figure 15-12.

Leser, H., Haas, H.-D., Mosimann, T., and Paesler, R., 1995, *Diercke Wörterbuch der Allgemeinen Geographie* [Diercke Dictionary of General Geography], 2nd ed. Munich, Germany: Deutscher Taschenbuch Verlag; and Braunschweig, Germany: Westermann, Vol. 2, 237.

Ley, D., 1983, *A Social Geography of the City.* New York, NY: Harper and Row, 73, Figure 3.7.

Lichtenberger, E., 1986, *Stadtgeographie. Band 1. Begriffe, Konzepte, Modelle, Prozesse* [Urban Geography, Vol. l. Ideas, Concepts, Models, Processes], Stuttgart, Germany: B. G. Teubner, 57, Figure 6.

_____ , *Stadtgeographie. Band 1. Begriffe, Konzepte, Modelle, Prozesse* [Urban Geography, Vol. l. Ideas, Concepts, Models, Processes], 2nd ed. Stuttgart, Germany: B. G. Teubner, 57, Figure 6.

_____ , 1997, Harris and Ullman's "The nature of cities": The paper's historical context and its impact on further research. *Urban Geography,* Vol. 18, 7–14.

Light, D., Jr., and Keller, S., 1975, *Sociology.* New York, NY: Alfred A. Knopf, Random House, 514, Figure 7-14.

Lloyd, J., Mack, R. W., and Pease, J., 1979, *Sociology and Social Life,* 6th ed. New York, NY: D. Van Nostrand, 349, Figure 14-6.

Lundberg, G. A., Schrag, C. C., and Larson, O. N., 1954, *Sociology.* New York, NY: Harper and Row, 151, Figure 32.

_____ , _____ , _____ , and Catton, W. R., Jr., 1968, *Sociology,* 4th ed. New York, NY: Harper and Row, 124, Figure 4.5.

Mabogunje, A. L., 1968, *Urbanization in Nigeria.* London, UK: London University Press, 178, Figure 22. (Distributed in U.S., New York, NY: Africana Publishing, 1969.)

MacIver, R. M. and Page, C. H., 1949, *Society: An Introductory Analysis.* New York, NY: Rinehart, 325, Chart XI.

_____ and _____ , 1962, *Sociology: An Introductory Analysis.* New York, NY: Holt, Rinehart, and Winston, 322–326.

Mack, R. W. and Pease, J., 1973, *Sociology and Social Life,* 5th ed. New York, NY: D. Van Nostrand, 257, Figure 10-3.

_____ and _____ , 1980, *Sociologia y Vida Social* [Sociology and Social Life]. Translated by Raúl Fernández Suárez. Mexico, DF: Union Tipografica Editorial Hispano-Americana (UTEHA), 249, Figure 10.3.

Mayer, H. M. and Kohn, C. F., editors, 1959, *Readings in Urban Geography.* Chicago, IL: University of Chicago Press, 277–286, Figures1–5.

McGahan, P., 1982, *Urban Sociology in Canada.* Toronto, Ontario, Canada: Butterworths, 183, Figure 10.

_____ , 1995, *Urban Sociology in Canada,* 3rd ed. Toronto, Ontario, Canada: Harcourt Brace, Canada, 152, Fig 7-1.

McKelvey, B., 1968, *The Emergence of Metropolitan America, 1915–1966.* New Brunswick, NJ: Rutgers Press, 111.

Morrill, R., Gaile, G. L., and Thrall, G. I., 1988, *Spatial Diffusion.* Scientific Geography Series, Vol. 10. Newbury Park, CA: Sage.

Murphy, R. E., 1966, *The American City: An Urban Geography.* New York, NY: McGraw-Hill, 215, Figure 12.4.

_____ 1974, *The American City: An Urban Geography,* 2nd ed. New York, NY: McGraw-Hill, 302, Figure 13.4.

Nader, G. A., 1975, *Cities of Canada: Theoretical, Historical and Planning Perspectives.* Toronto, Ontario, Canada: Macmillan of Canada, Vol. l, 52, Figure 2.7.

Nelson, H. J., 1969, The form and structure of cities: Urban growth patterns. *Journal of Geography,* Vol. 68, 201–204.

Nelson, R. E., Gabler, R. E., and Vining, J. W., 1995, *Human Geography: People, Cultures, and Landscapes.* Fort Worth, TX: Saunders College Publishing, Harcourt Brace College Publishers, 523, Figure 13.28.

Palm, R., 1981, *The Geography of American Cities.* New York, NY: Oxford University Press, 265, Figure 12.3.

Phillips, E. B., 1996, *City Lights: Urban-Suburban Life in the Global Society,* 2nd ed. New York, NY: Oxford University Press, 423–427, Figures 15-3, 15-4, and 15-5.

_____ and LeGates, R. T., 1981, *City Lights: An Introduction to Urban Studies.* New York, NY: Oxford University Press, 352–356, Figures 13-7, 13-8, and 13-9.

Platt, R. H., 1977, *Land Use Control: Interface of Law and Geography.* AAG Resource Paper, 75-1. Washington, DC: Association of American Geographers, 6, Figure 2.

Pokshishevskiy, V. V., editor, 1965, *Geografiya Gorodov* [Geography of Cities]. Translated into Russian by V. M. Gokhman. Moscow, Russia: Izdatel'stvo "Progress," 255–268, Figures1–5.

Pred, A. R., 1964, The intrametropolitan location of American manufacturing. *Annals of the Association of American Geographers,* Vol. 54, 171, Figure 1.

Putnam, R. G., Taylor, F. J., and Kettle, P. G., editors, 1970, *A Geography of Urban Places: Selected Readings.* Toronto, London, Sydney, Wellington: Methuen, 91–101, Figures 1–5.

Queen, S. A. and Carpenter, D. B., 1953, *The American City.* New York, NY: McGraw-Hill, 99, Figure 9. (Reprinted Westport, CT: Greenwood Press, 1972.)

Renwick, W. H. and Rubenstein, J. M., 1995, *An Introduction to Geography: People, Places, and Environment.* Englewood Cliffs, NJ: Prentice-Hall, 538–540, Figures 13-6, 13-7, and 13-8.

Rhind, D. W. and Hudson, R., 1980, *Land Use.* London, UK, and New York, NY: Methuen, 181, Figure 8.8.

Romanos, M. C., 1976, *Residential Spatial Structure.* Lexington, MA: Lexington, 46–53. Figures 3.5, 3.6, and 3.7.

Rose, A. J., 1967, *Patterns of Cities.* Melbourne and Sydney, Australia: Thomas Nelson (Australia), 161, Figure 9.1.

Rose, P. I., Glazer, M., and Glazer, P. M., 1977, *Sociology: Inquiring into Society.* San Francisco, CA: Canfield, 459.

_____ , Glazer, P. M., and Glazer, M. P., 1990, *Sociology: Understanding Society,* 3rd ed. Needham Heights, MA: Prentice-Hall, 283.

Rubenstein, J. M., 1996, *The Cultural Landscape: An Introduction to Human Geography,* 5th ed. Upper Saddle River, NJ: Prentice-Hall, 542–544, Figures 12.6, 12.7, and 12.8.

Rugg, D. S., 1971, *Spatial Foundations of Urbanism.* Dubuque, IA: Willam C. Brown, 106, Figure 4.2, and 182, Figure 5.12.

_____ , 1979, *Spatial Foundations of Urbanism,* 2nd ed. Dubuque, IA: Willam C. Brown, 124–129, Figure 4.10, and 215–222, Figure 8.1.

Schöller, P., editor, 1969, *Allgemeine Stadtgeographie* [General Urban Geography]. Darmstadt, Germany: Wissenschaftliche Buchgesellschaft, 220–237, Figures 1–5.

Schwab, W. A., 1982, *Urban Sociology: A Human Ecological Perspective.* Reading, MA: Addison-Wesley, 277, Figure 7.10.

_____ , 1992, *The Sociology of Cities,* 2nd ed. Englewood Cliffs, NJ: Prentice-Hall, 276, Figure 9.3.

Schwarz, G., 1966, *Allgemeine Siedlungsgeographie* [General Settlement Geography], 3rd ed. Berlin, Germany: Walter de Gruyter, 484, Figure 116.

_____ , 1989, *Allgemeine Siedlungsgeographie* [General Settlement Geography], 4th ed. Berlin, Germany: Walter de Gruyter, 728, Figure 127.

Schwirian, K. P., editor, 1974, *Comparative Urban Structure: Studies in the Ecology of Cities.* Lexington, MA: D. C. Heath, 217–226, Figures 1–5.

Shelley, F. M. and Clarke, A. E., 1994, *Human and Cultural Geography: A Global Perspective.* Madison, WI: Brown and Benchmark, 286–289, Figure 10-13.

Shepard, J. M., 1974, *Sociology: An Introduction.* New York, NY: Harper and Row.

_____ , 1981, *Sociology.* St. Paul, MN: West, 408, Figure 16.3.

_____ , 1993, *Sociology,* 5th ed. St. Paul, MN: West, Figure 19.1.

Small, J. and Witherick, M., 1995, *A Modern Dictionary of Geography,* 3rd ed. London, UK: Edward Arnold; and New York, NY: John Wiley, 163.

Smith, T. L. and McMahan, C. A., 1951, *The Sociology of Urban Life.* New York, NY: Dryden, 603, Figure 1.

Social Sciences Citation Index (SSCI), 1956– , *An International Multidisciplinary Index to Literature in the Social, Behavioral and Related Sciences*. Philadelphia, PA: Institute for Scientific Information.

Stoddard, R. H., Wishart, D. J., and Blouet, B. W., 1989, *Human Geography: People, Places, and Cultures*, 2nd ed. Englewood Cliffs, NJ: Prentice-Hall, 323–324. Figures 11-15, 11-16, and 11-17.

Sullivan, T. J., 1995, *Sociology: Concepts and Applications in a Diverse World*, 3rd ed. Boston, MA: Allyn and Bacon, 410, Figure 14.7.

_____ and Thompson, K. S., 1984, *Sociology: Concepts, Issues, and Applications*. New York, NY: John Wiley, 354, Figure 11.7.

Teevan, J. J., 1982, *Introduction to Sociology: A Canadian Focus*. Toronto, Ontario, Canada: Prentice-Hall of Canada, Figure 15.3.

_____ and Hewitt, W. E., 1995, *Introduction to Sociology: A Canadian Focus*, 5th ed. Scarborough, Ontario, Canada: Prentice-Hall of Canada, 504, Figure 16.2.

Thio, A., 1986, *Sociology*. New York, NY: Harper and Row, 480. Figure 21.1.

_____ , 1996, *Sociology*, 4th ed. New York, NY: HarperCollins, 512–513, Figure 22.2

Tidswell, V., 1976, *Pattern and Process in Human Geography*. London, UK: University Tutorial Press, 232, Figure 13.1.

Tischler, H. L., 1996, *Introduction to Sociology*, 5th ed. Fort Worth, TX: Harcourt Brace, 536–537, Figures 16.1, 16.2, and 16.3.

_____ , Whitten, P., and Hunter, D. E. K., 1983, *Introduction to Sociology*. New York, NY: Holt, Rinehart, and Winston, 206–207, Figures 7.1, 7.2, and 7.3.

Turner, B. L., II, 1988, Whether to publish in geography journals. *Professional Geographer*, Vol. 40, 15–18.

_____ and Meyer, W. B., 1985, The use of citation indices in comparing geography programs: An exploratory study. *Professional Geographer*, Vol. 37, 271–278.

Ullman, E. L., 1941, A theory of location for cities. *American Journal of Sociology*, Vol. 46, 853–864.

van der Merwe, I. J. and Nel, A., 1975, *Die Stad en sy Omgewing: 'n Studie in Nedersettingsgeografie* [The City in Its Region: A Study in Settlement Geography]. Stellenbosch, South Africa: Universiteits-Uitgewers en Boekhandelaars.

Vander Zanden, J. W., 1979, *Sociology*, 4th ed. New York, NY: John Wiley, 601–605, Figure 17.2.

_____ , 1996, *Sociology: The Core*, 4th ed. New York, NY: McGraw-Hill, 386, Figure 11.8.

Wehrwein, G. S., 1942, The rural-urban fringe. *Economic Geography*, Vol. 18, 219–220.

Wheeler, J. O. and Muller, P. O., 1981, *Economic Geography*. New York, NY: John Wiley, 135, Figure 6.6.

_____ and _____ , 1986, *Economic Geography*, 2nd ed. New York, NY: John Wiley, 144, Figure 6.6.

Whitehand, J. W. R., 1984, The impact of geographical journals: A look at the ISI data. *Area*, Vol. 16, 185–187.

_____ , 1985, Contributors to the recent development and influence of human geography: What citation analysis suggests. *Transactions, Institute of British Geographers*, Vol. 10, 222–234.

_____ , 1990, An assessment of "Progress," *Progress in Human Geography*, Vol. 14, 12–23.

_____ and Edmondson, P. M., 1977, Europe and America: The reorientation in geographical communication in the post-war period. *Professional Geographer*, Vol. 29, 278–282.

Wrigley, N. and Matthews, S. A., 1986, Citation classics and citation levels in geography. *Area*, Vol. 18, 185–194.

_____ and _____ , 1987, Citation classics in geography and the new centurions: A response to Haigh, Mead, and Whitehand. *Area*, Vol. 19, 279–284.

Yeates, M. H., 1990, *The North American City*, 4th ed. New York, NY: HarperCollins, 113, Figure 4.7.

_____ , 1998, *The North American City*, 5th ed. New York, NY: Longman, 209, Figure 7.6.

_____ and Garner, B., 1971, *The North American City*. New York, NY: Harper and Row, 62, Figure 3.1, and 247, Figure 9.6.

Zimolzak, C. E. and Stansfield, C. A., Jr., 1979, *The Human Landscape: Geography and Culture*. Columbus, OH: Charles E. Merrill, 245, Figure 8-8.

# CHAPTER 4

# Some Thoughts on the Development of Urban Geography in the United States during the 1950s and 1960s[1]

EDWARD J. TAAFFE

Urban geography evolved in the late 1940s from a mixture of traditional geography, Chicago School urban sociology, and city planning. In the 1950s, *Inventory and Prospect* first reflected a similar mixture in separate chapters; then the Mayer-Kohn readings pulled several strands together into a dominantly spatial framework which had considerable impact on the role of urban geography in the curriculum as did the later Murphy textbook organized along similar lines. During the 1960s, urban research focused on more explicitly spatial, quantitative, and theoretical concerns, which then merged with the mainstream model in the early 1970s as evidenced by the BASS committee report, the Berry-Horton reader, and the Yeates-Garner textbook. The resulting blend of spatial theory, traditional geography, and planning implications appears to have continued in the curriculum even though urban geographic research has become more divergent and pluralistic.

Although my own work has not been primarily in the field of urban geography, I have been involved in the field over the last forty years as a student, teacher, and rapporteur as well as in the study of intercity air transportation and commuting.

My first exposure to urban geography came in 1949 as a graduate student at the University of Chicago. At Chicago, I took Charles Colby's urban geography course and Albert Riess's urban sociology course. In addition, Jerome Fellman and I decided to take an evening course in urban planning, given by the Northwestern Geography Department and taught by the then-Director of Research at the Chicago Planning Commission—Harold Mayer. William Garrison, then a graduate student at Northwestern, was also in the class.

I had come into graduate work as a nongeography major, and it was interesting to get a look at this one subfield of geography from what turned out to be three quite different perspectives. My undergraduate degrees were in meteorology and in journalism and I had had virtually no exposure to geography. I now realize that the Colby course reflected his own Barrowsian man-land inclinations as well as a stress on regional study which was then dominant in the discipline. The physical bases for city location were emphasized, as were detailed discussions of individual cities, particularly their site characteristics. The main exceptions were those sections of the course dealing with Colby's own work on centripetal-centrifugal forces and with the work of his mentor, Mark Jefferson.

The sociology course was about 50% what we would now consider to be urban geography—patterns of land use, the urban hierarchy, etc. Reference was also made to Robert Dickinson and to Riess's sociology colleagues in the Chicago School of Sociology such as Park and Burgess. At the outset, Riess indicated that he was not going to cover the geography of the city—by which it turned out that he meant the *physical* geography of the city. Mayer also referred to Dickinson and to the Chicago School. His 1949 course was organized in a fashion similar to what I will identify for the purposes of this paper as a "mainstream" of urban geography. There were considerations of city classifications, city-size formulations, the three basic land-use theories, economic base ideas, central place, and a consideration of planning problems associated with land use in the city, approached somewhat more analytically than was true of the sociology course. In both the geography and the sociology courses, however, there was no absence of process concerns as later retrospective statements would have us believe.

In the early 1950s, Mayer came to the Chicago geography department, and I took a position in the Loyola University Department of Economics teaching geography, statistics, and economics. He served as my adviser on an air transport dissertation dealing with Chicago's hinterland/hierarchy relations, and I sat in on several of his classes. In 1954, *American Geography: Inventory and Prospect* (James and Jones, 1954) was published. Mayer chaired the committee writing the urban chapter, which was quite similar to his courses at Northwestern and Chicago. Other aspects of urban geography were treated in other chapters, however. In the settlement chapter, Clyde Kohn discussed Christaller; Edward Ullman discussed the gravity model in the transportation chapter; and Chauncy Harris discussed industrial location factors in the manufacturing chapter.

The Mayer-Kohn book of readings in 1959 marked a pulling-together of the separate strands of the mainstream noted above, establishing the basic structure of a university course in urban geography which, in my opinion, has persisted into the 1980s. It was clearly different in organization and in concepts treated from Eugene Van Cleef's *Trade Centers and Trade Routes* (1937) and Robert Dickinson's *City, Region and Regionalism* (1947), which had previously been used as urban-geography texts.

The Mayer-Kohn book and other selected texts are indicated in the diagram (Figure 4.1), and I will refer to them in the remainder of this discussion. Here I follow Thomas Kuhn's dictum (1962) that, at any given time, textbooks summarize the working paradigms of a discipline. Noting the concepts and ideas treated in textbooks as they develop through time is, therefore, one way to trace the intellectual history of a field. According to Kuhn, the development of a field is thus linearized in terms of its currently dominant paradigms and previous paradigms are rendered invisible. In the Mayer-Kohn book, concepts and topics were similar to the 1949 course and to the various *Inventory and Prospect* chapters. A rough separation was emerging, however, between those aspects of urban geography which were clearly *intra*-city and those which were not so confined, and there were several additions. These included expansions on economic-base ideas including Walter Isard's interregional input/output analysis; new classifications including CBD classifications; more varied transportation and hinterland studies, reflecting the increasingly strong heritage of Platt's functional region ideas; more urban fringe studies; and some early Berry-Garrison work involving preliminary attempts to revise Christaller's central-place concepts.

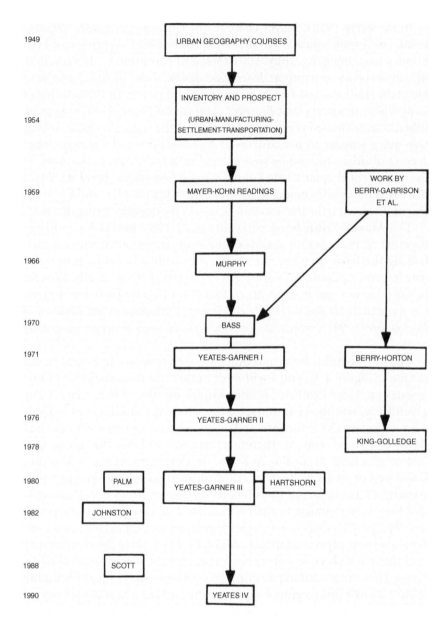

**Figure 4.1** This diagram represents one geographer's perception of the emergence of a "mainstream" in U.S. urban geography in the 1950s, 1960s, and 1970s, primarily as evidenced by textbooks.

In 1966, Raymond Murphy's *The American City* provided a further coalescence of urban geography. The mainstream structure and the set of articles relied upon were quite similar to Mayer-Kohn but now in the form of a textbook rather than a book of readings. The main differences were that topics such as central-place theory were given more extended treatment, and a number of local examples were added. In that connection it is interesting to observe that the link between urban and regional geography which had been noted in the Colby course was also evident in the frequent use by Murphy of Worcester examples and by Mayer-Kohn as well as by Berry of Chicago examples. As Murphy pointed out, this tendency also exemplified the idea of the region—in these cases, a city—as a laboratory.

Meanwhile, I had left Loyola and come to Northwestern University in 1956, initially in transportation. After Clyde Kohn left for Iowa, I added an urban emphasis to my work. During the Northwestern period, I worked on both urban and transportation questions with such talented graduate students as Barry Garner, Howard Gauthier, Peter Gould, Maurice Yeates, and others. The Mayer-Kohn book of readings was our basic text, but there were a number of other strong influences on all of us. The Lund studies, particularly those by Hägerstrand and Godlund, showed us the effectiveness of combined cartographic and quantitative analysis. The work of the newly formed Regional Science Association also had an influence. I had become interested in the work of Walter Isard during my five years in the Loyola Economics Department and, aided by such regionally inclined Northwestern economists as Charles Tiebout and Leon Moses, the graduate students and I made considerable use of the regional science publications. Perhaps the major influence, however, was the set of University of Washington discussion papers put out by William Garrison and such students as Brian Berry, Duane Marble, and Richard Morrill. These strongly analytical studies focused largely on urban topics. We first used them in the Northwestern field camp and were able to blend field-study empiricism and the theory associated with the early spatial models quite readily. Later, the Washington influence was intensified as Berry came to Chicago and first Morrill, then Garrison came to Northwestern. Finally, toward the end of my Northwestern stay, I directed two National Science Foundation quantitative institutes together with Berry, with Garner and Yeates as graduate assistants. The institutes brought together a number of the new approaches in a fashion which could readily be transmitted to the 60 or so participating geographers, many of whom were urban specialists. In 1963, I moved to Ohio State University; for several years, my involvement in urban geography consisted of discussions and seminar participation with Leslie King, Reginald Golledge, Howard Gauthier, and Lawrence Brown, each of whom represented a quite different perspective on the field.

In 1967, a role as rapporteur began with my appointment to the BASS Committee. The BASS or Behavioral and Social Science Committee was formed as a major collaborative, multidisciplinary effort by the National Science Foundation—Social Science Division, the Social Science Research Council, and the National Academy of Science-National Research Council, Behavioral Science Division. The charge was to develop reports on each of the social sciences indicating their recent research directions, with particular emphasis on the possible practical significance to society of their work. I was chairman of the geography panel, which included Ian Burton, Norton Ginsburg, Peter Gould, Fred Lukermann, and Philip Wagner. I met with them for two years as well as with the central committee of panel chairmen from the other social science disciplines. As was the case in the 1965 *Science of Geography* report, sponsored by the National Academy of Science-National Research Council, Earth Sciences Division, we had the thankless, not to say life-threatening, task of selecting certain fields which we felt had the greatest research promise, rather than of surveying the entire discipline as was true of the many committees involved in preparing *Inventory and Prospect* in 1954 or the recent *Geography in America* survey (Gaile and Wilmott, 1989). In 1965, the *Science of Geography,* confined to the sciences, selected physical geography, cultural geography, location theory, and political geography. In 1970, confined further to the social sciences, we selected locational analysis and cultural geography as major conceptual groupings and identified two fields in which those concepts were being actively applied: environmental and spatial behavior and urban geography (Figure 4.2). In a survey which we made in 1968 of all chairs of graduate departments of geography, urban geography was identified as the fastest growing subfield, and we felt it had particularly strong theoretical, practical, and curricular possibilities. Several of us felt that the increasing prominence of urban geography in the curriculum had important implications for geography as a discipline. The mainstream model exemplified by Mayer-Kohn (1959) and Murphy (1966) was providing a basis for the coalescence of an identifiable cluster of concepts and themes, similar to that provided in introductory economics by the Samuelson text (1948) and in introductory physical geography by the Finch and Trewartha (1936) and later Strahler (1978) texts, although the urban work was at a much-reduced scale and provided a less explicit curricular model.

Figure 4.2 also shows how we structured the urban discussion. The between-city, within-city dualism suggested by Mayer-Kohn and Murphy was by now being referred to as systems of cities versus the city as a

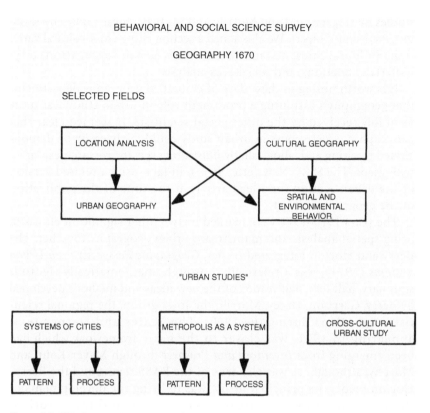

**Figure 4.2** This diagram shows the role of urban geography in the BASS report as well as the way in which the urban discussion was organized.

system. For each category we attempted to separate pattern and process, reflecting our belief that geographic research had been evolving from a pattern toward a process emphasis during the 1960s. In fact, the main definitional difference between the BASS report, *Geography* (Taaffe, 1970), and *The Science of Geography* (National Academy of Sciences-National Research Council, 1965) report lay in the more explicit addition of process to the former. The writing was aided greatly by position papers solicited from over 40 geographers who had made major recent contributions to geographic research. Many of the ideas in the previous studies were repeated, such as central place, size and functional studies, hinterland and linkage studies, and the land-use theories. Major additions were (1) the work of Brian Berry and his colleagues, notably in more conceptually ramified central-place studies, intra-city retail patterns, social-area factorial ecology, and commuting studies; (2) Borchert *et al.* and the Upper Midwest studies; (3) diffusion studies by Hägerstrand and Morrill; and (4) behavioral studies by Rushton, Pred, and Wolpert. We also added a section citing cross-cultural variations in hierarchies, distance decay, intra-urban segregation, city-hinterland relations, and social-area analysis.

It is worth noting in these days of embattled geography departments that geography's assuming a prominent role in urban study was quite favorably received by the other social scientists. It seemed clear that geographic effectiveness in urban study could fairly easily be demonstrated by citing and, above all, by *illustrating* specific research examples with maps. The SSRC Newsletter *Items,* in fact, used a revised version of the urban section for a lead article at the time of the publication of the geography report.

The years 1970–1971 were marked by two other significant efforts to bring spatial analysis and mainstream urban geography together. The Berry and Horton integrated reader, *Geographic Perspectives on Urban Systems* (1971), was a relatively thorough, mathematically rigorous summary, half text, half reader, of the new ideas and methods developed by Berry, Garrison, Dacey, Morrill, the Iowa group, the regional scientists, and others during the 1960s. The Yeates and Garner text, appearing in 1971, was closer to the basic framework which had been emerging from *Inventory and Prospect* through Mayer-Kohn and Murphy, although, as was also true of the BASS report and the Berry-Horton reader, its principal thrust was to bring the developments of the 1960s into this mainstream. The dichotomous framework of urban systems and internal structure was used, and a third section on problems, entitled "The Urban Dilemma," was added.

Thus, by 1971, there seemed to be a convergence of spatial analysis and a mainstream model as noted in three quite different books, a committee report, an introductory urban-geography textbook, and a collection of recent research articles. Since then, my own relatively cursory observations suggest that at least the textbook evolution of urban geography in the United States has reflected a continuation of this modified mainstream in the three subsequent Yeates-Garner and Hartshorn textbooks as well as the most recent Yeates book. The disciplinary context became more pluralistic, however, first in research and later in textbooks. A somewhat more quantitative line continued as represented by the King and Golledge (1978) and Cadwallader (1985) texts as well as by a large volume of research. David Harvey's *Social Justice in the City* (1973) represented an early Marxist interest in urban study and other approaches reflected strengthened economic, historical, and social-theoretic approaches. Risa Palm's *The Geography of American Cities* (1981) took an ecological-structuralist approach; R. J. Johnston's *The American Urban System* (1982) took a political-economy approach; Allen J. Scott's *Metropolis: From the Division of Labor to Urban Form* (1988) took an approach emphasizing the role of production and the division of labor in shaping cities in capitalist societies; other books focused separately on such things as social or behavioral approaches. In addition, later editions of the essentially mainstream texts showed further modifications in that model, this time in the direction of greater emphasis on historical and social concerns and less emphasis on some of the earlier topics—although most are still included.

It is of some interest to note the retrospective view of urban geography of the late 1950s and the 1960s which appears in many of the texts as well as in sections of *Geography in America*. In general, one might observe that such retrospective statements seem to follow a variant on Kuhn's dictum. Rather than render previous paradigms invisible, there is a tendency to convert them into straw men. The more recent statements announce an emphasis on concepts and theory as opposed to the descriptive approaches of the past. A glance at the preface of the Mayer-Kohn book, however, reveals that 30 years ago the authors announced, in effect, that they were emphasizing concepts and theories instead of the descriptive approaches of the past. The recent statements draw a rather sharp distinction between a concern in the 1960s with pattern and description almost to the exclusion of process and a concern in the later periods with process and theory. A common characterization of the 1960s is that they were an era dominated by morphological pattern description—initially verbal, then lent some temporary scientific stature with the infusion of a complex set of mathematical methods, most of which turned out to be conceptually sterile geometries. To those who had any involvement in the period, this greatly oversimplifies a complex and significant era in the evolution of U.S. geographic thought. As a pigeonhole label it seems to be taking its place, however, with the facile but misleading identification of Richard Hartshorne with the unique, first put forward by the quantitative geographers, or perhaps the equally misleading epitomization of Ellen Churchill Semple's entire outlook in the "dust-of-her-dust" quotation put forward by geographers of the 1930s and 1940s. It may also reflect the tendency to overstate the influence on the geographers of the 1960s of Fred Schaefer, who did indeed explicitly reject processes in 1953 in favor of morphological laws (Schaefer, 1953).

Although there was considerable mapping and measurement of empirical patterns in the 1960s and earlier, there was a growing concern throughout the period with the need to develop process statements, and there was little tendency to be satisfied with morphology whether verbally or mathematically described. In 1963, in fact, Ian Burton felt it was possible to declare the quantitative revolution over and state that it was time to emphasize the development of theory (Burton, 1963). Judging from our essentially retrospective BASS survey in the late 1960s, there was an increasing tendency through the 1960s to put forth various kinds of process statements in the sense of looking for the forces and behavior which seemed to be associated with observed patterns. At first these consisted of economic processes, planning policies, and historical trends. Later, the concepts of the types of behavior associated with spatial pattern broadened to include social, political, and psychological ideas. Most of the attempts in the 1960s were essentially inductive in nature, in line with geography's long empirical tradition, but deductive approaches were also present as early as 1953, with John Brush's applications of Christaller in southwest Wisconsin (Brush, 1953). There did seem to be a generally observable sequence during the 1950s and 1960s, however. First, a more explicit concern with spatial organization replaced a stress on areal differentiation. The emphasis was on a consideration of the spatial expression of a relatively small set of phenomena rather than an effort to attain a wider-ranging synthesis of a large number of characteristics of a particular urban place. Then, more precise mathematical descriptions of spatial patterns facilitated conceptualization and manipulation, which, in turn, led to attempts to identify processes. Attempts were made by some to link these processes to each other and to form middle-level generalizations or to make efforts to posit broader theories. Later this included the recognition that the process-pattern linkage is much more complex and inconsistent than was originally thought and that there was no generally agreed-upon meaning to a process/pattern dualism itself.

If it is felt useful or even possible to attach a philosophical label to such an era, the much-abused term "positivistic" is probably no worse than any other, as the above discussion suggests. The usefulness of such labeling may be questioned, however, on more grounds than the scope of this brief essay permits. Suffice it to say that the term as used by geographers seems to carry a wide variety of meanings ranging from Comte through Carnap to early Harvey. It also has been applied, with the all-purpose prefix "neo," to researchers who show little or no confidence in the possibility of value-free research, absolute standards of verification, the existence of true laws, or the truly objective nature of reality.

At any rate, it will be interesting to observe the path of urban geography in the 1990s, particularly in the curriculum. For a number of years the mainstream model provided a loose-knit consensus as to the ideas about cities that geography as a discipline wanted to convey to the general public. The model evolved to absorb first spatial analysis, then, to some extent, social and behavioral concerns, and it may be flexible enough to continue to absorb some of the pluralistic currents while retaining its identity and its place in the curriculum. Alternatively, there may be a breakup into a number of small courses and seminars, each representing a perspective on the subfield based on a different research cluster. Such a loss of focus would make a more general urban geography course more difficult and might even mean that the subfield would eventually find its main curricular expression in the world-cities variant of the introductory world survey course, thereby offering relatively limited opportunities for theoretical exposition but, in turn, strengthening the academic credentials of an increasingly popular type of course which meets a clear societal need.

## Notes

1. Assistance from W. Randy Smith, Howard L. Gauthier, and Thomas Klak is gratefully acknowledged as is Marilyn Raphael's work on the illustrations.

## Literature Cited

Berry, Brian J. L. and Horton, Frank, 1971, *Geographic Perspectives on Urban Systems*. Englewood Cliffs, NJ: Prentice Hall.

Brush, John E., 1953, The hierarchy of central places in southwestern Wisconsin. *Geographical Review*, Vol. 43, 380–402.

Burton, Ian, 1963, The quantitative revolution and theoretical geography. *The Canadian Geographer*, Vol. 7, 151–162.

Cadwallader, Martin, 1985, *Analytical Urban Geography*. Englewood Cliffs, NJ: Prentice Hall.

Dickinson, Robert E., 1947, *City, Region and Regionalism*. London.

Finch, Vernor C. and Trewartha, Glenn T., 1936, *The Elements of Geography*. First edition. New York: McGraw-Hill.

Gaile, Gary L. and Willmott, Cort J., 1989, *Geography in America*. Columbus: Merrill.

Hartshorn, Truman A., 1980, *Interpreting the City: An Urban Geography*. New York: Wiley.

Harvey, David, 1973, *Social Justice and the City*. London: Edward Arnold.

James, Preston E. and Jones, Clarence F., editors, 1954, *American Geography: Inventory and Prospect*. Syracuse, NY: Syracuse University Press.

Johnston, Ronald J., 1982, *The American Urban System: A Geographical Perspective*. New York: St. Martin's Press.

King, Leslie J. and Golledge, Reginald G., 1978, *Cities, Space and Behavior: The Elements of Urban Geography*. Englewood Cliffs, NJ: Prentice Hall.

Kuhn, Thomas, 1962, *The Structure of Scientific Revolutions*. Chicago: The University of Chicago Press.

Mayer, Harold M. and Kohn, Clyde R, editors, 1959, *Readings in Urban Geography*. Chicago: The University of Chicago Press.

Murphy, Raymond E., 1966, *The American City: An Urban Geography*. New York: McGraw-Hill.

National Academy of Sciences-National Research Council, 1965, *The Science of Geography*. Washington, DC.

Palm, Risa I., 1981, *The Geography of American Cities*. New York: Oxford University Press.

Samuelson, Paul A., 1948, *Economics: An Introductory Approach*. First edition. New York: McGraw-Hill.

Schaefer, Fred K., 1953, Exceptionalism in geography. *Annals of the Association of American Geographers*, Vol. 43, 226–249.

Scott, Allen J., 1988, *Metropolis: From the Division of Labor to Urban Form*. Berkeley: University of California Press.

Strahler, Arthur N., 1978, *Modern Physical Geography*. First edition. New York: Wiley.

Taaffe, Edward J., editor, 1970, *Geography: Report of the Behavioral and Social Sciences Survey*. Englewood Cliffs, NJ: Prentice Hall.

Van Cleef, Eugene, 1937, *Trade Centers and Trade Routes*. New York: Appleton-Century Crofts, Inc.

Yeates, Maurice and Garner, Barry, first edition, 1971, *The North American City*. New York: Harper and Row.

———, 1990, *The North American City*. New York: Harper and Row.

# II

# Urban Geography in the 1960s

# II

## Urban Geography in the 1960s

# CHAPTER 5

# Geography's Quantitative Revolution
## *Initial Conditions, 1954–1960.*
## *A Personal Memoir*

BRIAN J. L. BERRY

Humor me. I have avoided talking about the early days of geography's "quantitative revolution," but it is now close to 40 years since Henry Clifford Darby urged me to think about doing graduate work in the United States, and oral history does have value to those who write about disciplinary development. What I have to recount is a very personal view that will certainly differ from those of others. Think of the stories told by the participants in Kurosawa's *Rashomon* and accept my remarks for what they are, the selective recall of events now four decades old.

When Clifford Darby suggested that I go to the United States for graduate work, he did so because there were no graduate programs as such at the time in England. We talked about my interests in economic history and the theory of location, and he pointed me to programs and people he thought would meet my needs—to Chicago and Wisconsin, Minnesota and Washington,

*Urban Geography*, 1993, 14, 5, pp. 434–441.

and to Harris and Hartshorne and Ullman. Dutifully, early in 1954, I applied to all four universities. Chicago quickly rejected my application: "The faculty did not think I was well-suited to the program of studies the department offered." For the others I had to wait. In late fall I received a letter of admission and an offer of a teaching assistantship from G. Donald Hudson at the University of Washington. He asked for an immediate response and of course I said yes, overjoyed after the Chicago rejection. Little did I know about A.A.U.P. agreements. Donald consistently jumped the gun to get the students he wanted, and this had much to do with the pool of talent that was gathered in Seattle. It was another five months before Wisconsin wrote to say that I was on a waiting list for an assistantship if one should become available. Minnesota did offer me a fellowship, but I had long since made my commitment.

It was to be Washington and Ullman. Clifford Darby helped me obtain a Fulbright travel scholarship, and also arranged for me to attend a Fulbright summer school in Anglo-American studies at Oxford University in the summer of 1955. Robert D. Campbell, then at George Washington University, was the American geographer in residence, and he introduced me to geography in the United States via the newly published *American Geography. Inventory and Prospect* (Syracuse, NY: Syracuse University Press for the Association of American Geographers, 1954). One of the book's conclusions resonated to a young man whose undergraduate degree was the B.Sc. (Economics), not an arts degree in geography: "This inventory of the experience of the past few decades is undertaken for the purpose of setting up guidelines to point toward the frontiers. There is need that the advance toward those frontiers be based on sound theory and acceptable practice" (p. 17). I also felt very comfortable with the argument in the book that the "various kinds of duality which have been popular in the past, such as regional as opposed to topical geography, or physical as opposed to human geography, seem to have obscured ... the true nature of the discipline ... Actually, there is just one kind of geography" (p. 15).

These were articles of faith that I carried with me as I crossed the Atlantic in September 1955 on the *Queen Mary* and then took a continent-wide train ride from New York to Seattle via Chicago. In my suitcase I had the well-marked copy of *American Geography* that Bob Campbell had given to me, plus copies of Edgar Hoover's volume *The Location of Economic Activity* (New York: McGraw Hill, 1948) and the newly translated treatise by August Lösch, *The Economics of Location* (New Haven: Yale University Press, 1954). I already knew about von Thünen and Weber and distance-decay and hexagonal landscapes; Bill Mead and Brian Law had introduced them to me at University College.

When I arrived in Seattle I was met at the railroad station by my friend from London, Derrick Sewell, with whom I would share a dormitory room for the next two years, and by Duane Marble. Duane was the point man for Bill Garrison and was quick to tell me that Ullman was gone until Christmas and that I probably wouldn't want to study with him in any case. Bill Garrison, he said, was the person I should get to know. Donald Hudson, he said, was a man to be reckoned with, running things with an iron hand.

The next day Derrick took me to the department where I was to have an initial interview with Donald Hudson. On the way, I met other newly arrived students — Dick Morrill, John Nystuen, and others. Shepherded into Hudson's office, I was told that I would be a teaching assistant for Howard H. Martin's course in introductory economic geography, meeting lab sections in the early morning hours twice a week. He also said that since my undergraduate degree did many things other than geography, I needed to take some "real" geography courses. I demurred, told him what I had studied at University College, what I was interested in, and what I wanted to do. He gave me a very long hard stare and I went into a cold sweat. But then he leaned back in his chair, guffawed, slapped his thigh, and never again told me what to do. He readily concurred when I said I wanted to take Professor Garrison's brand new course in statistical inference for geographers—Duane Marble had urged this—and Professor Marion Marts's course on the planning and evaluation of water resource projects—Derrick Sewell's recommendation. Other students, I learned, were not so fortunate.

There began a process of acculturation; the natives *were* different. Howard Martin, an old-school give-'em-the-facts geographer, was the former department chair, replaced by and resentful of Hudson and the source of one set of tensions in the department. He had very poor eyesight, and I made a little extra pocket money grading examinations for his other regional and historical geography courses. Donald Hudson had created an archetypal department of the time—a course for each systematic field and each major world region, supported by introductory human and economic geography as required courses for the education and business schools, and physical geography to enable arts majors to satisfy physical science credits. The graduate students were the teaching assistants.

Our cohort entered the department as the last of the GI Bill students who had been admitted by Martin were receiving their Ph.Ds. We represented both a generational and an intellectual shift. We were housed in a very large room, dubbed by the ex-warriors "The Citadel," because Martin used to stride in there to give his folks their marching orders. It was there that a debate was already raging. The issue was one of nomothetic vs. idiographic geography, already articulated by Fred K. Schaefer in his criticism of Hartshorne's *Nature of Geography* ("Exceptionalism in Geography: A Methodological Examination," *Annals of the Association of American Geographers* Vol. 43 (1953), 226–49).

At the heart of the debate was Bill Garrison, recently returned from the University of Pennsylvania, where he had become involved with Walter Isard's call for a new field of Regional Science. Upon his return he began working with the civil engineers Robert Hennes and Edgar Horwood on funded highway research—Duane Marble was crunching numbers for him—so links between theory and practice were already being forged. The group that formed around him was well aware of quantitative stirrings elsewhere: John Weaver's use of descriptive statistics in agricultural classification, Howard Nelson's uses of similar statistics in urban classification, Chauncy Harris's work on population potentials, Bill Warntz's attempts to craft a macrogeography out of social physics, and Harold McCarty's flirtations with correlation coefficients as substitutes for visual map comparisons.

We were not alone: The sociology department at Washington had the most rabid of social physicists, Stuart Carter Dodd, whom I visited a number of times in the next two years. By then he was lonely, embittered, and largely ignored, yet he always gave me a royal welcome. Sociology also had the urban ecologist Calvin Schmid, who was interested in mapping the social geography of cities. A couple of years later I learned about the use of social area analysis from his student Maurice Van Arsdol in chance encounters at the university's computer lab.

It soon became clear that Garrison was the man with the ideas. His course on statistics was tough both because of his opaque teaching style and because he tried to go from means, medians, and modes to variance-covariance and multiple regression in a twelve-week academic quarter—theory, proofs, and applications! At the same time we learned machine-language programming on an IBM 604 computer, a year before the 32K IBM 1620 arrived complete with Fortran I. Most of the time, we crunched our numbers on desk calculators, inverting matrices by hand using the elimination method. We survived by developing a bootstrap help-each-other group that worked hard together, played hard together, and took great pride in seeking out analytic approaches to geographical problems—E.G. Ravenstein, G.K. Zipf, Colin Clark, etc.—and teaching each other about them. New discoveries were quick to circulate and were read with enthusiasm. We reached out wherever, whenever, and for whatever we could. Duane had spent a summer at an S.S.R.C. summer institute at Stanford on mathematics for social scientists and was working on an MA thesis that recast the Thünen agricultural location model using symbolic logic. He and Garrison subsequently published the results in the *Annals*. Duane had learned matrix algebra and taught it to the rest of us.

This is not to say that we were not competitive; we were, frequently intensely so. But we were welded into a community by a split among both faculty and students, with the differentiating feature the use of quantitative methods to solve geographic problems. There were heated debates about the pros and cons of both theory and statistics in geography, called familiarly the "battle of the mimeograph machine." Donald Hudson left the mimeo room open and the machine stacked with paper. The graduate students were encouraged to carry on their debates in writing as well as over coffee in the mornings, and continued at the Red Robin late into the night.

Ullman came back at Christmas 1955 and I took his course in urban geography, writing a long paper for him on urban location. I found the experience less than satisfactory, and was drawn more closely into the Garrison sphere. Ed was in love with himself and found graduate students a distraction. Other faculty made no bones about their disapproval of the "new geography," demanding that students "give them the facts" in oral examinations—none of this theoretical nonsense: Theory was determinism and geography had rejected determinism. This flew in the face of the admonitions I had taken to heart in 1954. Because of the opposition, our beliefs intensified. We came to believe in a cause that was new and that we knew was right. External threat is a powerful motivator.

The congealing interests of Garrison's "space cadets"—so dubbed by Joe Spencer at a Pacific Coast geographers' meeting—were reinforced in the third quarter of the 1955–1956 academic year by a visit of J. Ross Mackay from the University of British Columbia and the arrival on campus of a young new econometrician. Ross taught a course in statistical cartography, and we were enthusiastic consumers. The young econometrician was Arnold Zellner, and Marble, Morrill, Nystuen and I took his course along with one economics student. It was the first time Arnold had taught, and he went through Lawrence Klein's advanced text at breakneck speed. We were shell-shocked but survived; the economist didn't.

The year was soon over, and the battle lines were clearly drawn. But what to do in the summer? Contract research on a policy problem came to the rescue. Garrison and the civil engineers signed a contract to prepare a civil defense "survival plan" for the State of Washington, and our group spent the summer traveling the state gathering data. John Nystuen and I collaborated on the problem of emergency food supplies. By summer's end, a fat report had been prepared, and was for a while given a "Secret" classification. This was, after all, the height of the Cold War and the fear of nuclear Armageddon.

John and I wrote of the problem of people who might "survive a quick death from the bomb only to starve at leisure."

The experience did much for group cohesion, but I had a worry: What to do about a master's thesis? Duane Marble again came to the rescue. During one of Garrison's projects with the civil engineers, an attempt to assess the benefits of rural roads to rural property, they had surveyed the settlements in Snohomish County, just to the north of Seattle, inventorying types of business and numbers of stores in each of the county's nucleations. The project hadn't been able to make use of the data, Duane said, and since they were not to be used, I was welcome to them.

The paper that I had written for Ed Ullman suggested a problem: Was there a central-place hierarchy of the kind proposed by Walter Christaller? I decided that I must read Christaller's book, which had not yet been translated, although I knew of it from Lösch and from Ullman's article "A Theory of Location for Cities." Interlibrary loan found me a copy of the original German edition at the University of Chicago. When it arrived I discovered that the last person to check it out had been Ed Ullman more than sixteen years earlier!

Maybe I could devise a test of whether there was a central-place hierarchy. If there were hierarchies there would be persuasive support for a more nomothetic geography. The summer report finished, I tussled with ways to tackle the problem and came up with what, in hindsight, was a very primitive approach, but I didn't know any better at the time, and neither did anyone else. Three weeks of number crunching and writing and the thesis came together. Bill Garrison's help improved the draft. Bill found it hard to communicate with large groups of students in introductory undergraduate classes, but was at his very best, full of creative suggestions, in the one-on-one that is so important to the novice researcher.

I received the degree in the fall of 1956, and knowing that Garrison and Marble had an article forthcoming in the *Annals*, I proposed to Bill that we write a group of papers based on the thesis. In all, four were written. One, "Recent Developments of Central Place Theory," was read by Garrison at the 1957 Regional Science Association meetings and published in the *Papers*. Three, "The Distribution of City Sizes," "Functional Bases of the Central-Place Hierarchy," and "Central Place Theory and the Range of a Good," were submitted to the *Geographical Review* and promptly rejected: Wilma Fairchild said they were too mathematical—definitely "not geography." Suspicions about "us" versus "them" bubbled to the surface, but, pressing on, the three drafts were submitted to the *Annals*. Only the first was accepted, but the title had to change: "Distribution" meant spatial distribution, so Walter Kollmorgen insisted on substituting "Alternate Explanations of Urban Rank-Size Relationships." The two central-place papers were again rejected: They were "mathematics, not geography." Might things have been different if we had not pressed on and sent the Snohomish County papers to *Economic Geography*? Raymond Murphy accepted the first "as an experiment, to see whether his readers would object" to what he also called "mathematics, not geography." Because there were few complaints, he took the second. As a result, Ned Taaffe had his models for the Northwestern University field camps.

The experience with these papers epitomizes the geography of the time. Mainline geographers were suspicious, threatened, antagonistic; and we reciprocated. We felt we had to fight and fight we did, earning reputations for brashness and abrasiveness. So be it: If we had not been aggressive, geography would have rolled over us. Instead, we tried to roll geography: Remember *Rashomon*. We had chosen sides in a debate that by now had been joined by the discipline's luminaries. Stung by Fred Schaefer's critique, Richard Hartshorne's defense of exceptionalism, *Perspective on the Nature of Geography*, was already being discussed, ultimately to be published by the Association of American Geographers in 1959.

Our response was simply to fight harder, and for external support we reached out to embrace Regional Science and its battery of input-output analysis and other techniques, as well as to the received body of location theory: Thünen and Weber, Launhardt and Palander, Christaller and Lösch. The graduate student ranks were swelled by new entrants who joined the cause: Waldo Tobler, Michael Dacey, Arthur Getis, and, fresh from his battles with Hartshorne at Wisconsin, William Bunge. David Huff found his way to geography from the business school. There was now a critical mass that fed on itself. The mimeograph machine ran hot, and the students challenged, competed, and taught each other. The first Regional Science *Papers* arrived and were read with relish, as was Walter Isard's *Location and Space Economy* (New York: Wiley, 1956), which Garrison used as text in a new course. These were exciting times in which attention focused on the possibility of a *science* of geography with a theory of its own, tested and refined using statistical inference and mathematical models. To hone my own skills, in 1956 and 1957 I took courses in symbolic logic and learned elementary operations research, and, like the others, hung around the university's computer lab, bedazzled by the new machine. The group also was held together by Garrison's funded research—new highway impact studies as a consequence of the 1956 Interstate Highways Act. What was being crafted was quantitative and linked theory and practice: It was the antithesis of exceptionalism.

Funding was uncertain for the summer of 1957, however. That season's highway contracts had yet to be signed and I needed to be sure that I would eat. John Sherman, bless him, arranged a job for me in the city planning department of the City of Spokane. I spent the summer crafting an annexation policy for the city, but devoted my evenings and weekends to inventorying the city's retail and service business. I did not have a car and did it all on foot. I had my eye on a Ph.D dissertation: Perhaps I could repeat my master's research, applying central-place theory to the internal business structure of a large city. Returning to Seattle, I was assigned to teach the large introductory classes in economic and human geography. Art Getis was one of my TAs. In the time left over, I worked on the dissertation. To do the cluster analysis I needed a 49 × 49 correlation matrix, and the University of Washington's computer was unequal to the task. Large boxes of IBM cards went to the Western Data Processing Center in Los Angeles, and bundles of output came back. I did the cluster analysis by hand. Bill Garrison complained that I wrote the thesis faster than he could read it. As I was completing the dissertation early in 1958 I was interviewed for and offered a job at the University of Chicago. I received the degree in June, taught and worked on the highway studies for the summer, and left Seattle almost three years to the day from my arrival.

I also took advantage of Chicago's resources, retaking my statistics, seeking help on analytic problems from statisticians William Kruskal and David Wallace, developing a working relationship with sociologist Otis Dudley Duncan, whom I persuaded to title his new book *Statistical Geography*, and sitting in the back of Milton Friedman's courses in macroeconomics. Norton Ginsburg asked for my help with his *Atlas of Economic Development*, for which I designed a new kind of legend and taught myself principal components analysis to investigate his data set, learning the uncertain joys of programming direct factor analysis for the university's massive vacuum-tube Univac I, but developing a lively give-and-take with political scientist Duncan Macrae and educational psychologist Ben Wright, who were facing similar analytic problems. Gilbert White (who always put the fear of god into me) drew me into his floodplain studies, for which I devised procedures for spatial sampling, and had me serve on his dissertation committees. Harold Mayer took me under his wing and decided to include two of the Berry-Garrison papers in his *Readings in Urban Geography*, which did much to legitimate the quantitative cause. As a result, central-place studies came to play an important role in the early development of spatial analysis. I wanted to do a comprehensive series of comparative studies using new data that addressed the deficiencies of the Snohomish County work, and Harold worked with me on a proposal that was submitted to the NSF-precursor Office of Naval Research and was funded by Evelyn Pruitt.Because I had come from the University of Washington, which had John Sherman's famous cartography lab, Chicago assigned me to teach introductory cartography. Actually, it was not until after I had accepted the Chicago position that Gilbert White called to tell me what I would be expected to do. Someone had to teach cartography, and I was low man on the totem pole. I could hardly say no, but I had never taken a map-making course and was ham-fisted with ink and paper although I had benefitted from Ross MacKay's tutelage in statistical cartography. John Sherman came to the rescue and tutored Duane Marble and me during the summer of 1958. As a result, I survived, but worked to convert the offering into statistical cartography. I don't think I produced any cartographers!

The graduate students were another matter. Most were residual Robert Platt students—Plattaches—and they were quick to signal their disapproval of this brash young faculty member—younger by several years than most of them—and to engineer contention between the Sauerian philosophies of another arrival to the faculty, Marvin Mikesell, and what Bob Platt disparagingly called Brian Berry's "lab coat technicians." Fortunately, Marvin and I became and remain close friends, sharing common interests in the world of geographical ideas.

Around Christmas of 1958 I visited Northwestern and met Ned Taaffe and Charlie Tiebout, who had seen Ed Thomas through his Ph.D. Peter Gould, Bob Smith, and my undergraduate compatriot from London, Bruce Newling, were among Ned's students at the time. Ned and I developed a relationship that ultimately was to extend to summer institutes in quantitative methods for geographers, to joint work on the Ackerman Committee that produced *The Science of Geography*, and to a number of other ventures such as Barry Garner's doctoral dissertation.

Not long after, I read my first paper at the Pittsburgh AAG meetings, discussing the use of Poisson probabilities to assess the uniformity of spatial point patterns. There were barely a dozen people in attendance in what was a worst-time-and-place session. It took us quite a while to have our papers placed in the appropriate substantive sessions at Association meetings rather than a ghetto labeled "mathematical (or statistical) geography," although after the initial Berry-Garrison experience we were becoming more successful in getting papers accepted by the journals.

I returned to England in the summer of 1959 and gave a paper on the Washington group's highway impact studies to the geography section of the British Association. My paper was roundly criticized as being ungeographical by a number of old-school British geographers, and I was thrown out of one gathering because "only professional geographers are permitted to be present." Michael Wise and Dudley Stamp, whom I knew from my undergraduate days, rescued me—bless them! But it was painfully clear that there was no room in British geography in 1959 for a working-class kid with an East Midlands accent and an American Ph.D. who was doing all these new-fangled ungeographical things. Whatever the regrets, and there were many, an American career was indicated. I was pleased to get back to Chicago and resume my role as advocate, lightning rod, and spear catcher.

With the passage of only one more year, the important events had all occurred. Mine had been the first move from Seattle. Soon, Dick Morrill was at Northwestern, John Nystuen at Michigan, Duane Marble on his way to Pennsylvania, and Bill Bunge to join McCarty and Thomas at Iowa. Within five years of beginning their graduate studies, the initial cohort had positions in leading centers of geographic education. Bill Garrison and his brigade were all present in a series of seminal meetings organized in the summer of 1960 at Lund by Torsten Hägerstrand in connection with the XIXth International Geographical Congress—me through the good offices of Chauncy Harris. Walter Christaller was there. There were many battles to come, but it was clear that we had arrived. A new view of geography was on the move, one that believed that the frontiers were indeed to be found in sound theory and acceptable practice—in the application of the scientific method to geographic research and of geographic research to important issues of public policy.

# CHAPTER 6

## Yesterday as Tomorrow's Song
### *The Contribution of the 1960s "Chicago School" to Urban Geography*

MAURICE YEATES

## ABSTRACT

On December 17, 1971, a 33-column-inch article by Alan Wilson appeared on the front page of the "books" section of *The Times* (London, UK), titled "Big boost for new geography." In it, Wilson reviewed four textbooks: *Spatial Organization* (Abler, Adams, and Gould*); Industrial Location* (Smith); *Internal Structure of the City (ISC)* (edited by Bourne); and *The North American City (NAC)* (Yeates and Garner). Given that at least two of these books (*ISC* and *NAC*) are firmly in the realm of urban geography, what had happened in the field to warrant this banner treatment? Examination of ISC and NAC indicates that both are in large part rooted in research themes pursued in the Department of Geography (particularly by Brian Berry, and graduate students) at the University of Chicago from the late 1950s to the early 1970s. Some of these themes are as follows: the emphasis on process over place; the notion of the geographical matrix; the idea of systems within systems; the spatial topology of commercial structure; social theory and the internal structure of cities; the connection between social space and housing space in a race-determined society; the need for research relevance; and the importance of "professionalism" to the geographical research enterprise. The paper concludes by situating the collective work of this group within the proud tradition of the Chicago School of urban studies and by identifying the ways in which this efflorescence of activity continues to energize developments in urban geography and in the discipline at large. [Key words: urban geography, research themes, Chicago School, Brian Berry.]

On December 17, 1971, an article by Alan Wilson headlined the first page of the book review section of *The Times* (London, UK), titled "Big boost for new geography." This article covered 33 column inches and "Geography" blazoned the top of the first page! In it, Wilson stated: "Geography has undergone major transformation in the past decade. What has come to be called the "new geography" (from the mid-1960s) is now being taught in varying degrees in most universities." From a pedagogic point of view, the features of the "new geography" that were engaging the profession were its application to human geography of the scientific method, and problem orientation. One of the consequences of this application and orientation was that "teaching has not had an adequate backing of good text-books." However, this vacuum, Wilson noted, was being filled rapidly—"out of 70 or so "new geography" books on my own bookshelves, 40 have been published in the past two years, and most of these in the past 12 months." From this efflorescence of texts, Wilson selected four for review: *Spatial Organization* (Abler, Adams, and Gould); Industrial *Location* (Smith); *Internal Structure of the City (ISC)* (edited by Bourne); and *The North American City (NAC)* (Yeates and Garner).

Given that at least two of these books are firmly in the realm of urban geography, what had happened in the field to warrant this banner treatment? In the first place, it should be mentioned that what had happened had not happened, for the most part, in universities in the newspaper's country. Examination of the structure and contents of ISC and NAC indicates that both are rooted in research approaches/methodologies, theoretical developments, and empirical studies undertaken at major universities in the United States, particularly Washington, Chicago, North-western, Minnesota, Clark, Ohio State, UCLA, Pennsylvania State, and so forth. The research undertaken by faculty and graduate students at the University of Chicago, from the late 1950s to the early 1970s, was particularly important because the activity involved a number of themes that appear to have had a lasting influence not only on urban geography, but human geography in general.

The themes *embraced* by faculty and graduate students at Chicago included: an emphasis on process over place; the notion of the geographical matrix; the idea of layers of spatial systems; the spatial typology of urban commercial structure; social theory and the internal structure of cities; the connection between social space and housing space in a race-determined society; the need for research relevance; and the importance of "professionalism" in the geographic research enterprise. While researchers at other institutions also embraced some of these themes, the sheer volume and range of research emanating from Chicago during this period, as exemplified in the brave departmental Research Paper Series (all dissertations were published in the series!), in which 43 urban/spatial analysis studies were printed between 1961 and 1975[1] (inclusive), is impressive and commands attention.

## A Note on the "Chicago School"

The urban-oriented research themes that came to the fore in the Department of Geography[2] at the University of Chicago during the 1960s (referred to as the "Chicago School") were, of course, rooted in ideas emanating from a wide variety of sources—these are discussed in many texts, including those referred to previously. In addition to these external intellectual stimuli, faculty and students within the department were working in a university that had (and still has) a proud tradition of urban-oriented research. Although many cities in North America had grown rapidly and reached a metropolitan stage of development, few universities had developed by 1960 such a tradition of in-depth research on urban-oriented matters. For example, the range of publications derived from what is now referred to as the 1920s "Chicago School of Human Ecology" (Hawley, 1986) is well known. Less well known, perhaps, is the research undertaken by the residents of Hull House and their colleagues in Social Work during the same period (Sibley, 1995).

The pre-existence of such a well-recognized tradition of urban-oriented research in an institution can, of course, be a two-edged sword. On the one hand, researchers have the advantage of superior library resources and collections of original documents (both at the University and within the

City), and an array of stimulating colleagues in collateral departments and research centers. On the other hand, these same colleagues frequently have their own agendas, and can be quite critical of research approaches that may be incongruent with, or appear to replicate (or challenge) their own. In short, the urban-oriented researched that flourished in the department during the 1960s was undertaken in a tough and aggressive intellectual environment, and the faculty and graduate students who were involved had to be equal to the challenge.

## An Emphasis on Process over Place

Urban geography, as it would be recognized today, did not really exist until the late 1950s—urban research involved primarily analyses (in monograph form) of individual cities (Mayer, 1954; Harris, 1990). This focus on place made research in the area somewhat difficult—how can one develop an intellectual tradition in a field that is sketched on the basis of particularities? Furthermore, how can one attract graduate students to undertake research in a field more whimsy than reason? Thus, the publication by Harold Mayer, a faculty member in the Department (until 1967), and Clyde Kohn (University of Iowa) of *Readings in Urban Geography* in 1959 was important as it provided a path to the modern field (Taaffe, 1990). In the "Introduction" to *Readings,* the co-editors stated that "if urban geography is to be effective in its application to human affairs, it must rest upon a firm conceptual and theoretical base." In consequence, the articles selected for inclusion firmly emphasized process over place, though nearly all involved some kind of qualitative, and occasionally quantitative, empirical analysis.

The 54 articles included within *Readings* were grouped into 18 topics, with, for example, five relating to the economic base of cities, four concerning cities as central places, six on the commercial structure of cities, and four on the rural-urban fringe. Among the articles reprinted in the book, that have become classics in urban geography, were a number written by colleagues of Mayer at Chicago—Colby (1933), Harris (1943; Harris and Ullman, 1945), and, a new faculty appointment to the department, Berry (1958). The impact of *Readings* may be gauged from the fact that it was in its fourth impression by 1964, and, despite being a textbook, was increasingly referenced in the journal literature[3] up to 1969 (Figure 6.1), by which time other competitors, such as Murphy (1966), had entered the market.

The focus on process over place, but in a spatial context, had been emphasized by Harris (and Ullman, 1945) in the classic article "The

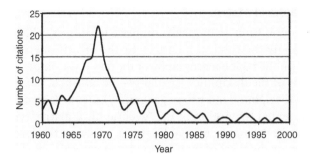

**Figure 6.1** Citations of *Readings in Urban Geography.*

Nature of Cities." The well-known multiple-nuclei model (or cartogram), which suggests that common processes lead to similar arrangements of land use among urban areas, is still cited and used as a framework for urban analysis (Figure 6.2)—with derivatives of the spatial model applied to the rapidly expanding metropolises of the less developed parts of the world. Mabogunge (1968, p. 179), for example, commented that "the idea of multiple nuclei is fundamental in understanding the nature of Nigerian cities." Harris (1997), in a review of the diffusion in use of the model, observed that it has been an object of study, or a point of departure, not only in geography, but also anthropology, sociology (its greatest use), economics, urban history, and communications. Its importance as an idea that one should know about is emphasized by the fact that requests for permissions for its use in books are still received by the journal in which it was first published (Figure 6.2).

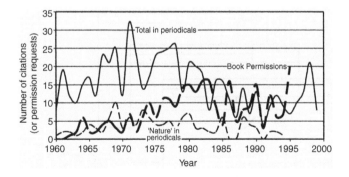

**Figure 6.2** Citations of Harris' "The Nature of Cities," and total citations in journals of all Harris publications. *Source:* Harris, 1997, for "Nature" and book permission counts.

## The Geographical Matrix

In this transition from an emphasis on place to process, which was interpreted (wrongly) by some as an attack by the practitioners of the "new geography" on the regional geography of the day, it was important that the central focus of the discipline—spatiality—be affirmed, but in a manner that led the discipline (and urban geography) forward. To do this, Berry (1964a) presented the notion of the "geographical matrix" (Figure 6.3). The columns in the matrix refer to locations, or places; the rows to characteristics defining the types of information being utilized. The characteristics in the diagram have been grouped into some of the main systematic branches of human geography—economic, social, political (others could be added). The intersection of a given column and row defines a cell which represents geo-referenced information.

Thus, a study involving cells in the whole, or in part, of a row amounts to studying the same characteristic as it occurs at a number of different locations—the emphasis being on the spatial variation of a given characteristic. On the other hand, a study of the whole or part of a single column amounts to looking at many different characteristics at the same place—the essence of regional geography, or regional analysis. Comparing information between rows amounts to the study of spatial associations. Comparison of two or more columns amounts to the comparative study of different places in terms of selected characteristics. Finally, and most important, if each matrix is viewed as representing a particular point in time, a number of slices for a number of time periods allows for analyses of temporal change in a geo-referenced information environment.

Now it is not too difficult, in this day and age when practically every faculty position advertised in both physical and human geography appears

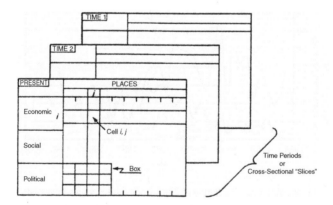

**Figure 6.3** The geographical matrix (Berry, 1964a, Figure 3).

to require (among other attributes) some facility with GIS, to recognize that the "geographical matrix" is a progenitor of GIS. The idea of the "geographical matrix" emphasized that geographical information needed to be stored and analysed within the context of geo-referenced relational databases—a "geographic information system" (Tomlinson, 1989; Coppock and Rhind, 1991). But, there is a big difference between being a progenitor and conceiving a practical means of achieving such a system. The solution had to wait for the advent of powerful micro-computers (about 1985), for prior to that time computing hardware and software involved fairly esoteric technologies, and computer time was quite expensive.

## Systems Within Systems

It may not be well known, but textbooks undergo pre-publication reviews, and user reviews (in the case of those being considered for subsequent editions), that are more numerous, detailed, and incisive (possibly because the reviewers are being paid), than most articles submitted to journals. A common characteristic of reviewers is that they would like the author to write the book that they wished they had written themselves. One welcome prepublication comment to the first edition of the NAC, however, was that its basic structure—the first part involving the city system, and the second part the internal structure of urban areas—reflected the "modern" approach (in 1971) accepted by most urban geographers. This duality became embedded in the course structure of many geography departments, though since about 1985 many city system courses have been deleted in favor of topics related to particular aspects of internal structure.

The city system/internal structure approach incorporates the notion of the "geographical matrix" applied to the urban realm (Figure 6.4); the concept of "cities as systems within systems of cities" (Berry, 1964b), which had its genesis in central place theory; and a way of looking at elements of internal structure, particularly those related to aspects of social structure and housing, and commercial structure, that had a spatial representation (Berry, 1965). This framework—combined with, on the city-system side, research into the evolution of American metropolitan centres by Pred (1962; 1965; 1971) and, later, Conzen (1977); and, on the internal-structure side, contributions from land rent/use theorists and social area analysts—provided a way of highlighting processes underlying urban spatial structure at national (Bourne and Simmons, 1978; Ray and Murdie, 1972), regional (Bourne and MacKinnon, 1972), and local (Bourne, 1968; Goheen, 1970) levels. These latter references also indicate the impact that the Chicago School was having on universities in Ontario, such as the University of Toronto.

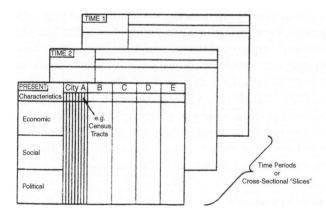

**Figure 6.4** The geographical matrix: Urban systems and the internal structure of cities (Berry, 1964b, Figure 3).

## A Spatial Typology of Commercial Structure

Building on the empirical work of earlier retail analysts and ideas derived from central place theory (referenced in: Berry and Pred, 1964, updated in 1966), Berry (1963) and some of his students (Berry and Tennant, 1963; Simmons, 1964; and Garner, 1966[4]) analyzed the complex spatial pattern of change in the major conformations of the commercial structure of Chicago. These conformations—defined as nucleations (shopping centers and unplanned); ribbons; and specialized areas—were identified on the basis of the *locational* requirements of different businesses within the urban area, not on their morphology or form. Because the conformations were rooted in locational requirements, the suggestion was that this spatial typology would be replicated in other North American metropolises, with variations in the consumer service structure depending on particularities of the urban areas concerned.

This suggestion (or hypothesis) appears not to have been examined in many locations, perhaps because the data requirements for a detailed analysis of the consumer service supply-side are so immense. Simmons (1966) undertook a study of Toronto, using City of Toronto and Metro Toronto planning data, which basically replicated the Chicago findings. Cohen (1972) examined the spread of shopping centers in the United States in the context of diffusion theory. Barnum (1966) undertook a regional study of consumer service delivery systems in southwestern Germany (adjacent to the area analyzed by Christaller, 1933). Other geographers in other universities took up the challenge, such as Morrill (1987), who examined the changing pattern of planned and unplanned nucleations in the Seattle metropolitan area; and Jones[5] (1984), who focused his analysis on special-ized retailing in inner-city Toronto. Potter (1982) tested the general hypothesis in the context of a number of urban areas in the United Kingdom, concluding that similar types of supply-side conformations either existed or were emerging in that part of the world.

Supply-side segmentation in the consumer service sector in North American urban areas is an outcome of highly competitive forces operat-ing in a weakly regulated planning environment. In such an environment, conformations will be in a continual state of flux. It is interesting, there-fore, to note the way in which these concepts have been adopted as norma-tive planning tools in many countries which have strong land-use controls, particularly in Europe and Japan (Guy, 1998). In this normative approach, a particular commercial structure may be preferred—invariably one that confers prominence to an historic downtown, and preserves commercial activity in traditional neighborhoods—and entrenched in a local (and perhaps national) regulative environment that controls supply-side com-petition (Yeates, 1998). The legacy of the Chicago School with respect to commercial structure, therefore, resonates today, particularly in more reg-ulated urban environments in which certain cultural values are embedded in planning regulations.

The activities of the 1960s Chicago School with respect to central place theory and its relation to urban system development, and the locational dynamics of supply-side competitive segmentation within metropolitan areas, were recapitulated in an integrated format, for a wider audience, in the monograph *Geography of Market Centers and Retail Distribution* (Berry, 1967b). This book had a significant impact on urban geography, and, in particular, the development of courses in *retail geography*. While sales figures are kept within the book trade, the impression is that they were quite significant. An indication of the lasting impact is indicated in the social science periodicals citation counts in Figure 6.5, which suggests that the major influence of this volume lasted for 15 years.

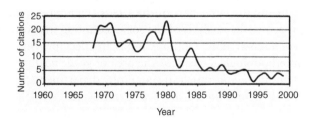

**Figure 6.5** Citations of *Geography of Market Centers and Retail Distribution.*

## Social Theory and the Internal Structure of Cities

There are many kinds of social theory, and that explored by faculty and graduate students on the Department of Geography at Chicago in the 1960s was (naturally, given the institution) in the human ecology framework pioneered by Park, Burgess, and McKenzie in Chicago the 1920s (Hawley, 1986), as reformulated and extended by the social area analysts in the 1950s (Shevky, Williams, and Bell, 1955). However, following the findings of Anderson and Egeland (1961), the Chicago School introduced to urban geography the intriguing notion that the social area constructs would tend to have particular spatial twists—the social construct, sectoral; the family construct, concentric; and a segregated (or clustered) construct, localized (Berry, 1965). Furthermore, rather than formulating the constructs from a conflation of pre-selected social and economic variables from the census (relating to aggregated areal units, such as census tracts), they decided to "test" the validity of the constructs using virtually all the socioeconomic information available in the census using factor analysis— which, in effect, provided a different way of achieving construct conflation.

The "tests" were undertaken using census tract data relating to Chicago (Rees, 1968) and Toronto (Murdie, 1969). The Chicago study not only identified the main factors, which were in accordance with the constructs suggested by the social area analysts, but also provided an imaginative model relating the main dimensions of social space (for households), housing space (for dwelling units), community space (for census tracts), to locational or physical space (for subareas)—providing, perhaps, a subliminal commentary on the modifiable areal unit problem, and the "ecological fallacy" blunderbuss (Amrhein and Reynolds, 1997). The Toronto study sought to establish whether the constructs, their spatial twist, and the processes that were involved, had some stability over time (they did). Research by Golant (1972) focused on the forces underlying the various patterns of residential clustering (national and local) that had emerged, and were developing, in the location of the elderly in Canada.

Although interest in urban ecology in general (Berry and Kasarda, 1977) faded during the latter part of the 1980s (Figure 6.6), the factorial ecology approach to the analysis of the socioeconomic structure of urban areas has proven remarkably obdurate. It has, in effect, provided the intellectual and methodological basis for the area of *geodemographics*, which, while it may not be taught much in geography departments, is a foundation of the whole field of *micro-marketing* in business schools. For those geography students adventurous enough to study geodemographics, along with GIS, and spatial analysis, there are exciting (and rewarding) opportunities in business geomatics in the "new information economy."

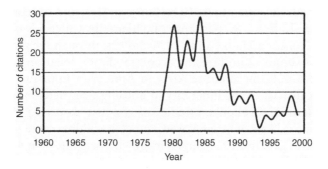

**Figure 6.6**  Citations of *Contemporary Urban Ecology.*

## Housing Location and Race

The 1960s were tumultuous times in many ways, not least because of federal (and, occasionally, local) attempts to eliminate the edifice of laws and practices that had been established to perpetuate racial segregation. The intentions and implications of the Fair Housing Act (1966), with subsequent reinforcing emendations, and the Civil Rights Act (1968), are still reverberating through American life. The Chicago School, with its natural concern for social justice, was as interested as many other groups of scholars with the progress that the Fair Housing Act, for example, was having on reducing discrimination in access to housing and achieving greater levels of spatial integration within metropolitan areas. Such research also proceeded naturally from the work in factorial ecology which had identified various manifestations of spatial segregation and pointed to some of its formative processes (Berry, 1971).

Thus, toward the end of the 1960s, some of the work in the Chicago School turned toward analyses of the ways in which African American urban households tended to concentrate within inner cities (Meyer, 1970); the impact on employment opportunities for some of these households of the decentralization of commercial and industrial activity which had been occurring at a rapid rate during the 1960s (Berry, 1968; Berry and Cohen, 1973); and the limited effect, by the early 1970s, of the Fair Housing Act on integration (Berry, et al., 1976). From a number of case studies of the reaction to the movement of African American households into some inner Chicago suburbs occupied by White households, it seemed apparent at that time that extensive integration of African-American and White households would be unlikely as long as race remained a status determining trait (Berry, 1979)—and so this has unfortunately proven, in large part, to be the case.

Parenthetically, one proactive response to this, and similar findings from other studies, was, perhaps coincidentally, Chicago's Gautreaux Program. This program began in 1976 and has, through the provision of initial rent support and housing accommodation location assistance, dispersed 5,000 low income African American families from the Gautreaux district of the inner city to housing elsewhere, mainly inner suburban locations. Follow-up studies suggest that "dispersed" household heads are more likely to be employed, and their children more likely to go to college and gain full-time employment with benefits, than households remaining in the inner-city district (Ramos, 1994).

## Research Relevance

One of the major tenets of the Chicago School was that research, while rooted in theory and geared toward the scholarly development of the discipline, should also have wider community relevance. Given that, by 1960, it was abundantly apparent that metropolitan areas were growing rapidly, and that this growth had given rise to new problems as well as exacerbating old ones, Mayer and Kohn (1959, p.1) emphasized that "... the field of urban geography has become more important both as an academic discipline and as one of the foundations for practical decision-making in government, business, and social affairs." Whether by intent or happenstance, research in the Department in the 1960s appears to have been directed entirely toward the formation of policy in the public sector.

There is, in fact, little recognition in any of the research emanating from the Department, or its funding, that the output was, or could be, tailored to the needs of decision-making in the private sector. The central importance of public sector research relevance is stated clearly in the opening paragraph of *Commercial Structure and Commercial Blight*: "This study was undertaken because of the need for "laying out the extent, location, nature, and trends of [commercial] blight and deterioration" in the City of Chicago, to provide one basis for determining alternative courses of action to alleviate and prevent commercial blight, and to facilitate evaluation of the benefits and costs of these alternatives" (Berry, 1963, p.1). The contract for the project was with the Community Renewal Program of the City of Chicago.

This focus on public policy relevance is also evident in the work of the Chicago School in the latter 1960s on metropolitan area definition. To many, metropolitan area definition may seem a humdrum pursuit. However, it is a serious issue in the United States because, apart from the widely accepted use of metropolitan areas as statistical reporting units, many federal entitlement (and other) programs define funding eligibility on the basis of "metropolitan area" definitions. The Office of Management and Budget has defined metropolitan areas, and required their use by all Federal agencies, since standard metropolitan areas (SMAs) were introduced for the 1950 census—when the need to incorporate spreading suburbanization had become evident (Dahmann and Fitzsimmons, 1995).[6]

If there is any place that urban geographers should be heard, it is in attempts to define the spatiality of the urban reality of the age. So when, by the mid-1960s, it had become clear that standard metropolitan statistical areas (SMSAs), as SMAs had by then become, were failing to capture the spatial nature of the urban system, it is not surprising that the Chicago School (not known for its lack of courage) was asked to review the definitional question. After considerable research, the team suggested that a metropolitan area should continue to consist of a nucleus (city or urbanized area) of a population of 50,000 or more, but to this nucleus should be added its functional economic region—demarcated on the basis of commuting patterns (Berry, 1967a; Berry, Goheen, and Goldstein, 1968). The effect of environmental factors on metropolitan reach was addressed subsequently in greater detail in Berry and Neils (1970).

This interest in the changing nature of metropolitan reach led, in the years prior to Berry's departure from Chicago in 1976, to considerations of the impact that urbanization was having on rural America (Lamb, 1975); identification of growth centers (Berry, 1976a); and, the counterurbanization hypothesis (Berry, 1976b). This hypothesis suggested that, by 1975, the United States had entered an era of low or declining population growth rates in the largest metropolitan areas, higher growth rates in mid-sized metropolitan areas, high growth rates in nonmetropolitan areas (particularly those adjacent to metropolitan areas), decreasing urban densities, and increasing community homogeneity with respect to class, race, age, or language (Berry and Gillard, 1977; Berry and Dahmann, 1977).

### "Professionalism" in the Research Enterprise

Graduate students in the Department of Geography at the University of Chicago in the 1960s, especially those in the urban field, were imbued with a "professional" research culture. The characteristics of this culture were: *Research was de-mythologized*—the activity was conducted openly in the department. Students could see faculty, and other graduate students, going about the research business—designing proposals, preparing grant applications, reviewing papers submitted for publication in journals, discussing research interests with guests (domestic and foreign), collecting and compiling information and data, tripping to and from the library and computing center, and, actually writing scripts for publication (invariably involving faculty and student coauthorship). The impression created was that, while research involved discipline and skills that had to be learned, it was an important and fulfilling daily activity that one could get into.

The research within the department was *methodologically open* to ideas and approaches developed in other disciplines, such as economics, sociology, politics, planning, psychology, history, architecture, transport engineering, statistics, operations research, mathematics, environmental studies, real estate, business, and so forth. For the most part, the *lingua franca* of this cross-fertilization was based on philosophical ideas that had emanated from the Vienna Circle of the 1920s and 1930s, commonly referred to as "logical positivism" (Achinstein and Barker, 1969), the basic tenets of which, during the late 1950s and 1960s, were being increasingly questioned by such philosophers as Quine (1953).

The various activities of the department, not just those in the urban field, were influenced by a *continental and global network* of professionals with similar interests which generated "value-added" in various research spheres. Networking is not a recent phenomenon of the mini-computer age—the most productive and ambitious researchers have always fostered research networks. In the Chicago case, these networks were strengthened by the contacts that were made through close involvement with the AAG and IGU,[7] and the stellar faculty and graduate student complement which attracted scholars visiting or on leave from their own institutions (who often participated in graduate courses within the department).

*The Urban Studies Center* became a means by which this professional research culture was enhanced through contacts with law, social work, and most of the social science departments—generating, for example, research into the social impacts of urban pollution (Berry et al., 1977). Like most university research centers, this was largely a "soft money" operation, which flourished only as long as it was involved with activities that other people (within the institution or outside) cared about and supported. Berry wrote the plan for the Center, organized its multidisciplinary training programs, and served as its Director from 1974 to 1976 (he also, by that time, held an endowed Chair within the University).

The high level of professionalism that was inculcated is evident in the career paths of many of the students who graduated from the Chicago School during the period discussed.

## Conclusion

The wave of creative energy, which drove urban geography in the Department of Geography at the University of Chicago in the 1960s and early 1970s, and has given rise to the appellation Chicago School had an extraordinary impact on the profession. The degree of impact is indicated by the aggregate journal citations[8] of publications authored or coauthored by Berry (Figure 6.7). The magnitude of the aggregate counts is staggering—the curve labelled "outstanding" simply refers, for comparative purposes, to another stellar urban geographer of the Chicago School (who has received numerous accolades). It is, for many reasons, difficult to sustain that type of research and writing intensity for a long period, and the move by Berry from Chicago to Harvard in 1976 could be interpreted as an attempt, through a change of research environment, to maintain that intensity.

The interesting feature, however, about the work of this period is that many of the themes that came to the fore at that time have prominence, albeit in slightly different guises, today. The emphasis on process over place is now firmly entrenched in urban geography. The notion of the "geographical matrix," with its anticipation of relational databases, has metamorphosed into GIS; and, coupled with remote sensing, GPS (global positioning systems), and spatial analysis, more broadly into a new field (and producer service industry) of geomatics (or "geographic information science"). Ideas discussed within the catchy phrase "cities as systems within systems of cities" are really attempts at addressing the "problem of market structure in the face of increasing returns" (Krugman, 1995, p.35)—that is, the nature and spatial consequences of compounding agglomerative forces in growing economies, sometimes referred to as "the new economic geography" (Martin, 1999). The detailed theoretical and empirical work on the commercial structure of cities is the basis for current research on supply-side competitive segmentation; and, it has also provided the theory underlying normative commercial planning, as well as its global vernacular.

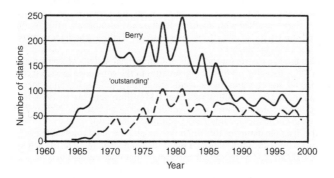

**Figure 6.7**  A wave of creative energy.

The work on the social structure of cities using the concepts and tools of factorial urban ecology underpins today's field of geodemographics, and the business of micromarketing. Studies of the impact of the Fair Housing Act (1966) on inner-suburban Chicago indicated that by 1976 African Americans were still largely liminal persons in American society— no longer legally "other," but still not quite accepted. Four decades later, race is now less a status determining trait, but it still, according to the 2000 U.S. Census, appears to have a strong influence on residential location. As far as research in urban geography is concerned, requirements for demonstrated relevance (and public and/or private sector partners) in research funding is now *sine qua non* in many agencies and jurisdictions. And the emphasis on the need for "professionalism" in the research enterprise is now universally entrenched as "good practice" in graduate studies and research. Yesterday, in many ways, is transmuting into tomorrow's song.

## Notes

1. 101 monographs, or Department of Geography research papers, were published in this 15-year period.
2. This paper focuses on urban geography. This was, of course, just one of many fields of faculty interest in the Department, such as: economic development and political geography (Norton Ginsberg), environmental analysis (Gilbert White), physical ecology (Karl Butzer), cultural geography (Marvin Mikesell), and so forth.
3. Citation counts in this paper have been gleaned from the *Social Sciences Citation Indexes* (Philadelphia, PA: Institute for Scientific Information), which go back to 1956. The annual counts refer to those in social science periodicals only, and the number of periodicals included in the surveys has *increased* with the passage of time. The journal *Urban Geography* was included only in 1988. *Thus, there are many reasons for interpreting these counts, especially in the social sciences, with caution.* Not all research in the social sciences is disseminated through peer reviewed journals (books and monographs are common), certain research groups go in for a lot of self-referencing, and many urban geographers publish in science (particularly those using spatial analytic methodologies) journals as well. The SSCI counts have been used recently in Rey and Anselin (2000). I thank Andy Charles for compiling the citation counts.
4. Although Garner worked with the Berry team on the commercial structure project, he was a doctoral student at Northwestern (supervised by Taaffe—who had, in turn, been supervised in the 1950s by Mayer).
5. Jones established a Centre for the Study of Commercial Activity at Ryerson University, Toronto, in 1993, focusing on business geomatics in the consumer service sector. It involves more than 60 industry and public sector supporters, and has external research, contract, and equipment grants totalling well over C$1million per year. In 1998, the Centre was included in the federally funded (C$7 million per year) National Network of Centres of Excellence in Geomatics for Informed Decisions" (GEOIDE).
6. Given the dynamic nature of the urban system, and the often conflicting needs of the various agencies using the information, it is not surprising that, by 1990, the definitions (MSAs, PMSAs, and CMSAs) were fraught with contradictions, and had become a thorny issue. "The result by 1990 was a system so arcane, so needlessly complex, so lacking in underlying principle, and so afflicted by ad hoc-ism ... as to be ludicrous." Berry, B. J. L., 1996, "Capturing Evolving Realities: Statistical Areas for the American Future" in Dahmann and Fitzsimmons (1995), pp. 86–87.

7. For example, Harris had been secretary of the AAG (1946-1948), vice president (1956–1957), president (1957–1958), and received its Honours award in 1976. Berry received an AAG Honours award in 1968, and was much involved with the AAG in the early 1970s, becoming president (1978–1979). Harris was vice president of the IGU (1956–1964) and its secretary/treasurer (1968–1976).

8. Note that the scale interval for the Y-axis in Figure 7 is in increments of 50 vs. increments of 5 as with the previous graphs.

# Literature Cited

Achinstein, P. and S. F. Barker, editors, 1969, *Legacy of Logical Positivism: Studies in the Philosophy of Science.* Baltimore, MD: The Johns Hopkins Press.

Amrhein C. and Reynolds, H., 1997, Using the Getis statistic to explore aggregation effects in metropolitan Toronto census data. *The Canadian Geographer,* Vol. 41, 137–149.

Anderson, T. A. and Egeland, J. A., 1961, Spatial aspects of social area analysis. *American Sociological Review,* Vol. 26, 392–398.

Barnum, H. G., 1966, *Market Centers and Hinterlands in Baden-Württemberg.* Chicago, IL: University of Chicago, Department of Geography, Research Paper 103.

Berry, B. J. L. and Garrison, W. L., 1958, The functional bases of the central place hierarchy. *Economic Geography,* Vol. 34, 145–154.

Berry, B. J. L., 1963, *Commercial Structure and Commercial Blight.* Chicago, IL: University of Chicago, Department of Geography, Research Paper 85.

Berry, B. J. L. and Tennant, R.J., 1963, *Chicago Commercial Reference Handbook.* Chicago, IL: University of Chicago, Department of Geography, Research Paper 86.

Berry, B. J. L., 1964a, Approaches to regional analysis: A synthesis. *Annals of the Association of American Geographers,* Vol. 54, 2–11.

Berry, B. J. L., 1964b, Cities as systems within systems of cities. *Papers and Proceedings of the Regional Science Association,* Vol. 13, 147–163.

Berry, B. J. L. and Pred, A., 1964, *Central Place Studies: A Bibliography of Theory and Applications.* Philadelphia, PA: Regional Science Association.

Berry, B. J. L., 1965, Internal structure of the city. *Law and Contemporary Problems,* Vol. 30, 111–119.

Berry, B. J. L., 1967a, *Functional Economic Areas and Consolidated Urban Regions of the United States.* Washington, DC: Bureau of the Census, Working Paper 26.

Berry, B. J. L., 1967b, *Geography of Market Centers and Retail Distribution.* Englewood Cliffs, NJ: Prentice Hall.

Berry, B. J. L., 1968, A Summary of Spatial Organization and Levels of Welfare: Degree of Metropolitan Labor Market Participation as a Variable in Economic Development" E.D.A. Research Review. Washington, DC: Department of Commerce, July, 1–6.

Berry, B. J. L., Goheen, P., and Goldstein, H., 1968, *Metropolitan Area Definition: A Re-evaluation of Concept and Statistical Practice.* Washington, DC: Bureau of the Census, Working Paper 28.

Berry, B. J. L. and Neils, E., 1970, Location, shape, and size of cities as influenced by environmental factors. In H. S. Perloff, editor, *The Quality of the Urban Environment.* Baltimore, MD: The Johns Hopkins University Press, 257–302.

Berry, B. J. L., 1971, Introduction: The logic and limitations of comparative factorial ecology. *Economic Geography,* Vol. 47, 209–219.

Berry, B. J. L. and Cohen, Y.S., 1973, Decentralization of commerce and industry: The restructuring of metropolitan America. *Urban Affairs Annual Reviews,* Vol. 7, 431–456.

Berry, B. J. L., 1976a, *Growth Centers in the American Urban System, 1960–1970.* Cambridge, MA: Ballinger.

Berry, B. J. L., 1976b, *Urbanization and Counterurbanization.* Beverly Hills, CA: Sage.

Berry, B. J. L., Goodwin, C. A., Lake, R. W., and Smith, K. B., 1976, Attitudes toward integration: The role of status in community response to social change. In B. Schwartz, editor, *The Changing Face of the Suburbs.* Chicago, IL: University of Chicago Press, 221–264.

Berry, B. J. L., Caris, S. (now Cutter), Gaskill, D., Kaplan, C. P., Piccinini, J., Planert, N., Rendall, J. H. III., and de Ste. Phalle, A., 1977, *The Social Burdens of Evironmental Pollution: A Comparative Metropolitan Data Source.* Cambridge, MA: Ballinger.

Berry, B. J. L. and Dahmann, D. C., 1977, Population redistribution in the United States in the 1970s. *Population and Development Review*, Vol. 3, 443–471.

Berry, B. J. L. and Gillard, Q., 1977, *The Changing Shape of Metropolitan America*. Cambridge, MA: Ballinger.

Berry, B. J. L. and Kasarda, J. D., 1977, *Contemporary Urban Ecology*. New York, NY: Macmillan.

Berry, B. J. L., 1979, *The Open Housing Question: Race and Housing in Chicago, 1966-1976*. Cambridge, MA: Ballinger.

Bourne, L. S., 1968, Market location and site selection in apartment construction. *The Canadian Geographer*, Vol. 12, No. 2, 211–226.

Bourne, L. S. and MacKinnon, R.D., editors, 1972, *Urban Systems Development in Central Canada*. Toronto, Canada: University of Toronto Press.

Bourne, L. S. and Simmons, J. W., editors, 1978, *Systems of Cities*. New York, NY: Oxford University Press.

Christaller, W., 1933, *Die Zentralen Orte in Süddeutschland [Central Places in Southern Germany]*. Jena, Germany: Gustav Fischer Verlag.

Cohen, Y. S., 1972, *Diffusion of an Innovation in an Urban System: The Spread of Planned Regional Shopping Centers in the United States, 1949–1968*. Chicago, IL: University of Chicago, Department of Geography, Research Paper 140.

Colby, C., 1933, Centrifugal and centripetal forces in urban geography. *Annals of the Association of American Geographers*, Vol. 23, 1–20.

Coppock, J. T. and Rhind, D. W., 1991, The history of GIS. In D. J. Maguire, M. F. Goodchild, and D. W. Rhind, editors, *Geographical Information Systems: Principles and Applications, Volume I*. New York, NY: Wiley, 21–43.

Dahmann, D. C. and Fitzsimmons, editors, 1995, *Metropolitan and Nonmetropolitan Areas: New Approaches to Geographical Definition*. Washington, DC: Population Division, Bureau of the Census, Working Paper 12.

Garner, B. J., 1966, *The Internal Structure of Retail Nucleations*. Evanston, IL: Northwestern University Press, Studies in Geography, No. 12.

Goheen, P., 1970, *Victorian Toronto, 1850 to 1900: Pattern and Process of Growth*. Chicago, IL: University of Chicago, Department of Geography, Research Paper 127.

Golant, S. M., 1972, *The Residential and Spatial Location of the Elderly: A Canadian Example*. Chicago, IL: University of Chicago, Department of Geography, Research Paper 143.

Guy, C. M., 1998, Controlling new retail spaces: The impress of planning policies in western Europe. *Urban Studies*, Vol. 35, 953–980.

Harris, C. D., 1943, Suburbs. *American Journal of Sociology*, Vol. 49, July, 1–13.

Harris, C. D., 1990, Urban geography in the United States: My experience of the formative years. *Urban Geography*, Vol. 11, 403–417.

Harris, C. D., 1997, "The nature of cities" and urban geography in the last half century. *Urban Geography*, Vol. 18, 15–35.

Harris, C. D. and Ullman, E. L., 1945, The nature of cities. *Annals of the American Academy of Political and Social Science*. Vol. 242, 7–17.

Hawley, A., 1986, *Human Ecology: A Theoretical Essay*. Chicago, IL: University of Chicago Press.

Jones, K., 1984, *Specialty Retailing in the Inner City*. Toronto, Canada: York University, Department of Geography, Geographical Monograph 15.

Krugman, P., 1995, *Development, Geography, and Economic Theory*. Cambridge MA: MIT Press.

Lamb, R., 1975, *Metropolitan Impacts on Rural America*. Chicago, IL: University of Chicago, Department of Geography, Research Paper 162.

Ley, D., 1983, *A Social Geography of the City*. New York, NY: Harper and Row.

Martin, R., 1999, The "new economic geography": Challenge or irrelevance? *Transactions of the Institute of British Geographers*, Vol. 24, 387–391.

Mayer, H. M., 1954, Urban geography. In P. E. James and C. F. Jones, editors, *American Geography: Inventory and Prospect*. Syracuse, NY: Syracuse University Press.

Meyer, D., 1970, *Spatial Variation of Black Urban Households*. Chicago, IL: University of Chicago, Department of Geography, Research Paper 129.

Morrill, R. L., 1987, The structure of shopping in a metropolis. *Urban Geography*, No. 2, 97–128.

Murphy, R. A., 1966, *The American City*. New York, NY: McGraw-Hill.

Murdie, R. A., 1969, *Factorial Ecology of Metropolitan Toronto, 1951/1961*. Chicago, IL: University of Chicago, Department of Geography, Research Paper 116.

Mabogunge, A., 1968, *Urbanization in Nigeria*. London, UK: London University Press.

Potter, R. B., 1982, *The Urban Retailing System*. Aldershot, UK: Gower.

Pred, A., 1962, *The External Relations of Cities During Industrial Revolution*. Chicago, IL: University of Chicago, Department of Geography, Research Paper 76.

Pred, A., 1965, Industrialisation, Initial Advantage, and American Metropolitan Growth. *Geographical Review*, Vol. 55, 165–180.

Pred, A., 1971, Urban systems development and the long-distance flow of information through preelectronic U.S. newspapers. *Economic Geography*, Vol. 47, 498–524.

Quine, W. V., 1953, *From a Logical Point of View: 9 Logico-Philosophical Essays*. Cambridge, MA: Harvard University Press.

Ramos, D., 1994, HUD-dled masses. *The New Republic*, March 14, pp. 12–16.

Ray, D. M. and Murdie, R. A., 1972, Canadian and American urban dimensions. In B. J. L. Berry, editor, *City Classification Handbook*. New York, NY: Wiley, 181–210.

Rees, P., 1968, The Factorial Ecology of Metropolitan Chicago, 1960. Master's thesis, Department of Geography, University of Chicago. [Largely reprinted in Berry, B. J. L. and Horton, F., 1970, *Geographic Perspectives on Urban Systems*. Englewood Cliffs, NJ: Prentice-Hall.]

Rey, S. G. and Anselin, L., 2000, Regional science publication patterns in the 1990s. *International Regional Science Review*, Vol. 23, 323–344.

Sibley, D., 1995, *Geographies of Exclusion: Society and Difference in the West*. London, UK, and New York, NY: Routledge.

Simmons, J. W., 1964, *The Changing Pattern of Retail Location*. Chicago, IL: University of Chicago, Department of Geography, Research Paper 92.

Simmons, J. W., 1966, *Toronto's Changing Retail Complex*. Chicago, IL: University of Chicago, Department of Geography, Research Paper 104.

Taaffe, E. J., 1990, Some thoughts on the development of urban geography in the United States during the 1950s and 1960s. *Urban Geography*, No. 4, 422–431.

Tomlinson, R., 1989, Geographic Information Systems and Geographers in the 1990s. *The Canadian Geographer*, Vol. 33, 290–298.

Yeates, M., editor, 1998, Metropolitan commercial structure and the globalization of consumer services. *Progress in Planning, Special Issue: Metropolitan Commercial Structure*, Vol. 50, 201–313.

CHAPTER 7

# The Quantitative Revolution in Urban Geography[1]

JOHN S. ADAMS

## ABSTRACT

The quantitative revolution in urban-economic geography flourished in the 1960s at a time when United States domestic policy focused on cities, problems of race and poverty, urban renewal and housing, land use and transportation, and environmental pollution. Theoretical work in previous decades in economic geography, spatial analysis, and quantitative social science had set the stage for the 1960s. After computers came along, psychology and economics moved briskly into theoretical areas, with application of quantitative analysis and statistical analysis, while applied mathematics vaguely promised to unify theoretical and applied realms of social science. Geography's main research agendas through the 1950s had been dominated by physical geography and regional analysis, but a new generation of scholars at Washington, Northwestern, Michigan,

Ohio State, Chicago, Minnesota, and Penn State pushed ahead with serious attention to theory development and quantification in human geography and statistical cartography. Scholars working on the fringes of other disciplines (e.g., Isard) as well as inside geography (e.g., Häger-strand, Berry, Garrison, Haggett, Chorley, Harvey) produced work attractive to the new geography. Mayer and Kohn assembled a stimulating reader illustrating theoretical and applied urban problems using the new approaches. Despite advances in the 1960s and early 1970s, the results of the quantitative revolution in human geography in general—and in urban geography in particular—fell well short of their promise. New ideas, approaches, methods, and problems were never fully exploited or pursued. The reasons are unclear—perhaps we should try to find out why. [Key words: theoretical geography, quantitative revolution in geography, history of geography, geographic thought, spatial analysis, quantitative geography.]

## The Quantitative Revolution in Urban Economic Geography

More than four decades have elapsed since the launching of what historians of our field refer to as geography's "quantitative revolution." It got underway in the late 1950s and flourished throughout the 1960s, so today our colleagues who worked at or near the center of those events are on the verge of retiring. A few, like Peter Gould and Barry Moriarity, have already passed away.

Geography's quantitative revolution occurred at a time when United States domestic policy focused on American cities, with serious national and local policy analysis and programmatic action devoted to problems of racial segregation and poverty, urban renewal and public housing, and neighborhood impacts of new highway construction. In succeeding decades, society's shortcomings and their vivid manifestations within cities persist, but new fashions of thought and inquiry in geography—and particularly in urban geography—have largely displaced the existing theoretical frameworks and robust analytical tools that were ushered in by the quantitative revolution in urban geography.

The quantitative revolution in urban geography occurred during the American involvement in the war in Vietnam. Perhaps when America turned its collective back on the war following the fall of Saigon, it also chose to leave behind some of the analytical frameworks and policy agendas of the 1960s. But changes in scientific methodologies are a regular feature of scholarly inquiry, and, for whatever reasons, urban geography had begun to move in new directions by the mid-1970s.

In preparing for the New York meetings of the Association of American Geographers, James O. Wheeler invited some of us who were involved in "the new urban geography" of the 1960s to reflect on what happened, speculating on why it happened, and assessing its effects. Time marches on, and apparently Wheeler hopes to record our reflections while we are still able to offer them.

I have organized my reflections into three parts. First is a summary of what was happening to me personally as the quantitative revolution heated up in the 1960s. Next, I offer a few comments on what had happened in quantification in the other social sciences before 1960. Finally, I summarize some events of the 1960s inside urban-economic geography that inspired a generation of us to try new ideas, and jump on the "quantitative analysis—theoretical urban-economic geography" bandwagon.

## Personal Background

I was in college in the late 1950s, majoring in economics with a minor in mathematics that included probability theory, statistical analysis, and matrix algebra, in addition to calculus and the mathematics of games, and an introduction to finite mathematical structures using an innovative text by John Kemeny and his colleagues (Kemeny et al., 1959).

From 1960 to 1962, I was at the University of Minnesota, studying economics and statistics in graduate school, while working as a research assistant on a regional economic analysis project titled "The Upper Midwest Economic Study." This was one of a half-dozen major regional studies under way around the United States at that time, including the Penn-Jersey Study, the Pittsburgh Study, the St. Louis Study, and the New York Metropolitan Region Study. Each of these regional studies assessed and interpreted the regional and structural changes in the American economy that occurred in the years following World War II, and described how national economic change played out within diverse metropolitan-centered regions.

Most of these projects were headed by regional economists, but, because this kind of inquiry was out of fashion, those economists were considered to be fringe types—doing interesting applied work, but no work that would get them called up to the "major leagues." Our Upper Midwest Study at Minnesota tried to develop an inter-industry input-output analysis of the Upper Midwest economy in order to track structural economic change across the region, and to show how various parts of the Ninth Federal Reserve District (Montana, North Dakota, South Dakota, Minnesota, Northwest Wisconsin, and Upper Michigan) were faring, given national and regional demographic shifts and economic change that followed the war (Henderson and Krueger, 1965).

A distinctive element of our regional study was the "Urban Research Project" of the Upper Midwest Study, headed by Minnesota geographer John Borchert, who tackled the job of portraying how regional economic and demographic transitions in the region were transforming the trade centers and trade areas of the Upper Midwest (Borchert and Adams, 1963).

But most young American geographers were otherwise occupied. In 1966, the year I received my Ph.D. from Minnesota, most of the Ph.D.s in geography awarded that year were in the area of physical geography. In my case, I moved away from graduate study in economics as a field too confining for purposes of comprehensive policy-oriented urban-economic analysis, and, from 1962 to 1966, I studied economic geography while continuing my work on the Upper Midwest Economic Study. I completed my studies in 1966 and then moved to The Pennsylvania State University, working with Peter R. Gould, Ronald F. Abler, and Anthony V. Williams. Our department chair, Allan Rodgers, and senior members of the department, to their great credit, encouraged us to reorganize the introductory graduate seminar in the spirit of the quantitative revolution in urban and economic geography, so we did (Abler et al., 1971).

A recent Ph.D. student of mine presented a paper at a transportation conference in Honolulu a few years ago. The paper was based on her dissertation, and after her presentation, a young member from the audience asked, "Who was your Ph.D. adviser?" She answered, "John Adams—at Minnesota." The questioner responded, with surprise, "Is he still alive?" Forty years is a long time for a person still in his/her twenties. Evidently, in urban geography, it represents two or three generations of thought and practice.

So, that's my personal story. What I would like to outline next are the other two parts of the story, namely what some of the other social sciences were doing before 1960, and what geography *was* and *was not* doing.

## Advances in Quantitative Social Science in the 1950s and 1960s

Much had been under way in economic geography and spatial analysis, and in quantitative social science, to set the stage for geography's advance in the 1960s. For example, psychology had helped during World War II, developing tests to sort recruits according to their skills and mental abilities, and I.Q. tests had been developed and came into use in the schools to assess pupils' varying abilities to do conventional schoolwork (Thorndike, 1968). Economics had made major theoretical strides with John Maynard Keynes' *The General Theory of Employment, Interest, and Money* (1936), which had provided a framework for national economic policy during the Depression years of the 1930s, as the cause of the Depression came to be diagnosed as flagging consumer demand and diminished national investment. At the National Bureau of Economic Research, founded in 1920, Victor Fuchs was measuring economic activity and analyzing economic census data (Fuchs, 1962). So, at the end of the 1950s, with electronic computers just coming onto the scene, psychology and economics were moving ahead briskly in theoretical areas and in the application of quantitative analysis and statistical methods, while the new areas of mathematical analysis such as the "theory of games" vaguely promised to unify several branches of theoretical and applied social science.

This work was enhanced by the exigencies of the Cold War, with the Office of Naval Research, other branches of the Department of Defense, and the National Science Foundation supplying research funding, while National Defense Education Act Fellowships attracted technically and scientifically adept college graduates into post-graduate study. When we got to graduate school, we found ourselves reading things like John von Neumann and Oscar Morgenstern's *Theory of Games and Economic Behavior* (1944). Walter Isard reviewed von Neumann's and Morgenstern's contributions to strategic analysis, which led to studies of conflict and defense, to two-person game theory, and, eventually, to applications in the locational behavior of firms (Isard, 1956, p. 165).

A stream of other works formed the reading list for many of us studying economic geography and urban geography in the early 1960s. They included Alfred Weber's *Theory of the Location of Industries* (1929), Edgar Hoover's *Location of Economic Activity* (1948), George Kingsley Zipf's *Human Behavior and the Principle of Least Effort* (1949), Wassily Leontief's *Studies in the Structure of the American Economy* (1953), and Edgar Dunn's *The Location of Agricultural Production* (1954), in which he credits Walter Isard, his doctoral adviser.

Isard's *Location and Space Economy* (1956) provided for many of us who were dismayed by the narrowness of conventional neoclassical microeconomics an exciting introduction to concepts and methods for spatial analysis of economic activity, plus an introduction that provided an insightful and comprehensive discussion of advances in theory and quantification in the social sciences. Isard reviewed the early efforts of Weber (1929) and von Thünen (1826) and showed how they found their way into the American social science literature in the 1920s and 1930s (Isard, 1956, pp. 27-29).

Then, there was Harvey Chernoff's and Leon Moses' *Elementary Decision Theory* (1959), and William Alonso's *Location and Land Use* (1964). The message of these works was that formal theory, testable with data, promised reproducible results, a solid basis for understanding and for policy action, and social science knowledge that was cumulative rather than simply additive.

Meanwhile, geography's principal research agendas in the 1950s had been heading in other directions. Physical geography and regional geography dominated, with practitioners asking mainly traditional questions and using traditional methodologies, approaches that were indicted by their brash Young Turk critics as "nontheoretical" and "nonquantitative." Environmental determinism had been discredited as a theoretical framework; Carl Sauer's *Morphology of Landscape* (1925) charted another course, and Hartshorne's *Nature of Geography* (1939) was required reading in some graduate programs. *The Nature of Geography* was highly influential, but argued an exceptionalist case for geography. Then, along came Fred Schaeffer's critical essay appearing in the *Annals of the Association of American Geographers,* (hereafter the *Annals*) entitled "Exceptionalism in Geography" (1953).

The growing divergence between mainstream geography and new developments in other areas of the social and physical sciences was highlighted in 1963 when Edward Ackerman, a Harvard-trained geographer (Ph.D., 1939) and executive officer at the Carnegie Institution in Washington, DC, published an article in the *Annals* titled, "Where is a Research Frontier?" (1963). When we read Ackerman's piece along with the National Research Council's *The Science of Geography: A Report* (1965), is seemed that we had received our marching orders to move the field of urban geography into new and more contemporary directions, using quantitative analysis. So, we marched with enthusiasm, with support from Washington agencies and Washington money.

## Urban Geography Joins the Quantitative Revolution in the Social Sciences

The expansion of quantitative social science included urban-economic geography in the 1960s. By mid-decade, geography was still behind psychology and economics in terms of quantitative analysis, but well ahead of anthropology and political science. Sociology was facilitated by the fact that demographers who had been trained mainly in sociology had been in charge at the Population Division of the Bureau Census since the late 19th century, and the structure and content of the decennial census and many other census surveys perfectly served their research needs.

As part of the effort to jump start quantitative approaches in the social sciences, the National Science Foundation sponsored a series of workshops on quantification. One such workshop for geographers and regional scientists that was organized by William Garrison led to a three-article series in the *Annals,* laying out the thinking behind these new approaches and illustrating the methods to be used in moving forward with quantitative urban and economic research in geography (Garrison, 1959, 1960).

With geography departments at Washington and Northwestern in the lead, soon followed by Michigan, Ohio State, Chicago, Minnesota, and Penn State, serious attention to theory and quantification in human geography moved ahead with alacrity, and no small amount of hubris. The books we used in graduate courses and seminars from the late 1950s through the 1960s—first as graduate students and then as young professors—illustrate what was happening.

One such book was Isard's *Interregional and Regional Input-Output Analysis: A Model of a Space Economy* (1951), which addressed problems with the economist's concept of equilibrium, and problems arising from difficulties dealing with space and the cost of distance. Another favorite was Hägerstrand's *The Propagation of Innovation Waves* (1952), which presented a creative attempt at modeling spatial processes using Monte Carlo simulation.

Mayer and Kohn's *Readings in Urban Geography* (1959) showed how several streams of investigation of the nature of cities and city systems were converging, although the convergence was still rough and uncoordinated. It summarized how scholarly attention had been slow to recognize the urban phenomenon that had spread over Europe, North America, and parts of the rest of the world during the previous century. Adna Weber had published *The Growth of Cities in the 19th Century* (1899) and a variety of studies classified cities according to how they earned their living. Others examined how cities interacted with their hinterlands through trade and migration links. But the most important and original work on cities originated at the University of Chicago where sociologists, geographers, and land economists were figuring out how a place that had been little more than a swamp in 1800 had emerged as one of the world's premier industrial metropolises by 1900.

Mayer and Kohn's collection illustrated how others actually beat geographers to the punch in producing a model of the spatial organization of Chicago (Park and Burgess, 1925), while Homer Hoyt's perceptive work explained the sectoral structure and market dynamics of residential neighborhoods in American cities (Hoyt, 1939). The fact that Chauncy Harris and Edward Ullman's classic article on the nature of cities was published in the *Annals of the American Academy of Political and Social Science* implies where the center of gravity of urban study probably resided at the end of the 1930s (Harris and Ullman, 1945).

The influential German school drew ideas from geomorphology that led to studies in Europe of urban morphology. The French school tried to adapt a method of studying the Franch *pais* to the study of cities as regions with a method defined by Blanchard (1922). When Raymond Murphy at

Clark University published the first urban geography text in 1966 (Murphy, 1966), students like James E. Vance were carrying out formal analysis of the central business districts of large cities while others in the United States and United Kingdom pursued studies of cities as functional centers of broader regions (Dickinson, 1947, 1964).

The urban geography symposium at Lund, Sweden, in 1960, in cooperation with the International Geographical Congress of that year, brought together a group that would include many of the leaders of the urban geography quantitative revolution that unfolded during the following decade (Norborg, 1960). Walter Christaller's original studies of central places in southern Germany and new analysis of the pioneering work of August Lösch stimulated a widening stream of influential theoretical and empirical studies—work that resided on the peripheries of geography and economics and exciting for the participants, but not fully embraced by the mainstream disciplines. Urban-economic geographers were developing formal models of spatial structure and process, testing them with census data and measures gathered from fieldwork; regional economists were incorporating spatial analysis into their work. They shared approaches and findings, and encouraged one another at national and regional meetings of the Regional Science Association. An important compendium of approaches was Isard's *Methods of Regional Analysis* (1960).

We saw the many ways to explore systems of cities when Brian Berry and Allan Pred published *Central Place Studies* (1961, 1965). When Alonso's *Location and Land Use* appeared (1964), we had a powerful theoretical framework for studying the internal residential structure of American cities. The most exciting single item to appear in the mid-1960s probably was Peter Haggett's *Locational Analysis in Human Geography* (1965) which argued a compelling case that these new ideas applied to a full range of topics and questions in human geography, and were not the sole preserve merely of urban and economic geography.

The following year, William Bunge gave us *Theoretical Geography* (1966), a brilliant, if eccentric and contentious, work. In the same year, Peter Hall published *Von Thünen's Isolated State* (1966), and Michael Chisholm published *Geography and Economics* (1966). The beat continued with Richard Chorley and Peter Haggett's *Models in Geography* (1967), which extended the arguments and approaches to a full range of inquiries in physical and human geography. They followed it up two years later with *Network Analysis in Geography* (1969). By the end of the 1960s the ideas and methods thought to be revolutionary a decade earlier were having a wide impact. Graduate and undergraduate departments were revising their ideas about curricula, and methods of analysis that seemed *avant-garde* in 1960 were becoming mainstream. Leslie King published *Statistical Analysis in Geography* in 1969, and David Harvey produced his *Explanation in Geography* in the same year. By the early 1970s, the *Spatial Organization* book that I did with Gould and Abler was readily accepted by many as a useful college and first-year graduate study text across the country and abroad, while Berry and Horton's *Geographic Perspectives on Urban Systems* enjoyed wide acceptance and use (Berry and Horton, 1970).

A number of advisers and mentors in geography contributed in subtle but powerful ways to the quantitative revolution by encouraging students to ask new questions and master new methods of quantitative analysis, to explore new theoretical frameworks, and to recast old inquiries in new ways. In this regard, the influence of Gus Caesar at Cambridge, Torsten Hägerstrand at Lund, Brian Berry at Chicago, Leslie Curry at Toronto, William Garrison at Washington and Northwestern, Waldo Tobler, John Nystuen and Gunnar Olsson at Michigan, Richard Morrill at Washington, Harold McCarty at Iowa, Edward Taaffe at Ohio State, Peter Gould at Penn State, and Fred Lukermann, Phil Porter and John Borchert at Minnesota stands out. There were others, as well, who helped—and some who did what they could to hurt—but the sociology of what happened during the quantitative revolution in urban-economic geography in the 1960s is another story for another time.

## What Did It All Mean? And What Followed?

By the early 1970s, physical geography was moving even faster than human geography in formal modeling and applications of quantitative analysis in geographical inquiry, while human geography was taking a fast turn toward perception and phenomenological inquiry. Traditional college-prep high school curriculum had largely collapsed by the end of the 1960s as college-going rates steadily rose and college and university education was increasingly democratized. The breakdown of the traditional college-prep curriculum accompanied a smorgasbord approach to high school course selection, meaning that students began arriving in our college and graduate geography courses with much less math and science, foreign language, literature, and philosophy than had been the case through the late 1950s.

At the same time that rigorous preparation waned, new ideas and fresh research approaches emerged that perhaps made fewer demands for prior preparation and analytical rigor. These approaches were influenced by efforts to model human perception, such as Kevin Lynch attempted in *The Image of the City* (1960), or as Kenneth Boulding theorized in *The Image* (1961). Good questions about the measurement of human perceptions and human behavior in the urban environment were asked, but they often led to scholarly work that was more impressionistic than analytical, more additive than cumulative.

Probably the most influential single book redirecting new thinking in urban geography was Yi-Fu Tuan's *Topophilia: A Study of Environmental Perception, Attitudes, and Values* (1974). And, if Tuan's influence were not enough to divert young minds away from the methodologies of the 1960s, we moved into a new realm with David Harvey's close reading of Marx and ended up with the influential *Social Justice and the City* (1973).

If I have one regret about all of this, it is that the promise of the quantitative revolution and its value to social science and public policy analysis never came to full fruition. For reasons I do not understand (and perhaps readers of *Urban Geography* might wish to suggest following this forum), as computers improved in capacity and speed, and as data collections were increasingly available in digital form, ability and interest in exploiting these new resources waned. In other words, it appears to me that the ideas, approaches, and methods were never fully employed, and that a subfield that was small to begin with simply failed to gain sufficient momentum. Perhaps if geography had been better represented in the National Academy of Sciences and in full-fledged departments of geography within the nation's major private research universities, the volume of research and its prominence and influence would have been greater, and things might have turned out differently. The possibility remains.

## Note

1. Presented at the Annual Meeting of the Association of American Geographers, New York City, February 28–March 3, 2001.

## Literature Cited

Abler, R. F., Adams, J. S., and Gould, P. R., 1971, *Spatial Organization: The Geographer's View of the World*. Englewood Cliffs, NJ: Prentice-Hall.

Ackerman, E. A., 1963, Where is a research frontier. *Annals of the Association of American Geographers*, Vol. 53, 429–440.

Alonso, W., 1964, *Location and Land Use*. Cambridge, MA: Harvard University Press.

Berry, B. J. L. and Pred, A., 1961, 1965, *Central Place Studies: A Bibliography of Theory and Applications*, Bibliography Series 1, with supplement. Philadelphia, PA: Regional Science Research Institute.

Blanchard, R., 1922, Une methodé de graphie urbaine [A method of urban geography]. *La Vie Urbaine*, Vol. 4, 301–319.

Borchert, J. R. and Adams, R. B., 1963, *Trade Centers and Trade Areas of the Upper Midwest*. Minneapolis, MN: University of Minnesota, Upper Midwest Economic Study.

Boulding, K., 1961, *The Image*. Ann Arbor, MI: University of Michigan Press.

Bunge, W., 1966, *Theoretical Geography*. Lund, Sweden: C. W. K. Gleerup.

Chernoff, H. and Moses, L. E., 1959, *Elementary Decision Theory*. New York, NY: John Wiley.

Chorley, R. J. and Haggett, P., editors, 1967, *Models in Geography.* London, UK: Methuen and Company.

Dickinson, R. E., 1947, *City, Region, and Regionalism: A Geographical Contribution to Human Ecology.* London, UK: Routledge and Kegan Paul.

Dickinson, R. E., 1964, *City and Region: A Geographical Interpretation.* London, UK: Routledge and Kegan Paul.

Dunn, E., 1954, *The Location of Agricultural Production.* Gainesville, FL: University of Florida Press.

Fuchs, V. R., 1962, *Changes in the Location of Manufacturing in the United States Since 1929.* New Haven, CT: Yale University Press.

Garrison, W. L., 1959, Spatial structure of the economy. *Annals of the Association of American Geographers,* Vol. 49, 232-239, 471–482.

Garrison, W. L., 1960, Spatial structure of the economy. *Annals of the Association of American Geographers,* Vol. 50, 357–373.

Hägerstrand, T., 1952, *The Propagation of Innovation Waves.* Lund Studies in Geography: Series B., No. 4, Lund, Sweden: Gleerup.

Haggett, P., 1965, *Locational Analysis in Human Geography,* London, UK: Edward Arnold.

Haggett, P. and Chorley, R. J., 1969, *Network Analysis in Geography.* London, UK: Edward Arnold.

Hall, P., editor, 1966, *Von Thünen's Isolated State.* Oxford, UK: Pergamon.

Harris, C. D. and Ullman, E. L., 1945, The nature of cities. *Annals of the American Academy of Political and Social Science,* Vol. 142, 7–17.

Hartshorne, R., 1939, *The Nature of Geography: A Critical Survey of Current Thought in the Light of the Past.* Lancaster, PA: Association of American Geographers.

Harvey, D., 1969, *Explanation in Geography.* London, UK: Edward Arnold.

Harvey, D., 1973, *Social Justice and the City.* Baltimore, MD: The Johns Hopkins University Press.

Henderson, J. M. and A. O. Krueger, 1965, *National Growth and Economic Change in the Upper Midwest.* Minneapolis, MN: University of Minnesota Press.

Hoover, E. M., 1948, *Location of Economic Activity.* New York, NY: McGraw-Hill.

Hoyt, H., 1939, *The Structure and Growth of Residential Neighborhoods in American Cities.* Washington, DC: Federal Housing Administration.

Isard, W., 1951, Interregional and regional input-output analysis: A model of a space-economy. *Review of Economics and Statistics,* Vol. 33, 318–328.

Isard, W., 1956, *Location and Space Economy: A General Theory Relating to Industrial Location, Market Areas, Land Use, Trade, and Urban Structure.* Cambridge, MA: MIT Press.

Isard, W., 1960, *Methods of Regional Analysis,* Cambridge, MA: MIT Press.

Kemeny, J. G., Mirkil, H., Snell, J. L., and Thompson, G. L., 1959, *Finite Mathematical Structures.* Englewood Cliffs, NJ: Prentice-Hall.

Keynes, J. M., 1936, *The General Theory of Employment, Interest, and Money.* London, UK: Macmillan and Company.

King, L. J., 1969, *Statistical Analysis in Geography.* Englewood Cliffs, NJ: Prentice-Hall.

Leontief, W. W., Chenery, H. B., Clark, P. G., Duesenberry, J. S., Ferguseń, A. R., Grosse, A. P., Grosse, R. N., Holzman, M., Isard, W., and Kistin, H., 1953, *Studies in the Structure of the American Economy: Theoretical and Empirical Explorations in Input-Output Analysis.* New York, NY: Oxford University Press.

Lynch, K., 1960, *The Image of the City.* Cambridge, MA: MIT Press.

Mayer, H. M. and Kohn, C. F., editors, 1959, *Readings in Urban Geography.* Chicago, IL: University of Chicago Press.

Murphy, R. E., 1966, *The American City: An Urban Geography.* New York, NY: McGraw-Hill.

National Research Council, Ad Hoc Committee on Geography, 1965, *The Science of Geography: A Report.* NRC Report No. 1277. Washington, DC: National Academy of Sciences, National Research Council.

Norborg, K., editor, 1960, *Proceedings of the IGU Symposium in Urban Geography.* Lund, Sweden: Gleerup.

Park, R. E. and Burgess, E. W., editors, 1925, *The City.* Chicago, IL: University of Chicago Press.

Sauer, C. O., 1925, Morphology of landscape. *University of California Publications in Geography,* Vol. 2, No. 2, 19–53.

Schaeffer, F. K., 1953, Exceptionalism in geography: A methodological examination. *Annals of the Association of American Geographers,* Vol. 43, 226–249.

Thorndike, R. L., 1968, Intelligence and intelligence testing. In Sills, D. L., editor, *International Encyclopedia of the Social Sciences*. New York, NY: Macmillan, Vol. 7, 421–429.

Tuan, Y., 1974, *Topophilia: A Study of Environmental Perception, Attitudes, and Values*. Englewood Cliffs, NJ: Prentice-Hall.

von Neumann, J. and Morgenstern, O., 1944, *Theory of Games and Economic Behavior*. Princeton, NJ: Princeton University Press.

von Thünen, J. H., 1826, *Der Isolierte Staat [The Isolated State]*. Hamburg, Germany: Perthes.

Weber, A., 1929, *Alfred Weber's Theory of the Location of Industries*. Chicago, IL: University of Chicago Press.

Weber, A. F., 1899, *The Growth of Cities in the Nineteenth Century*. New York, NY: Columbia University Press.

Zipf, G. K., 1949, *Human Behavior and the Principle of Least Effort*. Cambridge, MA: Addison-Wesley Press.

# CHAPTER 8

# Pacific Views of Urban Geography in the 1960s

WILLIAM A. V. CLARK

## ABSTRACT

The 1960s were times of considerable change both within the discipline of geography and in the geography of urban areas. Urban geography came of age in a time when suburbanization and major population decentralization were transforming both inner cities and the outer reaches of the newly expanding cities. Geographers joined planners to investigate the nature and significance of these changes and moved from largely descriptive interpretations of cities and city sites to attempts to theorize the urban transformation. The move to theory and conceptualization placed urban geography at the heart of the changing concept of the discipline and it remains central even in the face of increasing splintering of human geography. The influence of a small group of geographers at the University of Chicago and the University of Washington resonated across the U.S., and the Pacific, through the edited collection of articles in *Readings in Urban Geography* (Mayer and Kohn) and *Studies of Highway*

*Development and Geographic Change* (Garrison, Berry, Marble, Nystuen and Morrill). Those volumes contained theoretical insights and rich empirical analysis, hallmarks of the creativity of urban geographers in the 1960s. [Key words: urban geography, University of Chicago, University of Washington.]

Looking back is more than an exercise in nostalgia; it is an important way of understanding how and why we are at the present juncture in our studies of urban geography. The changes in the two decades of the 1950s and 1960s were central in initiating new and creative ways of looking at cities and the urban process. The outcomes of those initiatives still infuse the work of urban geographers and associated work in spatial demography and urban planning. While the quantitative emphasis of the research in the 1960s and 1970s has been overlain with a complex set of methodologies and contexts, the basic themes which were set in place by the research at the Universities of Washington, Chicago, and Iowa still resonate in the field. I examine these changes but I add the dimension of considering the changes from an antipodean and West Coast perspective. To a great extent the changes which were to affect the Pacific Coast and Australian and New Zealand geography were a direct outcome of the spatial diffusion of ideas from Washington, Chicago and Northwestern.

Before the development of quantitative analyses of cities and towns most of the work in urban geography approached the discussion of cities through descriptive interpretations of townscapes and urban land use. British geographers had developed a rich literature which interpreted and discussed the location of towns and cities in the context of the cultural and historical evolution of European society. These studies ranged from individualistic discussions of significant trading centers to more complex analytic interpretations of the pattern of English villages. These and the studies of American central business districts, the nature of commuting patterns and urban service areas, and the interpretations of the economic contribution of cities to the life of the state all hinted at the possibility for more structured studies of urban phenomena. That flowering came quite quickly in the mid-to-late 1950s with the development of strong urban research programs at Washington and Chicago.

## The Contemporary Background

Two compendiums of research (Mayer and Kohn, 1959; Norborg, 1962) and one collaborative research volume (Garrison, Berry, Marble, Nystuen and Morrill, 1959) defined the initiation of the new research in urban geography but they were not independent of the rich literature, which began to emerge during the 1950s. The research, which emerged at Washington, Northwestern and Chicago in the late 1950s, had a solid basis in papers which ranged from Christaller's work in the 1930s in Germany to Ullman's attempt to define a theory of location for cities in 1941. The papers in Mayer and Kohn and Norborg show just how much the work which emerged in the 1960s and 1970s was an outgrowth of quantitative research and thinking which had been going on for two decades. There is much more *evolution* than revolution in the urban geographic literature of the postwar period in North America. Indeed, any examination of the citations in the articles such as those by Stewart in *Geographical Review* or Harris and Ullman's 1945 article on the nature of cities, only serves to remind us that the flowering of quantitative research in the 1960s was solidly based on a long tradition of earlier research.

The research of the 1960s can be seen reaching back to papers that wrestled with analytical issues of classification and location. The study of classification, always a precursor to the development of more general concepts, was set in place by the Harris (1943) formulation of a functional classification of United States cities and the papers by Ullman (1941) and Harris and Ullman (1945) which examined both central place theory and the internal structure of cities. These early analytic papers set a context for developments in the 1950s and 1960s which are extensions and developments of specific issues of location and classification. The research that was created following these early investigations, which, it is good to remember, were published in both geography and sociology journals, built on the ideas of classification, location theory and the processes of urban change.

The compendium of papers in Mayer and Kohn, including excerpts from Harris, and Harris and Ullman, was notable for three important aspects of the contemporary background of urban geography at the beginning of the 1960s. The papers in Mayer and Kohn included a wide range of regional scientists, sociologists, economists and planners in addition to geographers. Jean Gottmann, Kingsley Davis, Homer Hoyt, Walter Isard and Richard Ratcliff were all anthologized in this book of readings. Second, the book included a mix of statistical and qualitative approaches to understanding city structure, and, third, it emphasized social problems as well as more abstract questions about urban spatial structure. As the first book of readings in urban geography it set a tone by its eclectic but in-depth concern with urban issues. That tone was to be continued in *Geographic Perspectives on Urban Systems* (Berry and Horton, 1970) published ten years later.

*Geographic Perspectives*, published at the end of the 1960s, reflected the changes which had occurred during the 1960s and the important work which grew out of the developments in urban geography at the University of Chicago. During the 1960s work by Berry and his students set out important contributions on policy issues and census administrative decisions. The studies of blight and urban renewal, of functional urban regions and definitions of metropolitan areas were significant pieces of research which connected academic investigations with public policy concerns. The studies at Chicago explicitly examined the urban framework and the impact of policy decisions. It was a part of the wider concern with the application of urban research to pressing urban problems.

## The Demographic Context

The new research was not independent of the fundamental changes occurring in United States society. It is a well worn cliché, but worth repeating, that our social science research is necessarily related to the real world changes that are going on about us. This is particularly true of the link between urban change and urban research in the 1960s and 1970s.

In 1950 the United States population was 154 million; 20 years later it was 208 million, a growth of more than 50 million. In addition to raw population growth there were three important administrative/spatial changes which were critical components in stimulating urban research. In 1950 the U.S. Bureau of the Census defined urbanized areas for the first time. It was the initial recognition of a geographic unit rather than a political or administrative unit. Second, the Census defined a county-based measure of metropolitan structure, the Standard Metropolitan Statistical Area (SMSA). And third, there was a fundamental change in the distribution of the urban population itself. In 1950, 58.5% of the metropolitan population lived in central cities. In 1970, 31.2% of the metropolitan population lived in central cities. In a short 20-year period the "suburbs" had more than doubled in population. The spatial implications for services, land-use planning and new statistical analyses of these changes were enormous, and I think it can be argued it was an implicit if not explicit push for the creation of new urban geographical research.

Urban research at Chicago was focused solidly on the changes that were occurring in urban structure and urban society. The themes which urban geography embraced in this period were very much concerned with the internal structure of the city and of the organization of political, social and economic spaces within the city. The studies ranged from investigations of the changing racial patterns to attempts to classify and analyze the retailing structure of the city, certainly a wide range of theoretical and practical concerns.

## Geographical Science and Urban Geography

There has been an extensive discussion of the role of positivism, much of it highly critical, and little of it reflecting how the adoption of a hard science paradigm created a new way of thinking about spatial phenomena and one which has continued to inform our approaches to understanding cities and city relationships. There is no question that some of the research involved rather uninteresting observations on urban structure but on the other hand the development of serious thinking about space and spatial relations came out of those changes.

As I have suggested elsewhere, we need to exercise some caution before we abandon the approaches and findings from past analytic work on cities. For example, there is a central debate about whether "the new urban structure" is different from past structures or simply a continuation of processes set in place with the emergence of the car as the principal means of urban transport. One should view with caution postmodernist calls for flexism and keno capitalism and an abandonment of current thinking about polycentric urbanism (Clark, 2000). The research on the slow change from monocentric density gradients to a complex structure of multinodal density surfaces only serves to emphasize how adaptable the *models* of the city have been.

Other commentators have discussed the evolution of the quantitative approach in geography in general, and it is more important here to empha- size the role of specific research initiatives at Washington, Northwestern and Iowa, which were summarized in the International Geographical Union (IGU) Symposium at Lund in 1960, and published in 1962. That volume brought together a wide variety of statistical studies of urban phe- nomena, including a range of theoretical works on central place theory. The publication of the IGU Symposium in Urban Geography, (the Lund Symposium of 1960) provided one of the legs of a new and reformulated urban geography (if we view Mayer and Kohn as a first-leg and Garrison et al., as the other leg of the stool). The participants at the IGU sympo- sium represented the younger "urban" faculty—Berry, Marble, Morrill, Dacey, and Thomas among others, as well as their mentors, McCarthy, Garrison, and Mayer. But in addition the meeting was a link to European urbanists and to a wide circle of faculty interested in the broad spectrum of urban problems. Indeed the first half of the book is closely focused on studies of central places and urban hierarchies, where new methods of spa- tial analysis were introduced to examine the nature and distribution of towns and service centers within towns. Interestingly, the second half of the Lund symposium included more traditional studies of the internal structure of the city. Also interesting were the conversations and discussions on philosophical issues which were an integral part of the emergence of an analytical urban geography.

As new graduate students began their studies across the university campuses of the Midwest, the publication of the Mayer and Kohn *Readings in Urban Geography* summarized much of the research initiatives then underway in a far flung set of contexts and set out a single source for new research endeavors. In the same year, Garrison and his students at the University of Washington published, *Studies of Highway Development and Geographic Change*, a detailed set of studies of the nature of highway-related business activities in the metropolitan areas of Washington State. These volumes, all published in a three-year period, firmly established the structure of urban geography and gave a central role to Washington and Chicago and later to Iowa and Northwestern.

The Garrison volume was spatially focused, theoretically informed and analytically quantitative. The theoretical perspective used the notions of rent theory from urban economists, notably Richard Ratcliff, and the economics of transport inputs from Walter Isard. The book was avowedly practical and concerned with one of the pressing problems of the day, the nature of transportation arteries and the associated impacts on businesses and land use. Even so, the book was also concerned to provide a spatial order to the structure of business, and, most importantly, to provide the means to predict future changes in the order and structure of land-use arrangements and the order of business activities. Beyond examining land use changes and the nature and evolution of highway-related business, the research initiated studies of trip making and to a lesser extent of trip making behavior. The analysis employed simple regression, clustering and some straightforward gravity models. It was also important to the extent that it gathered and processed survey data on consumer travel behavior, a topic that was to become more central in the evolution of the geography of behavior in urban areas.

## Coastal and Antipodean Contexts and Viewpoints

The changes in urban geography accelerated with the infusion of students into the geography programs at Washington, Northwestern, Chicago and Iowa, and soon by extension to Penn State, Ohio State, Wisconsin, and other Midwestern universities. The development of a modern urban geography was slower to flower in the universities on the Pacific Coast, which is perhaps puzzling given the early quantitative research at Washington, and the appointment of Nelson at the University of California, Los Angeles (UCLA) and later Pred and Vance at Berkeley. In part it was related to substantial changes in personnel, as senior faculty moved from Washington, though Ullman and Thomas stayed on, and there were strong programs in cultural and regional geography competing for scarce faculty resources within the University of California systems. Still, the context for change was there and the establishment of a new program in Geography at Santa Barbara in the early 1970s with the appointments of Simonett, Tobler and Golledge was notable and created a significantly different intellectual perspective on analytic work in geography.

The creation of an extended Chicago School was a slow process and was related to the diffusion of students from Chicago. The appointment of Howard Nelson at UCLA and Allen Pred at Berkeley provided different paths in urban geography on the Pacific Coast. Nelson brought the perspectives of Harris and classification while Pred brought the central place and locational and process perspectives of Berry. They were both representatives of the Chicago School but they are illustrations of the broad and encompassing role of that school. That they both developed their own and different traditions speaks to the strengths of a training that was not rigidly based on a particular paradigm or methodological tradition. The emphasis on process rather than place, a strength of the Chicago School, offered multiple paths for its students (see Yeates in this volume for more detailed comments). Berkelely and UCLA established urban programs which reflected the Chicago ideas and paradigms but they took on local and often place-based foci, perhaps not surprising in two urban regions which were to become, 20 years later, emblematic of the changes in urban regions in general. The emergence of Silicon Valley in the Bay area and the first truly multiethnic metropolis in Southern California provided a basis for both place-based research and new conceptualizations of urban change. Although notions of a Los Angeles School are rather inflated, they do emphasize how the notions of urban analysis are transformed in new locations.

On the other side of the Pacific, Australia and New Zealand have always had strong links to North America, and it was those links that created and exported the changes in urban geography to New Zealand and Australia. Prior to 1960 there was little urban research in New Zealand geography. The focus in New Zealand geography was regional, physical and historical. However, the links which Leslie Pownall (lecturer and later professor at the University of Canterbury) created from his visits to the United States, and especially with Homer Hoyt, led to a course in Urban Geography. This course, at the then-Canterbury College of the University of New Zealand in 1960, with Mayer and Kohn as the central book, brought the ideas, principals and theoretical conceptualizations of the Chicago School to New Zealand. The decision to use Mayer and Kohn was a decision to bring quantitative, process-oriented thinking to a program that was still largely regional and historical in focus. At the same time the program was introduced onto fertile ground. Physical geography in Australia and New Zealand had a long tradition of process-oriented analysis, and the strong connection between human and physical geography provided a platform for a process-oriented human geography.

Leslie Pownall's work on the New Zealand town attempted to apply and extend the techniques of urban analysis being developed in North America. Pownall's early paper on the functions of New Zealand towns, published in the *Annals of the Association of American Geographers* (Pownall, 1953) was the outgrowth of his reading in the North American literature. His attempts to develop a method of classifying New Zealand towns struggled with the same issues that were central to Harris, and later work by Steigenga (1955), Nelson (1955) and others. The paper was a true extension of the work by Harris and Ullman and in a sense is a classic example of the diffusion of ideas—the application and modification of a technique developed elsewhere to a local problem. Thus, the early work at Chicago was extended across the Pacific to new realms and colonial urban structures.

The links between the New Zealand and the Chicago School were strengthened with the Fulbright visit of Harold Mayer (co-editor with Clyde Kohn of *Readings in Urban Geography*), to Auckland University in 1961, and of Harold McCarthy from Iowa, to the University of Canterbury in 1962. McCarthy's visit to Canterbury where Leslie King had returned from his Ph.D. work in Iowa further strengthened the ideas and approaches of the Midwestern quantitative approaches to urban geography. The appointment of Reginald Golledge from Australia who then followed McCarthy to Iowa in turn created a set of links that flowed into the 1970s with a continuing stream of students traveling to North America for study and of visits from North American academics to New Zealand.

The publication of a Festschrift (McCaskill, 1962), for a distinguished New Zealand Geographer, George Jobberns, included papers on changes in the functional structure of New Zealand towns and central place theory and the spacing of towns in the United States. The paper by Pownall cited Harris' and Nelson's pioneering work on the functional classification of towns and King's paper begins with references to Christaller's classic work. We must remember that this Festchrift, published in 1962, in the same year as the Lund Symposium and only two years after Mayer and Kohn, is notable for the publication of two papers which were direct links to the development of an analytic urban geography. While King's paper on central place spacing is an extension of his dissertation work at Iowa and a direct extension of the attempts to provide a coherent set of explanations for the rise and function of urban places, Pownall's extension of his studies of classification is notable for the attempts to straddle both past idiosyncratic explanations of urban structure and more "modern" attempts to provide a general explanation for urban structure. King concluded that (paraphrasing loosely) the application of statistical techniques will provide a basis for a generalized theory of town location (King, 1962), while Pownall (1962) concluded that detailed field studies would be needed to understand idiosyncratic change in the urban functions of the small towns of New Zealand. Both authors, from very different perspectives, were struggling with the issue of explanation and coherence. They both reflected the fundamental changes that were occurring in North American geography, translated to the Antipodean urban realm.

A similar story links Australia and North America, though the linkages take different and perhaps less direct forms with respect to urban geography. Australian geography in fact had a stronger tradition of economic and industrial geography than urban geography, but Herbert King on the faculty at one of the new Australian universities—Armidale—was teaching Christaller and Lösch at about the time Garrison and his students were pursuing locational analysis at Washington. John Holmes, later at Queensland, was writing about rural settlement patterns and related questions about service areas.

Two individual cases illustrate the linkages and the way in which the linkages extended beyond North America and the antipodes. R. J. Johnston did his doctoral work in Melbourne followed by an appointment in New Zealand in the late 1960s before returning to the United Kingdom. Eric Moore, was trained in Queensland in the early 1960s before joining Northwestern University where he influenced a wide range of students who took up positions both in North America and Canada. In reverse, New Zealand geographers who traveled to Canada or the United States returned and took up positions in Australian universities and developed strong urban programs which invariably were influenced by their training by Chicago graduates. These antipodean programs drew heavily on the research and teaching in place at Iowa, Minnesota, Wisconsin, Chicago and Ohio State, and, in Canada, at Toronto.

## New Paradigms and Connections

The entry of a large number of new young geographers expanded and enriched the study of urban geography and took it in new directions. Studies of social spaces, of neighborhood change and ethnic patterns brought geography into closer contact with colleagues in sociology who were exploring the same issues from a social perspective. Studies of human behavior in shopping and travel brought geographers into contact with transportation planners and marketing professionals, and studies of residential mobility and migration provided linkages to economists interested in population change and the outcomes of these flows. The transformation of modes of investigation and the spread of these ideas to New Zealand and Australia have been important to the continuing strength of geography. In Austrialia and New Zealand, geography is central and respected both within the academy and outside, more central than in the American academy where there is a continuing struggle for position within universities. By strengthening the linkages between the United States and the antipodes the field as a whole was enriched during the 1960s and that enrichment continues.

## Observations

Not surprisingly in a retrospective such as this, the outlook is likely to be particular and selective. At the same time by examining the particular ways in which changes in the practice of urban geography in the antipodes was based on the new research in the U.S. which was then strengthened with the flows and interconnections of students and teachers between the U.S. and Australasia, is a reminder of how powerful those changes occurring in social science in general were, across space and through time. The changes in urban geography today, while more eclectic than 40 years ago, should still be based on an analytic concern with discovering new knowledge and integrating that knowledge into the wider geographic practice.

## Literature Cited

Berry, B. J. L. and Horton, F. E., 1970, *Geographic Perspectives on Urban Systems*. New York, NY: Prentice Hall.

Berry, B. J. L., 1963, *Commercial Structure and Commercial Blight*. Chicago, IL: University of Chicago, Department of Geography Research Paper 103.

Berry, B. J. L., 1964, Cities as systems within systems of cities. *Papers and Proceedings of the Regional Science Association*, Vol. 13, 147–163.

Berry, B. J. L., 1967, *Functional Economic Areas and Consolidated Urban Regions of the United States*. Washington, DC: Bureau of the Census.

Berry, B. J. L., Goheen, P., and Goldstein, H., 1968, *Metropolitan Area Definition: A Re-evaluation of Concept and Statistical Practice*. Washington, DC: Bureau of the Census.

Clark, W. A. V., 2000, Monocentric to polycentric: New urban forms and old paradigms. In Gary Bridge and Sophie Watson, *A Companion to the City*. Oxford, UK: Basil Blackwell.

Garrison, W. L., Berry, B. J. L., Marble, D. F., Nystuen, J., and Morrill, R., 1959, *Studies of Highway Development and Geographic Change*. Seattle, WA: University of Washington.

Harris, C., 1943, A functional classification of cities in the United States. *Geographical Review*, Vol. 33, 86–99.

Harris, C. D., and Ullman, E. L., 1945, The nature of cities. *Annals of the American Academy of Political and Social Science*, Vol. 242, 7–17.

King, L. J., 1962, Central place theory and the spacing of towns in the United States. In M. McCaskill, editor, *Land and Livelihood: Geographic Essays in Honour of George Jobberns*. Christchurch, New Zealand: New Zealand Geographical Society.

Mayer, H., and Kohn, C., 1959, *Readings in Urban Geography*. Chicago, IL: University of Chicago Press.

McCaskill, M., editor, 1962, *Land and Livelihood: Geographic Essays in Honour of George Jobberns*. Christchurch, New Zealand: New Zealand Geographical Society.

Nelson, H. J., 1955, A service classification of American cities. *Economic Geography*, Vol. 31, 189–210.

Norborg, K., 1962, *Proceedings of the IGU Symposium in Urban Geography Lund, 1960*. Lund, UK: C. W. K. Gleerup Publishers.

Pownall, L. L., 1962, Changes in the functional structure of New Zealand towns. In M. McCaskill, editor, *Land and Livelihood: Geographic Essays in Honour of George Jobberns*. Christchurch, New Zealand: New Zealand Geographical Society.

Pownall, L. L., 1953, The functions of New Zealand towns. *Annals Association of American Geographers*, Vol. 43: 332–350.

Stewart, C. T. Jr., 1958, The size and spacing of cities, *Geographical Review*, Vol. 48, 222–245.

Steingenga, W., 1955, A comparative analysis and a classification of Netherlands towns. *Tijdschrift voor Economishe en Sociale Geographie*, Vol. 46, 105–119.

Ullman, E., 1941, A theory of location for cities. *American Journal of Sociology*, Vol. 46, 853–864.

# CHAPTER 9

# Assessing the Role of Spatial Analysis in Urban Geography in the 1960s[1]

JAMES O. WHEELER

Bare ruin'd choirs, where late the sweet birds sang.

William Shakespeare

## ABSTRACT

Using content analysis, this study examines four leading geography journals and two regional science outlets for the number and kinds of urban geography articles published during the 1960s. Whereas the 1960s are typically portrayed as the heyday of the theoretical and quantitative revolution, a surprisingly large number of urban geography articles published during this time reflected a pre-1958 research mode. *The Professional Geographer*, for example, published 386 articles during the 1960s, most of which were quite short. A total of 61 of these articles, or 15.8%, were classified as urban—broadly defined. Of these urban geography articles, only 28, or 45.9%, were classified as following the spatial analysis format—that is, they were considered quantitative-theoretical. This study calls into question the dominance of spatial analysis in urban geography during the 1960s. In hindsight, however, it is evident that it was those relatively few articles following the spatial tradition that transformed pre-1970 urban geography from a weakly structured subdiscipline into a more mature and expansive field of investigation. [Key words: content analysis, spatial analysis, 1960s, urban geography.]

Apparently, it is widely believed in geography that the 1960s constituted a decade dominated by quantitative-theoretical geography. Martin and James (1993, p. 227), for example, declared that "the geography of the 1960s is referred to as spatial science." Likewise Holt-Jensen (1988, p. 76) wrote, "In retrospect it may be said that the 1960s [in geography] can be characterized as an era of 'hard science.'" Mikesell (1984, p. 192) observed that "the 1960s was a decade of remarkable turbulence and heated debate. That this turbulence receded and debate ceased to be heated within four or five years suggests that the controversy about "old" and "new" geography was both initiated and terminated during the 1960s." Johnston (1987, p. 74) asserted that "by the mid-1960s, the changes seem to have been widely accepted, and the regional approach had certainly been ousted from its prime position in the publications of human geographers." Taaffe (1974, p. 1) also believed that "the sixties were dominated by the spatial view...." While it is certainly true that "the 1960s ushered in the theoretical and quantitative revolutions in geography" (Martin and James, 1993, p. 226), the degree to which this movement—often miscalled the "quantitative revolution"—actually held ascendancy in American geography during this period is questionable. As Taaffe (1979, p. 133) has reminded us, at the end of the 1950s, "the changes were extremely concentrated...at just a few universities, notably Washington, Iowa, Chicago, and Northwestern."

In this study the term "spatial analysis" is used to characterize this fresh approach to the discipline, which in fact emerged in the mid-to-late 1950s, first at the University of Washington (Berry, 1993). The spatial view constituted one of Pattison's (1964) four traditions in geography, a theme elaborated upon by Taaffe (1974, p. 1) in focusing on its "strengths and weaknesses during the 1960s." To assess the role of spatial analysis in urban geography in the 1960s, this study relies upon content analysis to examine four leading geography journals and two regional science journals of that era.

The concentration on urban geography in this study appears to be appropriate, given the emphasis of spatial analysis on urban topics during the 1960s. "Within human geography in the United States, the initial development of systematic studies using the positivist scientific method was very largely focused on economic and the associated economic aspects of urban geography" (Johnston, 1987, p. 74). "The initial focus of the effort, in urban and economic geography, produced heavy impact on two of the most populated fields of American geography" (Mikesell, 1984, p. 193). Therefore, the goal of this study is to address the question of the degree to which leading geography and regional science journals published, during the 1960s (1) urban articles, and (2) urban articles that followed the spatial analysis approach. Just how paramount was the "new" urban geography during this decade?

## Urban Geography in the Context of the 1960s

A number of developments and issues are pertinent to understanding the field of urban geography in the 1960s. Few would disagree with Kohn (1970, p. 211) that "the emergence of geography as an abstract, theoretical science appears to have been the most overriding development in geographic research during the 1960s." As just noted, in addition to urban geography being a relatively large subfield of geography, this subfield was at the forefront in applying theoretical constructs and employing statistical methods and mathematical modeling. In 1957, the Soviet Union's launching of Sputnik, the world's first space satellite, generated a great interest in the United States in scientific and technological issues. Bulky mainframe computers became increasingly available on college and university campuses for faculty and graduate-student use, as computer-center administrators were eager to purchase ever larger and ever faster machines to handle punch cards and magnetic tapes. Data availability skyrocketed. Beginning with the 1960 United States Census, much more detailed spatial data became available from the Bureau of the Census, particularly for metropolitan areas. Federal-government-mandated metropolitan planning led to large urban data inventories made accessible on boxes and boxes of punch cards, a primitive technology compared with the vast amount of information that can be stored on the tiny computer disk today. Graduate geography classes in computer programming became commonplace. Elementary computer-assisted cartography and digitizing became widespread. The Symap system, for example, came increasingly into use in the late 1960s to portray choropleth and isoline maps (a great breakthrough at the time but ugly by today's standards). These and many other technological developments had an impact on urban geography during this period.

It was also during the 1960s that baby boomers entered college, creating a great swell in undergraduate and, later, graduate enrollments. Urban geography attracted its share of this burgeoning student population. Membership in the Association of American Geographers (AAG) more than tripled, from 2004 in 1960 to 6866 a decade later. Demand for college and university teaching was so great that many with only M.A. degrees became professors. Since urban geography was so strongly aligned with spatial analysis, many—though not all—of its graduate students of the 1960s not only were exposed to thinking in terms of theoretical geography, that is, location theory, but also were required to take at least one course in quantitative methods in geography (Lavalle et al., 1967). Since these quantitative courses typically involved applied statistics, students learned to keypunch, use card-sorting machines, run basic programs (again using punch cards placed before the punch-card data and formatted to identify, for example, where the decimal points were located in the data fields). Many graduate students studying urban geography, therefore, held little in common with those investigating cultural geography (farms and fences, house and barn types) or, say, the regional, historical geography of Madison County, Georgia. Although Ian Burton's (1963) early declaration that the so-called "quantitative revolution" was over, the use of quantification and the cogitating of theoretical geography had not penetrated into all areas of geography (nor had the methods been thoroughly embraced even in urban geography by 1963).

## Data Selection

Four geography journals and two regional science outlets were selected for content analysis to shed light on the status of urban geography, and particularly the role of spatial analysis, in the 1960s. In addition to the two major serials published by the Association of American Geographers (*Annals of the Association of American Geographers* [hereafter referred to as the *Annals*] and *The Professional Geographer*), *Economic Geography* and *Geographical Review* were selected for study. Prior to the establishment of *Urban Geography* in 1980, *Economic Geography* was the principal outlet for research in urban geography. Although younger geographers today may think it quaint to include *Geographical Review* in this investigation, the fact is that this journal—published by the long-established and prestigious American Geographical Society—was generally viewed as one of the three or four highest quality geography journals during the 1960s, before the subsequent spawning of the many specialty journals we have today. *Journal of Regional Science*, published by the Regional Science Association, first appeared in 1958 and was an important outlet for urban research by economists, geographers, and others. *Papers of the Regional Science Association*[2] was the published proceedings of various regional science meetings. The

Regional Science Association, founded in 1954 by Walter Isard, was vital, especially in the early 1960s, in sustaining and amplifying research in urban geography.

Content analysis of these six publications consisted merely of counting (1) the total number of articles published in each issue, (2) classifying the articles as urban or nonurban, and (3) identifying those urban-focused articles as following or not following the spatial analysis approach. An inclusive interpretation as to what constituted an urban article was adopted. To be classified as falling within the spatial analysis sphere, the article had to embrace a theoretical and/or quantitative perspective. My interpretations of course are highly personal, and others might find certain differences in their explication of this same undertaking—though I believe little change in this study's overall results would be found.

## Article Evidence

Approximately 21% of all articles published in these six journals treated urban topics during the 1960s (Table 9.1), though many articles published in *Journal of Regional Science* and *Papers of the Regional Science Association* were written by economists. Only about 11% of the total articles published treated urban topics from the spatial analysis viewpoint. (Of course a number of articles used quantitative-theoretical approaches in nonurban areas of study, especially in transportation and economic geography.) The outlet with the highest percentage of total articles using spatial approaches to urban studies was *Journal of Regional Science*, with more than 26%,

**TABLE 9.1** Urban Geography Articles Published in Selected Journals, 1960–1969

| Journal | Total Number of Articles | Number of Urban Articles[a] | Number of Urban Articles Using Spatial Analysis[a] |
|---|---|---|---|
| *Annals* | 329 | 59 (17.9) | 26 (7.9) |
| *Economic Geography* | 224 | 76 (33.9) | 23 (10.3) |
| *Geographical Review* | 250 | 35 (14.0) | 15 (6.0) |
| *Journal of Regional Science* | 125 | 36 (28.8) | 33 (26.4) |
| *Papers of the Regional Science Association* | 235 | 62 (26.4) | 51 (21.7) |
| *The Professional Geographer* | 386 | 61 (15.8) | 28 (7.4) |
| Total | 1549 | 329 (21.2) | 176 (11.4) |

[a] Number of articles as percentage of total number of articles is given in parentheses.
*Source*: Compiled by author.

followed by *Papers of the Regional Science Association*, with over 21%. The lowest was *Geographical Review*, with just 6%, closely followed by *The Professional Geographer* (7.4%), the *Annals* (7.9%), and *Economic Geography* (10.3%), though *Economic Geography* published the largest number of urban articles (76) of any of the six journals. *Geographical Review* published the lowest number of urban studies (35) as the journal was viewed then as now as focusing more on historical, regional, and cultural topics, but the mere 15 urban articles the journal did publish following the spatial analysis tradition were, however, some of the most-cited of the 1960s. Although overall the spatial analysis proportion of the total number of publications was not large, it was enormous compared to the proportion of spatial analysts in the profession at the time.

*Journal of Regional Science* had the highest percentage of urban articles using the spatial analysis viewpoint (92%), followed by *Papers of the Regional Science Association* (82%), *The Professional Geographer* (46%), the *Annals* (44%), and *Geographical Review* (43%) (Table 9.2). The journal with the lowest percentage was *Economic Geography* (30%), the serial with the highest total number of urban articles. The overall mean among the six journals was 53.5%.

The data were broken down into two five-year time periods, 1960–1964 and 1965–1969, to determine if there were any notable differences between them (Table 9.3). Not unexpectedly, there was an increase in the number of urban articles published in the later time period (35%), as well as a corresponding increase in the number of urban articles using spatial analysis (35%). The former increase—largely attributable to the *Annals, The Professional Geographer, Journal of Regional Science,* and *Geographical Review*—relates, no doubt, to the growth of the discipline and to the attraction of the subfield of urban geography. Increases in the number of urban articles

**TABLE 9.2** Percentage of Urban Articles Using Spatial Analysis, Selected Journals, 1960–1969

| Journal | Percent |
|---|---|
| *Annals* | 44.1 |
| *Economic Geography* | 30.3 |
| *Geographical Review* | 42.9 |
| *Journal of Regional Science* | 91.7 |
| *Papers of the Regional Science Association* | 82.3 |
| *The Professional Geographer* | 45.9 |
| Overall Mean | 53.5 |

*Source*: Compiled by author.

**TABLE 9.3** Breakdown of Data by Two Five-Year Periods

| Journal | Number of Urban Geography Articles | | Number of Urban Geography Articles Using Spatial Analysis | |
|---|---|---|---|---|
| | 1960–1964 | 1965–1969 | 1960–1964 | 1965–1969 |
| *Annals* | 22 | 37 | 10 | 16 |
| *Economic Geography* | 39 | 37 | 12 | 11 |
| *Geographical Review* | 13 | 22 | 6 | 9 |
| *Journal of Regional Science* | 13 | 23 | 13 | 20 |
| *Papers of the Regional Science Association* | 29 | 33 | 21 | 30 |
| *The Professional Geographer* | 27 | 41 | 13 | 15 |
| Totals | 143 | 193 | 75 | 101 |

*Source*: Compiled by author.

classified as following the spatial analysis approach occurred principally with the two regional science journals and to a lesser extent with the *Annals*. No outlet showed a notable decrease in this category, indicating greater acceptance of and training in the spatial tradition in the later 1960s, despite the developing wave of Marxism and radical geography late in the decade (e.g., *Antipode: A Radical Journal of Geography* began publication at Clark University in 1969).

## Editorships

The four geography journals were all edited during the 1960s by leading scholars in the field, but their graduate education was in the traditional historical, regional, and cultural areas of geography. Wilma Fairchild, long-time and highly respected editor of *Geographical Review*, served throughout the 1960s, as did Raymond Murphy, who edited *Economic Geography* during the entire decade of the1960s. Professor Murphy, author of the first actual textbook in urban geography (*The American City*, 1966), had obtained his Ph.D. in 1930 from the University of Wisconsin. The *Annals* was served during the 1960s by Walter M. Kollmorgen (1960), Ph.D., 1950, Columbia University; Robert Platt (1961–1964; see note to Table 9.5), Ph.D., 1930, University of Chicago; Norton Ginsberg (1961–1962), Ph.D., 1949, University of Chicago; and Joseph Spencer (1964–1969), Ph.D., 1936, under Carl O. Sauer, University of California, Berkeley. *The Professional Geographer* was edited during this period by Phyllis R. Griess (1960–1962), Ph.D., 1948, Pennsylvania State University, and Hallock Raup (1963–1971), Ph.D., 1935, also under Carl Sauer, University of California, Berkeley. The point of this detail is to indicate that these editors were indeed educated in a different era—long before spatial analysis entered geography in a significant way—and were viewed as not especially enthusiastic toward quantitative and theoretical approaches to geography. In fact, the general feeling that spatial analysis was not getting its fair share of space in the leading geography journals led to the establishment of *Geographical Analysis: An International Journal of Theoretical Geography* in 1969 at Ohio State University (Golledge, 1979).

The two journals published by the Association of American Geographers, which appoints editors, were examined to determine if there were any notable differences among editors in the publication rates of urban spatial analysis articles (Tables 9.4 and 9.5). (As already noted, *Economic Geography* and *Geographical Review* maintained the same editors throughout the 1960s, and the two regional science journals, though having changing editorships, retained editors extremely sympathetic to spatial analysis.)

In the case of *The Professional Geographer*, during the 1960s, Phyllis Griess was editor for three years (1960–1962) and H. F. Raup for seven (1963–1969). Table 9.4 shows a noticeable contrast between the two editors, with Raup publishing a higher percentage of urban spatial analysis articles and averaging over three and one half per year compared to one per year for Griess. Since Raup's term came later, one wonders to what degree the difference reflects the editors' attitudes toward spatial analysis and to what extent it reflects the changes taking place in the field of urban geography. In any case, very few urban spatial analyses (28) were published during the 1960s in *The Professional Geographer*.

The *Annals* published even fewer urban spatial analysis articles (26), but the case of the *Annals* is more complex, with four different editors serving during the 1960s. Walter Kollmorgen, editor from 1955 through

**TABLE 9.4** Article Analysis, *Professional Geographer,* by Editor, 1960–1969

| The Professional Geographer | Griess, 1960–1962 | Raup, 1963–1969 |
|---|---|---|
| Total number of articles | 118 | 268 |
| Total number of urban articles | 11 | 50 |
| Total number of urban spatial articles | 3 | 25 |
| Urban spatial articles as percentage of total articles | 2.5 | 9.3 |
| Urban spatial articles as percentage of total urban articles | 27.3 | 50.0 |

*Source:* Compiled by author.

**TABLE 9.5** Article Analysis, Annals, by Editor, 1960–1969

| Annals | Kollmorgen, 1960 | Platt,[a] 1961–1964 | Ginsburg,[a] 1961–1962 | Spencer, 1964–1969 |
|---|---|---|---|---|
| Total number of articles | 19 | 77 | 18 | 215 |
| Total number of urban articles | 3 | 13 | 1 | 42 |
| Total number of urban spatial articles | 2 | 5 | 0 | 19 |
| Urban spatial articles as percentage of total articles | 10.5 | 6.5 | 0.0 | 8.8 |
| Urban spatial articles as percentage of total urban articles | 66.7 | 38.5 | 0.0 | 45.2 |

[a] Platt's term as editor was interrupted due to field work in Pakistan. Norton Ginsburg was acting editor for the December 1961, March 1962, and June 1962 issues. Platt returned as editor for the September 1962 issue and continued through the March 1964 issue, whereupon Joseph Spencer became editor as of the June 1964 issue.
*Source*: Compiled by author.

1960, actually published the highest percentage of urban spatial articles to total articles (10.5%) during the period covered by the study, even though the absolute number was only two (Table 9.5). This percentage is likely an anomaly, since his editorship, for purposes of this study, covers only one year (1960). Robert Platt published only five urban spatial articles during his interrupted editorship from 1961 to 1964 (see note to Table 9.5). Since Norton Ginsburg was acting editor for only three issues, his record of publishing no urban spatial articles is probably not significant. Joseph Spencer, editor for six years (1964–1969), published 19 urban spatial articles, an average of more than three per year, a figure higher than that for any of the three other editors. All in all, it appears, however, that it is impossible to separate editorial attitude toward urban spatial analysis from the growth of urban spatial analysis during the decade, though one might argue that Spencer should have published more analytical articles given the higher level of interest in spatial analysis in the late 1960s, as implied on certain anecdotal grounds.

## Significance of Articles

Even a study not limited to six journals could not fully assess the role of urban spatial analysis by number counts alone. Somehow the significance of the articles published must be taken into account, however subjectively. Wrigley and Matthews (1986, 1987) identified most-cited articles based on

**TABLE 9.6** Most-Cited Urban Geography Articles Following the Spatial Analysis Tradition, 1958–1969

| Author | Date | Journal |
|---|---|---|
| B. J. L. Berry and W. Garrison | 1958 | *Economic Geography* |
| B. J. L. Berry and W. Garrison | 1958 | *Papers and Proceedings of the Regional Science Association* |
| P. R. Gould | 1963 | *Annals* |
| E. J. Taaffe et al. | 1963 | *Geographical Review* |
| J. Wolpert | 1964 | *Annals* |
| R. L. Morrill | 1965 | *Geographical Review* |
| J. Wolpert | 1965 | *Papers and Proceedings of the Regional Science Association* |
| R. L. Morrill | 1967 | *Annals* |
| A. G. Wilson | 1967 | *Transportation Research* |
| G. Rushton | 1969 | *Annals* |

*Source:* Wrigley and Matthews (1986, 1987).

the Social Science Citation Index. Table 9.6 shows the ten most-cited urban geography articles that used spatial analysis from 1958 through 1969, some of which continue to be included in introductory textbooks and in contemporary texts in urban geography. It is clear that not all urban articles following spatial analysis were of equal significance.

## Concluding Comments

This study indicates that, while spatial analysis no doubt reached its apex in the 1960s, it was not the dominant force perceived by scholars such as those quoted early in this paper. It represented overall only about one-half of the urban studies among the six journals examined here. Among the four geography outlets, spatial analysis contributed to less than 40% of all urban articles. Although this study based on article counts excludes books published and papers presented at professional meetings, it is evident, in hindsight, that spatial analysis never had the overarching power that some attributed to it. It was, however, that small number of articles in the urban spatial tradition that transformed urban geography from an essentially regional-study approach to a topical, analytic one that linked urban geography more closely with the social sciences and remains central to urban geographic research today.

Several other points have been made. First, although articles in urban spatial analysis represented a small share of total articles published during the 1960s, the number was huge compared with the small number of geographers using spatial analysis at the time. Second, urban geography was a large subfield within the profession in the 1960s, as it is today. Third, the growth of geography was enormous, with the number in the discipline more than tripling during the decade. Fourth, this overall growth was also reflected in the gradual expansion of urban spatial analysis during the decade. Fifth, the role of regional science was essential in stimulating and sustaining urban spatial analysis, particularly in the early 1960s, when there were so few geographers knowledgeable in the spatial analysis approach. Sixth, while anecdotal evidence suggests that editors of the four geography journals examined here were not especially supportive of spatial analysis, it is difficult to separate an editor's attitude toward this new approach from the growth and evolution of spatial analysis throughout the decade. We will never know if more sympathetic editors might have been willing to publish a greater volume of research from spatial analysts. Finally, simple number counts can overlook the quality and influence of particular articles, as indicated, for example, by the most-cited articles identified by Wrigley and Matthews (1986, 1987).

## Notes

1. I wish to thank Dr. Brian J. L. Berry for suggestions for improving an earlier draft of this chapter and to acknowledge with appreciation Jodie Traylor Guy for many helpful editorial suggestions.
2. This publication has also appeared under the title *Papers and Proceedings of the Regional Science Association.*

## Literature Cited

Berry, B. J. L., 1993, Geography's quantitative revolution: Initial conditions, 1954–1960. A personal memoir. *Urban Geography,* Vol. 14, 434–441.

Burton, I., 1963, The quantitative revolution and theoretical geography. *Canadian Geographer,* Vol. 7, 151–162.

Golledge, R. G., 1979, The development of *Geographical Analysis. Annals of the Association of American Geographers,* Vol. 69, 154–155.

Holt-Jensen, A., 1988, *Geography: History and Concepts* (2nd ed.). Totowa, NJ: Barnes and Noble.

Johnston, R. J., 1987, *Geography and Geographers.* New York, NY: Arnold.

Kohn, C. F., 1970, The 1960s: A decade of progress in geographical research and instruction. *Annals of the Association of American Geographers,* Vol. 60, 211–219.

Lavalle, P., McConnell, H., and Brown, R. G., 1967, Certain aspects of the expansion of quantitative methodology in American geography. *Annals of the Association of American Geographers,* Vol. 67, 423–436.

Martin, G. J. and James, P. E., 1993, *All Possible Worlds: A History of Geographical Ideas.* New York, NY: John Wiley & Sons.

Mikesell, M., 1984, North America. In R. J. Johnston and P. Claval, editors, *Geography since the Second World War.* Totowa, NJ: Barnes and Noble, 185–213.

Pattison, W., 1964, The four traditions in geography. *Journal of Geography,* Vol. 63, 211–216.

Taaffe, E. J., 1974, The spatial view in context. *Annals of the Association of American Geographers,* Vol. 64, 1–16.

Taaffe, E. J., 1979, Geographies of the sixties in the Chicago area. *Annals of the Association of American Geographers,* Vol. 69, 133–138.

Wrigley, N. and Matthews, S., 1986, Citation classics and citation levels in geography. *Area*, Vol. 18, 185–194.

Wrigley, N. and Matthews, S., 1987, Citation classics in geography and the new centurions: A response to Haigh, Mead, and Whitehand. *Area,* Vol. 19, 279–284.

# III
## Urban Geography in the 1970s

# III

Urban Geography in the 1970s

# CHAPTER 10

# A Decade of Methodological and Philosophical Exploration

MARTIN CADWALLADER[1]

## ABSTRACT

The 1970s were a time of considerable theoretical debate in urban geography. The once dominant paradigm of spatial analysis came under increasing scrutiny from behavioral, Marxist, and realist perspectives. A decade that had opened with a largely unexamined confidence in the scientific method was to close with an uneasy sense of pluralism. This paper examines the major contours of that debate from an autobiographical perspective. My own academic journey through the 1970s involved an early engagement with the behavioral approach, especially within the contexts of residential mobility, migration, and consumer behavior.

[1]Correspondence concerning this chapter should be addressed to Martin Cadwallader, Vice Chancellor for Research and Dean of the Graduate School, The Graduate School, University of Wisconsin, Bascom Hall, 500 Lincoln Drive, Madison, WI 53706; telephone: 608-262-1044; fax: 608-262-5134; e-mail: cadwallader@mail.bascom.wisc.edu

131

Toward the end of the decade, however, I was pursuing a more methodological focus. In particular, I had begun to explore the role that structural
equation models might play in our understanding of spatial patterns and
the processes that both produce them and are constrained by them.
These structural equation models ultimately included causal models,
path analysis, simultaneous equations, and the use of latent, or unmeasured variables. [Key words: spatial analysis, behavioral approach, Marxist
political economy, structural equation models.]

The 1960s had been an era of theoretical consensus. Urban geography
was dominated by the positivist tradition, an empirically based approach
to social science that involved generous doses of hypothesis-testing and
model-building. The goal was to articulate the mutual interrelationships
between socioeconomic processes and their spatial manifestations (Berry,
1967). Form and process was the mantra of the day. The analysis of form
required a preoccupation with the geometry of spatial patterns and a
vocabulary involving points, networks, surfaces, and regions (Haggett,
1965). By contrast, the causal processes responsible for generating those
patterns were often borrowed from cognate disciplines, especially economics and sociology. Neoclassical economics and human ecology, for
example, proved fertile theories for extracting concepts with which to
explain patterns of urban land use and residential differentiation (Yeates
and Garner, 1971).

Much of the excitement associated with this positivist tradition, and
the associated scientific method, was captured in David Harvey's seminal
*Explanation in Geography* (Harvey, 1969). Indeed, it was this book that
first captured my imagination when, in 1969, I joined the department of
geography at the University of California at Los Angeles to pursue my
graduate education. I was excited to join the still relatively small community of urban geographers who were expecting to rewrite the subdiscipline
in terms of hypotheses, generalizations, models, and theories. We were
utterly confident that abstract modes of thought, especially statistics and
mathematics, could be invoked as a means of describing, explaining, and
predicting the phenomena of interest. We were as yet unencumbered by
any doubts concerning the following assumptions: that social science can
generate scientific laws that are well-confirmed empirical regularities; that
scientific discourse can be value free or neutral; and that peculiarly spatial
laws can be envisaged in which space is somehow independent of time and
matter.

## The Behavioral Turn

Some of us, however, were becoming increasingly disenchanted with the rather unsatisfactory concept of "economic man" that had been borrowed from neoclassical economics. In particular, the associated assumptions of profit maximization and perfect information led us to seek a more realistic model of human behavior that would embrace the principles of satisficing behavior, bounded rationality, and incomplete information. It was from these beginnings that the behavioral approach began to be formulated. A central tenant was that a person's behavior was based on his or her perception of the environment, and not on the environment as it actually exists. Learning, attitude formation, and decision-making became part of the urban geographer's lexicon. In retrospect, however, the break from the conventional analytical framework was only relatively minor (Golledge, 1981). Such concepts as residential stress, place preferences, and cognitive distance were used in addition to, rather than as a replacement for, more traditional measures of physical and socioeconomic environments. The search for empirical generalizations continued unabated, although individual rather than aggregate level data were used as input for the various statistical manipulations. Also, psychology, rather than economics or sociology, was the preferred hunting ground when searching for suitable concepts to borrow and adapt.

Models of spatial choice to describe how people choose between locationally specific alternatives, such as supermarkets or shopping centers, were of increasing interest. In this context, the revealed preference approach was widely used to examine observed behavior in order to uncover the underlying preference structure. In particular, a series of indifference curves could be derived, thus identifying the trade-offs between various combinations of attributes in given situations (Rushton, 1981). However, as the revealed preference approach only enables one to deduce a preferential ordering for the range of opportunities available in the study area, some researchers began to use experimental designs to generate a set of hypothetical alternatives that are independent of any particular spatial structure. For example, information integration theory, which involves constructing an algebraic model of information processing, allows attribute values to be manipulated such that a variety of abstract combinations can be presented to the subjects. Similarly, the conjoint measurement approach also required subjects to evaluate multiattribute alternatives by comparing predetermined levels of supposedly important variables.

With encouragement from my major advisor, Bill Clark, I began to explore the phenomenon of residential mobility from a behavioral perspective. For my master's thesis, I developed a concept of residential stress in which the decision to seek a new residence was treated as a behavioral response to the stress created by the difference between a household's current level of satisfaction and the level of satisfaction it believes it can attain elsewhere (Clark and Cadwallader, 1973). Within the context of cognitive mapping, I also began to explore the relationship between cognitive and physical distance. As part of my doctoral dissertation, I was able to show that cognitive distance was more successful than physical distance as a predictor of consumer behavior, thus empirically demonstrating the hypothesized relationship between cognition and behavior (Cadwallader, 1975). In later work of this genre, I investigated the process of neighborhood choice (Cadwallader, 1979a).

Our initially high expectations for the behavioral approach were ultimately dampened, however, by a series of potentially intractable problems. Having postulated the importance of environmental images, it proved difficult to capture the salient properties of those images. For example, noncommutativity and intransitivity in cognitive distance estimates suggested that people do not possess internal representations of cities that can be portrayed in terms of Euclidean geometry (Cadwallader, 1979b). Indeed, the use of such terms as mental and cognitive map were often taken too literally. The implied assumption of subject-object separation also proved problematic. It is overly simplistic to think that the world can be neatly divided into an objective world of material objects and a subjective world of the mind, thus allowing a theoretical distinction between the observer and the observed.

## Marxist Political Economy

Discussions about the relative strengths and weaknesses of the behavioral approach were soon displaced, however, by the increasingly tense debates associated with the growing popularity of Marxist political economy. Although Marx himself was not especially interested in urbanism per se, Marxist theories of political economy were used to provide a framework for analyzing urban phenomena. A pivotal advocate, of course, was David Harvey, who summarized many of the issues in *Limits to Capital* (1982). In particular, Harvey (1978) identified three circuits of capital: a primary circuit involving the production of commodities; a secondary circuit involving fixed capital investment in the urban built environment; and a tertiary sector comprising investment in research and technology. Thus the suburbanization process could be conceptualized as an example of capital switching from the primary circuit to the secondary circuit as a result of the underconsumption problems of the 1930s.

For those of us who were comfortably numb within the confines of the spatial analytical tradition, the dialectic approach of Marx and his followers was something of a challenge. We were exhorted to penetrate the realm of appearances in order to uncover the underlying relations that give rise to those appearances. A new vocabulary involving mode of production, surplus value, capital accumulation, and social reproduction had to be understood and communicated to students. Perhaps most challenging of all were the intricacies of dialectical analysis, with its perpetual resolution of opposites in which each resolution generates its own contradiction. For those of us used to the building-block approach, which isolates individual components of the overall system before making them fixed foundations for further enquiry, the idea of reinterpreting concepts as they are juxtaposed with other concepts came as something of a shock.

As with most paradigmatic shifts, however, a counterveiling wave of criticism was soon unleashed. Charges of economic determinancy became increasingly commonplace, as researchers argued that reducing the range of social experience to the surface manifestations of some underlying economic structure presents an excessively impoverished view of the social, cultural, and political realms of experience (Duncan and Ley, 1982). That is to say, these other realms of experience are not uniquely determined by the prevailing mode of production. Rather, they assume particular configurations in specific historical situations. Similarly, it was increasingly posited that Marxist political economy is class reductionist, that capitalism involves a more complex social structure than that implied by the simple distinction between the bourgeoisie and the proletariat. For example, during the 20th century there was a growing interpenetration of the capitalist and labor classes, as evidenced by worker-owned companies. Ultimately, the rather passive model of human agency invoked by structural Marxism began to be replaced by the more complex version contained in structuration theory. Structuration theory emphasized the recursive relationship between individuals and society; social structure is not merely a barrier to action, but is actively reproduced by that action. In this way, the theory provided a mode of analysis for transcending the seemingly endless debate about whether the social formation or individual human agency should be the basis of explanation.

## Structural Equation Models

Toward the end of the decade my own interests began to reflect the increasing focus on structural equation models that was sweeping through the social sciences (Duncan, 1975). These models integrated a number of originally disparate research traditions in economics, psychology, and sociology, and provided a powerful approach for investigating the interrelationships among a set of variables. In particular, while econometricians had traditionally favored simultaneous equation models, involving nonrecursive relationships among a set of variables that contain negligible measurement error, psychometricians had tended to emphasize the problems of measurement error and had thus concentrated on various types of factor analysis. Meanwhile, work on path analysis in sociology showed that identification could be achieved in the presence of both measurement error and simultaneous relationships. In this way, causal models, path analysis, and systems of simultaneous equations, with or without unmeasured variables, can all be subsumed under the general heading of structural equation models (Cadwallader, 1986).

Originally, most causal models in urban geography were of the single-equation variety, as exemplified by the plethora of multiple regression models, but gradually more complex causal structures were generated that included both direct and indirect effects (Mercer, 1975). Path analysis represents a logical extension of the causal modeling framework, as it involves estimating the magnitude of linkages among variables, rather than merely assessing their presence or absence. My own efforts along these lines involved using causal models and path analysis to articulate the interrelationships between housing patterns, social patterns, and residential mobility (Cadwallader, 1981). In particular, the aim was to construct a model that linked housing demand, as expressed by different social groups, housing supply, as represented by different types and quality of housing, and residential mobility. A further refinement involved the use of simultaneous equations, thus allowing for two-way, or reciprocal causation (Cadwallader, 1982).

Perhaps the most exciting advance, however, involved the integration of unmeasured, or latent variables within these models. For example, the advent of the first generation of so-called LISREL models allowed us to think in terms of two components: a measurement model and a structural equation model. The measurement model linked the observed variables to a set of unobserved or latent variables through a factor analysis model, while the causal relationships among the latent variables themselves were specified in terms of a structural equation model. In this way, we were able to combine the previously disparate attributes of factor analysis and regression analysis into one overall methodological framework. The individual parameters were usually estimated via maximum likelihood techniques and a chi-square statistic was used to determine goodness-of-fit.

## Summary

In summary, the 1970s were a decade of methodological and philosophical experimentation in urban geography. Various paradigms and approaches were enthusiastically embraced and then often discarded. Each of these approaches, however, allowed us to ask different types of questions and encouraged interaction with other disciplines in the social sciences. For myself, the 1970s represented a decade of intellectual exploration, at the end of which I had returned to my methodological roots in model-building and statistical analysis.

## References

Berry, B. J. L., 1967, *Geography of Market Centers and Retail Distribution*. Englewood Cliffs, NJ: Prentice-Hall.

Cadwallader, M., 1975, A behavioral model of consumer spatial decision making. *Economic Geography*, Vol. 51, 339–349.

Cadwallader, M., 1979a, Neighborhood evaluation in residential mobility. *Environment and Planning A*, Vol. 11, 393–401.

Cadwallader, M., 1979b, Problems in cognitive distance: implications for cognitive mapping. *Environment and Behavior*, Vol. 11, 559–576.

Cadwallader, M., 1981, A unified model of urban housing patterns, social patterns, and residential mobility. *Urban Geography*, Vol. 2, 115–130.

Cadwallader, M., 1982, Urban residential mobility: A simultaneous equations approach. *Transactions of the Institute of British Geographers*, new series, Vol. 7, 458–473.

Cadwallader, M., 1986, Structural equation models in human geography. *Progress in Human Geography*, Vol. 10, 24–47.

Clark, W. and Cadwallader, M., 1973, Locational stress and residential mobility. *Environment and Behavior*, Vol. 5, 29–41.

Duncan, J. and Ley, D., 1982, Structural Marxism and human geography: A critical assessment. *Annals of the Association of American Geographers*, Vol. 72, 30–59.

Duncan, O., 1975, *Introduction to Structural Equation Models*. New York, NY: Academic Press.

Golledge, R., 1981, Misconceptions, misinterpretations, and misrepresentations of behavioral approaches in human geography. *Environment and Planning A*, Vol. 13, 1325–1344.

Haggett, P., 1965, *Locational Analysis in Human Geography*. London, UK: Edward Arnold.

Harvey, D., 1969, *Explanation in Geography*. London, UK: Edward Arnold.

Harvey, D., 1978, The urban process under capitalism: A framework for analysis. *International Journal of Urban and Regional Research*, Vol. 2, 101–131.

Harvey, D., 1982, *The Limits to Capital*. Chicago, IL: University of Chicago Press.

Mercer, J., 1975, Metropolitan housing quality and an application of causal modeling. *Geographical Analysis*, Vol. 7, 295–302.

Rushton, G., 1981, The scaling of locational preferences. In K. Cox and R. Golledge, editors, *Behavioral Problems in Geography Revisited*. New York, NY: Methuen, 67–92.

Yeates, M. H. and Garner, B. J., 1971, *The North American City*. New York, NY: Harper & Row.

## Summary

In summary, the 1970s were a decade of methodological and philosophical examination in urban geography. Various paradigms and approaches were enthusiastically embraced and then often discarded. The methodological movement, however, allowed us to ask different types of questions and encouraged interaction with other disciplines. In this respect, the 1970s represented a decade of methodological examination and ...

# CHAPTER 11
# A Personal History

RISA PALM[1]

## ABSTRACT

In the 1970s, urban geography in the United States was primarily concerned with understanding the development of our system of cities and also with discerning and modeling patterns in the internal structure of the metropolitan area. Hypothesis testing and the application of some rather complex multivariate statistics to census and other survey data were common. This paper outlines the author's personal history in this subfield through this decade. [Key words: American cities, urban geography, modeling.]

[1]Correspondence concerning this chapter should be addressed to Risa Palm, Executive Vice Chancellor and Provost, Office of Academic Affairs, Louisiana State University, Baton Rouge, LA 70803; telephone: 225-578-8863; fax: 225-578-5980; e-mail: rpalm@lsu.edu

I first learned about urban geography when I became a graduate student at Minnesota in the fall of 1967. Unlike many other graduate students, I had not as a child had the slightest interest in patterns of railroad lines, nor had I pondered maps and imagined far-away places. But I was interested in people—in housing—in communities and neighborhoods. And so at that time, the study of the geography of cities sounded fascinating to me, and the field seemed like the most dynamic and "scientific" part of geography.

My professors at Minnesota had an immense impact on the way I understood urban geography. My first urban course was from John W. Webb, and we used Robert E. Dickinson's (1964) *City and Region* as one of our texts. Focusing a great deal of attention on "site and situation," we studied the history of cities, the initial locus, and relative location within a local region of the urban area.

I then took a two-course sequence on the American city from John Borchert. Borchert had recently published (1967) his landmark article on the development of the American urban system and was also deeply engaged in the Lake Shore Development project (1970) (known locally as the LSD project—a term that meant more to me then than now). John Borchert's American city courses were popular and thought-provoking because they combined empirical information with a simple theoretical perspective that made sense of both the city system and local urban structure.

Borchert's American cities course was organized in the same way that several texts in the 1970s followed: the first half on the development of the system of cities, and the second on the internal structure of the city. The course contained a field experience in one of the "big cities" near Minneapolis—that is, Chicago, Kansas City, or St. Louis. These field trips were unforgettable experiences—partly in their preparation, and partly in their execution. In preparation, the students had to plot 1960s census information on tract maps (income, race, etc.) so that they could then "see" patterns on the ground. In execution, Borchert was a superb field teacher—knowledgeable yet setting students up for an adventure.

The primary questions being asked and "answered" during this period were concerned with where cities were located, how "systems" of cities developed, how the city itself could be seen as a "system," and what generalizations could be made about spatial patterns and the processes that underlie these patterns.[2] Although Raymond Murphy's (1966) book on American cities was influential, students were far more interested in Berry and Horton's (1970) and also Yeates and Garner's (1971) approaches to

[2]Urban geography had turned to works such as B. J. Garner, 1967, Models of urban geography and settlement location, in R. J. Chorley and P. Haggett, editors, *Models in Geography,* London, UK: Methuen, 303-360, and D. Harvey, 1969, *Explanation in Geography,* London, UK: Edward Arnold.

urban geography and the Abler, Adams, and Gould (1971) perspective on the field as a whole.

The mode of inquiry was clearly quasi-scientific: hypotheses were developed and tested using multivariate statistical methods. In fact, the more complex the method, the better. Graduate students delighted in learning multiple regression and factor analysis, and they applied multivariate methods to anything that moved (or did not). Whether our work showed much at all was of less importance than the pride we took in applying a method that someone else had not yet applied to a data set. The paper that Doug Caruso and I did on "factor labeling in factorial ecology" (Palm and Caruso, 1972)—our first *Annals* paper—was done as an exercise to teach ourselves principal components analysis. Again, I can recall spending night after night in the basement of the Social Sciences building waiting for a hoped-for "thick" printout (as opposed to the thin printout one got if one missed a punctuation mark in the command line). Doug and I did believe we were striking a blow for a more humanistic analysis of cities, although I am not certain in retrospect that our work made much difference.

Fred Lukermann (1961) had taught us that geography was defined by two questions: the spatial organization of phenomena—questions of location—and the relationship between people/societies and their environment. In the pedagogical collection of readings that Dave Lanegran and I had put together (Lanegran and Palm, 1972), we simplified Fred's definition to say that "the concern of geography is with places, spatial analysis, and the relationships between people and land" (p. 12). However, it was clear that the brand of urban geography I practiced was *solely* spatial in emphasis. When I interviewed at Berkeley, my lecture was based on my dissertation—an empirical look at how the regions or "communities" defined by factorial ecology resembled regions that might be defined based on telephone call volume or common subscription to magazines and newspapers. I can recall Ann MacPherson asking me a very sensible question—whether urban communities had anything to do with their physical setting—and I can recall finding the question almost bizarre. Although I could imagine that "view" or "air pollution" or local weather (in the Bay area) might affect land value, none of these had been factors in the "paradigmatic" Midwestern American city. Only much later, after I had moved to Colorado, did I even consider the physical geography of the city—at least its natural hazard patterns—as in any way related to land value or housing choice.

My own work in the 1970s focused on, as John Borchert kindly called it, "the real estate agent as geography teacher." Since the real estate agent controlled a great deal of information about housing vacancies, one could draw a parallel between the selection and communication of such information to buyers and the way a teacher selects, edits, and conveys information to students. My first work (begun with Doug Caruso when we were graduate students) focused on the process that we believed might lie behind the empirical finding of "directional" or "sectoral" bias in intra-urban migration described by John Adams (1969), Larry Brown (Moore and Brown, 1970), Bill Clark (1970), Eric Moore (Moore and Brown, 1970), Ron Johnston (1969), and others. Our working hypothesis was that it was the spatial bias in information conveyed to prospective buyers which then influenced move patterns: that if real estate agents only knew about limited portions of the metropolitan area, they would only "teach" these areas to prospective buyers (Palm, (1976).

I also developed an interest in the pattern of housing prices that was not unrelated to the focus on real estate agents. I suspected that the strong bias we had found in real estate agent knowledge and preferences might also have a marginal impact on demand for houses by neighborhoods, with a concomitant influence on relative house prices. I did a series of hedonic analyses in the San Francisco Bay area, trying to identify the factors that would result in relative differences in house price inflation (Palm, 1977, 1978, 1979).

My Oxford University Press book (1981), completed in May of 1980, summarized my perspective of urban geography in the 1970s and also marked my departure from this specialty and my transition to the study of natural hazards (especially earthquake hazards) as integrated into the housing market. In the introductory notes, I acknowledged the influence that Fred Lukermann, John Borchert, John Adams, Gilbert White, Jay Vance, Allan Pred, and Ron Johnston had exerted on my thoughts on the structure of urban geography.

I tried to set this book within what I understood to be current methodological frameworks. I stated in the book that future research directions for urban geography would have to supplement hypothesis testing and model building with the study of history, politics, economy, and society in order to explain the processes shaping and changing metropolitan areas. I also felt that geographers needed more emphasis in deciphering subjective meanings in the environment.

I also tried to relate cities to their physical environment, describing some of the work on the physical geography of the city and arguing for the need for a greater integration of physical geography into studies of urban geography. Although early urban geography emphasized geomorphic characteristics as related to "site," I argued that modern urban geography needed to look at the physically differentiated environment of the city, and the way in which settlement and environment interact. I imagine that this interest was partly due to my move from the Midwest to more variegated settings in the San Francisco Bay area and later in the foothills of the Rocky Mountains, as well as the influence that both Mel Marcus (1979) and David Greenland had on my work.

My vision of urban geography in the 1970s was summarized in the outline of the American cities textbook I completed in the spring of 1980 (Appendix). As I mentioned earlier, I left the field of urban geography in the 1980s and focused my work on the human dimensions of earthquake hazard response: the disclosure of hazards by real estate agents, the impacts of known hazards on land values, integration of hazards into valuation and practice by real property appraisers and the insurance industry, and the interpretation of hazards within cultural contexts that stress the individual as opposed to the collective.

Quite obviously, urban geography has greatly evolved and advanced in the years since the 1970s. Yet, because of my personal involvement in the field, I still view this time as the halcyon days of urban geography.

## References

Abler, R., Adams, J. S., and Gould, P., 1971, *Spatial Organization: The Geographer's View of the World.* Englewood Cliffs, NJ: Prentice-Hall.

Adams, J. S., 1969, Directional bias in intra-urban migration. *Economic Geography,* Vol. 45, 302–323.

Berry, B. J. L. and Horton, F., 1970, *Geographic Perspectives on Urban Systems with Integrated Readings.* Englewood Cliffs, NJ: Prentice-Hall.

Borchert, J. R., 1967, American metropolitan evolution. *Geographical Review,* Vol. 57, 301–332.

Borchert, J. R., 1970, *Minnesota's Lakeshore.* With George W. Orning, Les Maki, and Joseph Stinchfield. Minneapolis, MN: University of Minnesota Department of Geography.

Clark, W. A. V., 1970, Measurement and explanation in intra-urban residential mobility, *Tijdschrift voor Economische en Sociale Geografie,* Vol. 61, 49–57.

Dickinson, R. E., 1964, *City and Region: A Geographic Interpretation.* London, UK: Routledge & Kegan Paul.

Johnston, R. J., 1969, Some tests of a model of intra-urban population mobility: Melbourne, Australia. *Urban Studies,* Vol. 6, 34–37.

Lanegran, D. and Palm, E., editors, 1972, *Invitation to Geography.* New York, NY: McGraw-Hill.

Lukermann, F., 1961, The role of theory in geographical inquiry. *Professional Geographer,* Vol. 13, 1–6.

Marcus, M. G., 1979, Coming full circle: Physical geography in the twentieth century. *Annals of the Association of American Geographers,* Vol. 69, 421–432.

Moore E. G. and Brown, L. A., 1970, Urban acquaintance fields: An evaluation of spatial models. *Environment and Planning A,* Vol. 2, 443–454.

Murphy, R. E., 1966, *The American City: An Urban Geography.* New York, NY: McGraw-Hill.

Palm, R., 1976, Real estate agents and geographical information. *Geographical Review,* Vol. 66, 266–288.

Palm, R., 1977, Homeownership cost trends. *Environment and Planning A,* Vol. 9, 795–804.

Palm, R., 1978, Spatial segmentation of the urban housing market. *Economic Geography,* Vol. 54, 210–221.

Palm, R., 1979, Financial and real estate institutions in the housing market: A study of recent house price changes in the San Francisco Bay Area. In D. T. Herbert and R. J. Johnston, editors, *Geography and the Urban Environment,* Vol. 2. Chichester, UK: John Wiley, 83–123.

Palm, R., 1981, *Geography of American Cities*. New York, NY: Oxford University Press.
Palm, R. and Caruso, D., 1972, Factor labelling in factorial ecology. *Annals of the Association of American Geographers*, Vol. 62, 122–133.
Yeates, M. H. and Garner, B. J., 1971, *The North American City*. New York, NY: Harper & Row.

# Appendix
## Outline of *Geography of American Cities* textbook

    1.  Introduction
  I.  Background to the Study of American Cities
    2.  The City and the Countryside: Attitudes toward Urban Life
    3.  Ethnicity in the American City
    4.  Political Economy and Urban Structure
  II.  Urban Development and the Physical Environment
    5.  Resources and Urban Growth
    6.  Cities and their Physical Environments
    7.  Political Economy, Attitudes, and the Physical Environment of Cities
  III.  The System of Cities in the United States
    8.  The Development of the City System in the United States
    9.  The Changing City System
  IV.  The Metropolitan Area and its Spatial Organization
    10. Land Values in the City
    11. Location Decisions in the Private and Public Sectors of the Economy
    12. The Residential Structure of American Cities
    13. Mobility within the City
    14. The Changing Metropolitan Area

# CHAPTER 12
# Competing Visions of the City

PETER G. GOHEEN[1]

## ABSTRACT

For urban geography the decade of the 1970s was a period of critical reflection on the positivist social science that had flourished in the 1960s. Two challenges to existing practice were of special importance. One, identified with humanistic geography, sought to broaden the idea of science and dissociate it from positive statistical analysis. The other, espousing a Marxist analysis, challenged the legitimacy of the science practiced by the profession. The former served to diversify the nature of the research contributed to the field. The latter sought to supplant it. The decade ended in discord, with urban geography suffering a loss of coherence in the absence of a productive discussion of the issues which confronted its practice. [Key words: positivism, humanism, Marxism, 1970s.]

[1]Correspondence concerning this chapter should be addressed to Peter G. Goheen, Department of Geography, Queen's University, Kingston, ON K7L 3N6, Canada; telephone: 613-533-6044; fax: 613-533-6122; e-mail: goheenp@qsilver.queensu.ca

This paper offers a retrospective glance at the research and writing of urban geography in the 1970s, work that was fresh when my career was young. In rereading the literature my intention has been to discover a sense of shape to the period, to see it as a unit. This was, needless to say, not my perspective as the decade unfolded, and I have paid as little attention as possible to what I may remember of my earliest response to the work discussed here. I have approached the 1970s from the vantage point of the present, aware that my perspective bears the influence of scholarly trends since that time. I offer a personal view of what was a vital endeavor: urban geography was at the creative center of a vibrant and disputatious human geography in the 1970s.

Urban geography entered the 1970s as a favored standard bearer on behalf of the ongoing, enthusiastic campaign to renovate the practice of human geography. Throughout the decade its leading practitioners continued to include many who remained committed to this social science approach. It attracted the attention of others who vigorously championed alternatives to the positive science that had become widely accepted as the authorized approach to research. Humanism and Marxism, respectively, offered as reformist and revolutionary alternatives, were the principal vehicles for challenging the new orthodoxy. The city became a battleground for contending parties who proved to be unable to sustain a constructive debate over their divergent visions of what constituted quality research. The published record of the decade shows precious little evidence of any constructive conversation among the protagonists. The contenders were going their own ways. They defined their positions with reference to thinking outside geography while neglecting to address the arguments of their geographic colleagues with which they disagreed. The confusion which arose from their unwillingness or inability to address what were thought to be profound disagreements about the conduct and even the purposes of research can be described in David Livingstone's words as "situated messiness" (Livingstone, 1992). A decade which witnessed an impressive growth in the number of researchers and in the volume of published work in urban geography came to be dominated by the discord of unresolved disputes about the conduct of the enterprise.

As the decade opened the new direction of the field was strikingly articulated in two texts which defined the structure and illustrated the fruits of the new thinking (Berry and Horton, 1970; Yeates and Garner, 1971). Urban geography, as codified in these books and in the preponderance of published articles, was conceived as a contribution to the new science which had recently been defined as involving "extensive application of mathematical methods to facilitate refinement of theory" (National Academy of Sciences-National Research Council, 1965, p. 44). Urban geographers approached their subject prepared to examine any questions that they could address through mathematical and statistical modeling. Often the presentation focused more on the nature of the model than on the urban place. Given the urbanized nature of modern western society there was an almost boundless field for urban geography so conceived (Hudson, 1972; Webber, 1972).

The leading journals, most of which were long established generalist publications, were joined by new serials which served the new and specialized markets created by the growing number of researchers committed to the field. Nevertheless, the space which editors of the most prestigious generalist journals devoted to mathematical and statistical modeling of urban phenomena was significant. The significance of this work to their journals can be seen in the accommodations which editors made to the style of presentation favored by this brand of science. Authors often eschewed interpretive essays in favor of articles intended primarily as theoretical statements; these contributions often contained little or no discussion either of data or its contextual meaning (Clark and Burt, 1980; Curry, 1976; Mumphrey and Wolpert, 1973). The prominence and character of urban geography, especially in the early 1970s, is well illustrated in the March 1970 issue of the *Annals of the Association of American Geographers.* Five of the ten articles published in that issue are devoted to urban geography. Four of the five, on topics as diverse as the formation of the Black ghetto in Milwaukee and traffic generation by nonresidential land use in Perth, Scotland, share a central purpose in building and testing statistical models (Rose, 1970; Eliot Hurst, 1970). These authors adopted a popular style of presentation in their field, focusing on a demonstration of innovative scientific methodology rather than on an interpretation of their results. So infrequently did an author choose to focus the interpretation on the meaning of the findings for understanding particular urban places that when one did so it could attract comment (Murdie, 1971, reviewing Robson, 1969).

The exuberant expectations which fueled the reformed urban geography were perhaps nowhere better expressed than in a project of unprecedented scope in which urban geographers collectively engaged during the first half of the decade. The Comparative Metropolitan Analysis Project was a collaborative venture which aimed to demonstrate the analytical capacity of urban geographers to contribute to an understanding of the condition of urban America circa 1970. In the words of Brian Berry, it aimed to assess where we stood in 1970 with respect to a range of stated national urban-policy goals; [to provide] a consistent comparative documentation of a variety of social, economic, political, and physical aspects of the nation's twenty largest urban regions, containing close to half of the total U.S. population; [to offer] a set of skillfully drawn portraits of each region highlighting both common problems and local individuality; [to contribute] an assessment of the management and performance of these urban regions in twelve major policy areas—in short, a record of what has and has not been accomplished as perceived by a particular research community... (Berry, 1976, p. v).

The project was completed with the publication in 1976 of three volumes. Volume 1, entitled *Contemporary Metropolitan America* and consisting of four books, was edited by John Adams and comprised 20 vignettes of the leading metropolitan areas (Adams, 1976a). Thirteen of these extended essays saw their way into print as separate monographs. Volume 2 concerned *Urban Policymaking and Metropolitan Dynamics* (Adams, 1976b). Volume 3, by Ronald Abler and John Adams, was an atlas of the nation's 20 largest metropolitan regions (Abler and Adams, 1976). At a time when public interest in America had become increasingly focused on social and economic trends in its major cities, urban geographers were seeking to make visible, in both academic and policy circles, the nature of what they could contribute.

The 13 published vignettes collectively represented the most significant contribution American geographers had made to understanding the geography of their leading cities. There was nothing to rival this series in the geographical literature. These studies provided the first broadly comparative look at the geography of America's most important cities. They illustrated a historical and interpretive tradition in urban geography that had deep roots but bore little relation to the modern science which had come to dominate urban geography at this time. The volumes reflected both the geographical imaginations of their individual authors and the particularities of place. They were the most accessible of the written products of this massive undertaking.

Volumes 2 and 3, like the entire project, are distinguished by the immense volume of research that the many contributors undertook. They dealt with the natural environments of the cities, with housing, transportation, population, education, public health, ethnic spatial segregation, gender and employment, crime, governance, and poverty among other topics. The economic aspects of these issues were more fully documented in the wealth of available evidence than were the social concomitants, and the published work reflected this bias. The sources on which the researchers relied provided far more quantitative than qualitative evidence; the questions that could be addressed reflected this bias. What may impress the reader is the great range in subject matter concerning life in the largest urban regions that these volumes seek to address. The reader who wishes to distill the lessons of the massive volumes 2 and 3 faces a daunting challenge. The authors have presented a compendium; they have not attempted to integrate the findings of their many investigations, and offer no guidance to the reader wishing to understand the relations of the many separate issues investigated. This presentation did little to render the results of this massive research enterprise amenable to policy makers who might be interested in the work. It remains unclear what if any identifiable influence this great investment of talent and work exerted on the subsequent practice of urban geography.

The 1970s saw the appearance of landmark interpretive works by Paul Wheatley and James E. Vance, Jr. Wheatley's interpretation of the origins and character of the ancient Chinese city, *The Pivot of the Four Quarters* (1971) remains a noble and enduring monument from the decade. His forceful argument and his interest in understanding the Chinese city in comparative theoretical terms made his work a widely fashionable reference among urban geographers with diverse interests. James Vance saw into print his imaginative interpretation of the role and structure of the city in the geography of western civilization, *This Scene of Man* (Vance, 1977). His deep historical imagination and broad geographical canvas appealed especially to those who yearned for an alternate to the economic logic and statistical theory of positive social science. Both works were highly ambitious undertakings; they are scholarly works of significant originality and mature reflection. We should also recognize that both remained eccentric to the main currents of thought in urban geography, even during a period when urban geographers had begun to acknowledge research and publications on the historical nature of cities (Ward, 1971). They are among the most impressive statements of the decade, yet little evidence can be found to indicate that they deflected the magnetic orientation of the field that was urban geography.

A book of less weight but more heft in pointing to an alternate course for urban geography was David Ley's somewhat tentative exploration of the American Black inner city, or to state his declared intentions more accurately, the Black neighborhood of Philadelphia that he called Monroe (Ley, 1974). His approach, which he identified as "behavioral geography," did not proceed "from an examination of spatial form to a deduction of spatial process." He wanted instead to "understand a decision-making world," and to do so he needed a paradigm "which treats variables because they are salient and not just convenient, which proceeds from the individual to the aggregate, and which recognizes inductive understanding as no less appropriate than deductive prediction" (Ley, 1974, pp. 8–9). His experiential world, he announced:

> is a complex entity consisting of a potent internal or behavioral environment and the strategies and resources of adaptively rational men. The world is subjective, socially and culturally enacted, and perceived through distinct cognitive filters. Moreover, distortion is not only a feature of perception within the world. It is strongly perpetuated between the world of the inner city and the world of mainstream America, by powerful images, or stereotypes, which constrain incoming information to a form harmonious with their own survival. (Ley, 1974, p. 10)

He organized his presentation under two main headings: "The Shadows," where he discussed theories and models, and "The Real," where he examined the daily lives of the Black youths themselves as a basis for evaluating the model he proposed. The reader cannot fail to grasp the distance separating this work from that of mainstream urban geography of the period. The argument he wished to join was not about the need for theory and models but rather about aims, appropriate theory, the place of experience, and the nature of interpretation.

Humanistic geography came to embody the approach that Ley outlined in his 1974 book. Yi-Fu Tuan (1976, p. 266) defined it as an inclusive endeavor, belonging with both the humanities and the social sciences "to the extent that they all share the hope of providing an accurate picture of the human world." Humanism, he urged, offers an expansive view encompassing rather than denying the scientific perspective on man. Humanistic geography "tries to understand how geographical activities and phenomena reveal the quality of human awareness" (p. 267) while critically building on scientific knowledge. Nicholas Entrikin (1976, p. 616) identified its appeal as providing an alternative to what its proponents "believe to be an overly objective, narrow, mechanistic and deterministic view of man presented in much of the contemporary research in the human sciences." Humanist geographers, he wrote, "hold that the study of human behavior cannot be modeled after the physical sciences. They reject the positivist claim of the isomorphism of social and physical science…" (p. 625). The discussion continued in a series of essays appearing throughout the rest of the decade. Ley, writing in 1977, urged an alternative to "a too-severe positivism which in dissociating subject from object, had removed the human context from behaviour, and even from life" (Ley, 1977, p. 502). In introducing a book of essays exploring the nature of humanistic geography, the editors wrote that there was "one central, irreducible message: a principal aim of modern humanism in geography is the reconciliation of social science and man, to accommodate understanding and wisdom, objectivity and subjectivity, and materialism and idealism" (Ley and Samuels, 1978, p. 1).

Humanistic geographers were not inventing a new perspective. They were refining ideas long associated with the discipline, and offering a sounder understanding of the relation of their thinking to broad currents within the larger academic world. The geographic literature on cities had always included work dealing with such concepts as place and landscape. The issue for the 1970s was whether urban geography, after more than a decade of close identification with the new science of geography, would be regarded as a sufficiently inclusive field to encompass such a broad range of practice. As the decade progressed the answer was clearly affirmative. Scholars working under the rubric of urban geography explored a wide range of social practices and institutions and the values they embodied. Authors studying urban landscapes and exploring the concept of place self-consciously identified themselves as social scientists exploring alternatives to the physical science model of research they had rejected. In the journals their studies discussed such topics as the aesthetics of Black ghetto environments, the experiential impact of mental health policies on inner-city neighborhoods, urban squatting as a mechanism of adjusting to stress over housing, and the spatial knowledge biases of real estate agents (Palm, 1976; Dear, 1977; Ford and Griffin, 1979; Kearns, 1979). A mark of acceptance by urban geographers of the argument for a broadened concept of scientific practice within urban geography came with the publication of edited collections which offered representative samples of several styles of research (Gale and Moore, 1975; Herbert and Johnston, 1978).

It would misrepresent the decade of the 1970s to attribute the debate about urban geography exclusively to the discussion of humanism. Urban geography was central to an increasingly critical scrutiny of the new scientific geography. In particular, questions were raised about the adequacy of the statistical methods being employed and about their contribution to science. Peter Gould challenged the suitability of commonly used statistics, arguing that "it is difficult not to think of traditional non-spatial inferential statistics as totally irrelevant as correspondence rules in geographic inquiry" (Gould, 1970, p. 446). Leslie King a few years later questioned research practice in urban and economic geography in different terms. He argued for the need to focus on issues of social change and social policy rather than on matters of technique, the matter which in his judgment had been the principal concern of those involved in the so-called "quantitative revolution" (King, 1976, p. 293). He argued that neither the positivism to which many urban geographers were committed nor the Marxism espoused by others would suffice. He wished to promote research "that will be concerned explicitly with social change and policy, and that will recognize the ethical content of all its analyses and freely admit the biases which are inherent in them as a result" (King, 1976, p. 294). Values and goals as well as facts, he suggested, can become the subject of analysis that tries "'to discern patterns but…seeks general principles that take account of the context and comingle facts and values'" (Rein quoted in King, 1976, p. 307).

More directly, in 1973, David Harvey, in *Social Justice and the City*, challenged urban geographers to redefine their goals and to rethink their relation to their object of study: the city. He identified what he called counter-revolutionary and revolutionary theory; the distinction between them is fundamental. Counter-revolutionary theory

> is based on…man's ability…to manipulate and control human activity and social phenomena…Immediately the question arises as to who is going to contol whom, in whose interest is the con-trolling going to be exercised, and if control is exercised in the interest of all, who is going to take it upon himself to define that public interest? We are thus forced to confront directly…the social bases and implications of control and manipulation. We would be extraordinarily foolish to presuppose that these bases are equitably distributed throughout society. These groups may be benevolent or exploitive with respect to other groups. This, however, is not the issue. The point is that social science formu-lates concepts, categories, relationships and methods which are not independent of the existing social relationships. As such, the concepts are the product of the very phenomena they are designed to describe. A revolutionary theory upon which a new paradigm is based will gain general acceptance only if the nature of the social relationships embodied in the theory are actualized in the real world. A counter-revolutionary theory is one which is delib-erately proposed to deal with a revolutionary theory in such a manner that the threatened social changes which general accep-tance of the revolutionary theory would generate are, either by cooptation of subversion, prevented from being realized. (Harvey, 1973, p. 125)

Harvey was not challenging colleagues to a debate. He was trying to revolutionize geographers' approach to their subject.

To my reading, Harvey finds the city to be an awkward object of study. He began his book by devoting three chapters to criticizing what he calls liberal formulations. Here he discovered many problems with the way social science has conceptualized the city. In the chapter on social justice and spatial systems the city fails to figure in his discussion: it is simply not analytically or conceptually important. The problem does not disappear when, in the second part of his work, he discussed socialist formulations. In his interpretive essay, "Urbanism and the City," "the main problem" is to elucidate "the general proposition that some sort of relationship exists between the form and functioning of urbanism (and in particular the various forms of town-country relationship) and the dominant mode of production..." (Harvey, 1973, p. 205). The city fades from focus, to reappear only when absolutely necessary to the purposes of implementing his socialist manifesto.

In light of Harvey's disapproval of research which aims to understand the city as constituted at present it is not surprising that he had little to say in this book about an urban research agenda. Indeed, he was occasionally dismissive of the matter, as when he discussed the problem of ghetto housing. A "rigorous and critical examination" of the matter, he urges, "does not entail yet another empirical investigation of the social conditions in the ghettos....There is already enough information...Nor does it lie in what can only be termed 'moral masturbation' of the sort which accompanies the masochistic assemblage of some huge dossier on the daily injustices to the populace of the ghetto...." (Harvey, 1973, p. 145)

*Social Justice and the City* attracted much interest; it is undoubtedly one of the most widely recognized titles from the 1970s. Perhaps paradoxically, it provoked little published debate within urban geography. The most celebrated exchange occurred in the pages of the self-proclaimed radical journal, *Antipode,* whose editor published a review of the book by Brian Berry. The exchange between author and reviewer demonstrated the incompatibility of their assumptions and approaches to the subject matter, each talking past the other and unable to find agreed common ground on which to conduct a constructive conversation (Berry, 1974; Harvey and Berry, 1974).

The search for paradigms deriving from Marxist thinking attracted a growing following, and inasmuch as interest focused on contemporary western societies it often dealt with urban societies. The city occasionally emerged as a conceptually significant object of study, but imprecise references to urban were more common (Harvey, 1979; Walker, 1978). A fuller discussion of Marx and the significance of his thinking to the study of urban geography would await the 1980s. When the 1970s ended there were only the first signals of what would come to be a larger stream of contributions to the geographical literature on cities and urbanization that would qualify by Harvey's standard as socialist formulations.

The impasse separating Harvey's agenda from that of urban geographers who saw themselves as social scientists was different in nature to the distinctions which marked the humanists from the positivists. Perhaps what impresses the observer at this remove is not the fact of distinct views but rather the absence of engaged discussion about them. Urban geographers, members of a fairly small community, occupied several solitudes. Was their segregation part of a general crisis of confidence that gripped the discipline, as Wilbur Zelinsky suggested? He used the opportunity of speaking to the profession as President of the Association of American Geographers to announce "[t]he failure of science, most conspicuously social science..." (Zelinsky, 1975, p. 124). Brian Berry, in his presidential address to the same Association at the end of the decade, suggested that geography had "succumbed to a new tribalism" (Berry, 1980, p. 451). Berry pleaded for a reintegration for which in 1980 there was little prospect in urban geography. The lines of division represented a fundamental disagreement that was to become a central feature of the scholarly urban landscape.

## References

Abler, R. and Adams, J. S., 1976, *A Comparative Atlas of America's Great Cities*. Minneapolis, MN: University of Minnesota Press.

Adams, J. S., 1976a, *Contemporary Metropolitan America*. Four volumes. Cambridge, MA: Ballinger.

Adams, J. S., 1976b, *Urban Policymaking and Metropolitan Dynamics*. Cambridge, MA: Ballinger.

Berry, B. J. L., 1974, David Harvey: Social justice and the city. *Antipode*, Vol. 6, 142–145.

Berry, B. J. L., 1976, Foreword. In R. Abler and J. S. Adams, *A Comparative Atlas of America's Great Cities*. Minneapolis, MN: University of Minnesota Press.

Berry, B. J. L., 1980, Creating future geographies. *Annals of the Association of American Geographers*, Vol. 70, 449–458.

Berry, B. J. L. and Horton, F., 1970, *Geographic Perspectives on Urban Systems*. Englewood Cliffs, NJ: Prentice-Hall.

Clark, W. A. V. and Burt, J. E., 1980, The impact of workplace on residential relocation. *Annals of the Association of American Geographers*, Vol. 70, 59–67.

Curry, L., 1976, Location theoretic style and urban policy. *Economic Geography*, Vol. 52, 11–23.

Dear, M., 1977, Psychiatric patients and the inner city. *Annals of the Association of American Geographers*, Vol. 67, 588–594.

Eliot Hurst, M. E., 1970, Nonresidential traffic land use generation. *Annals of the Association of American Geographer*, Vol. 60, 153–173.

Entrikin, J. N., 1976, Contemporary humanism in geography. *Annals of the Association of American Geographers*, Vol. 66, 615–632.

Ford, L. and Griffin, E., 1979, The ghettoization of paradise. *Geographical Review*, Vol. 69, 140–158.

Gale, S. and Moore, E. G., editors, 1975, *The Manipulated City*. Chicago, IL: Maaroufa .

Gould, P., 1970, Is *Statistix Inferens* the geographical name for a wild goose? *Economic Geography*, Vol. 46, 439–448.

Harvey, D., 1973, *Social Justice and the City*. Baltimore, MD: Johns Hopkins University Press.

Harvey, D., 1979, Monument and myth. *Annals of the Association of American Geographers*, Vol. 69, 362–381.

Harvey, D. and Berry, B. J. L., 1974, Discussion. *Antipode*, Vol. 6, 145–149.

Herbert, D. T. and Johnston, R. J., editors, 1978, *Social Areas in Cities: Process, Patterns and Problems*. New York, NY: Wiley.

Hudson, J. C., 1972, *Geographical Diffusion Theory*. Evanston, IL: Northwestern University Department of Geography, Studies in Geography, No. 19.

Kearns, K. C., 1979, Intraurban squatting in London. *Annals of the Association of American Geographers*, Vol. 69, 589–598.

King, L. J., 1976, Alternatives to a positive economic geography. *Annals of the Association of American Geographers*, Vol. 66, 615–632.

Ley, D., 1974, *The Black Inner City as Frontier Outpost*. Washington, DC: Association of American Geographers, Monograph Series, No. 7.

Ley, D., 1977, Social geography and the taken-for-granted world. *Transactions, Institute of British Geographers*, n.s., Vol. 7, 498–512.

Ley, D. and Samuels, M. S., editors, 1978, *Humanistic Geography*. Chicago, IL: Maaroufa.

Livingstone, D., 1992, *The Geographical Tradition*. Oxford, UK: Blackwell.

Mumphrey, A. J. and Wolpert, J., 1973, Equity considerations and concessions in siting public facilities. *Economic Geography*, Vol. 49, 109–121.

Murdie, R. A., 1971, Review of Urban Analysis by B. T. Robson [book review]. *Geographical Analysis*, Vol. 3, 103–105.

National Academy of Sciences - National Research Council, 1965, *The Science of Geography*. Washington, DC: NAS-NRC, Publication No. 1277.

Palm, R., 1976, Real estate agents and geographical information. *Geographical Review*, Vol. 66, 266–280.

Robson, B. T., 1969, *Urban Analysis*. New York, NY: Cambridge University Press.

Rose, H. M., 1970, The development of an urban subsystem: The case of the Negro ghetto. *Annals of the Association of American Geographers*, Vol. 60, 1–17.

Tuan, Y. F., 1976, Humanistic geography. *Annals of the Association of American Geographers*, Vol. 66, 266-276.

Vance, J. E., Jr., 1977, *This Scene of Man*. New York, NY: Harpers College Press.

Walker, R. A., 1978, The transformation of urban structure in the nineteenth century and the beginnings of suburbanization. In K. R. Cox, editor, *Urbanization and Conflict in Market Societies*. Chicago, IL: Maaroufa, 165–212.

Ward, D., 1971, *Cities and Immigrants*. New York, NY: Oxford University Press.

Webber, M. J., 1972, *Impact of Uncertainty on Location*. Cambridge, MA: MIT Press.

Wheatley, P., 1971, *The Pivot of the Four Quarters*. Chicago, IL: Aldine.

Yeates, M. H. and Garner, B. J., 1971, *The North American City*. New York, NY: Harper & Row.

Zelinsky, W., 1975, The demigod's dilemma. *Annals of the Association of American Geographers*, Vol. 65, 123–143.

# CHAPTER 13
# The Comparative Metropolitan Analysis Project

PATRICIA GOBER[1]

## ABSTRACT

The rise in spatial analysis, numerical methods, and theory testing of the 1960s thrust the field of urban geography into a period of reductionist research, one that assumed the urban system was no more than the sum of its parts and that these parts could be studied without reference to the system itself. Amid this reductionist approach to urban inquiry came the Comparative Metropolitan Analysis Project of the 1970s, which sought to use results of the 1970 Census to assess progress on a range of urban-policy issues, including race, class, poverty, and housing in the nation's 20 largest metropolitan areas. The project produced (1) a comparative atlas of the 20 urban regions, (2) a book assessing management

[1]Correspondence concerning this chapter should be addressed to Patricia Gober, Department of Geography, Arizona State University, Tempe, AZ, 85287-0104; telephone: 480-965-7533; fax: 480-965-8313; e-mail: gober@asu.edu

and performance in 12 major policy areas, and (3) a collection of 20 short monographs outlining the physical, social, and economic make-up of urban areas, highlighting both common problems and local individuality. This essay reviews the history, intellectual context, and significance of the project with an eye toward the fundamental tension between its integrative and outward-looking aspirations and disciplinary trends of fragmentation, specialization, and insularity. Due to this tension, the project quickly disappeared from our disciplinary consciousness. [Key words: metropolitan vignette, synthesis, comparative atlas, urban policy.]

I look upon the field of urban geography in the 1970s as something of an outsider. My graduate education and research at Ohio State centered on the study of regional-scale population and economic processes, and it was only after I found myself trying to make sense of these processes in Phoenix—the dynamic urban region in which I lived—that my interests became more urban in scale. As I looked for materials that would be useful to me, one of my colleagues suggested that I examine the Comparative Metropolitan Analysis Project, an ambitious effort to map America's 20 largest urban regions, to identify the social problems and solutions that cut across them, and to describe their individual character and personality (Abler, 1976; Adams, 1976a, 1976b, 1976c, 1976d, 1976e). I discovered that I was not alone as a geographer in trying to make sense of a complicated urban environment and in struggling with the delicate balance between its unique social history and geographic setting and the more general forces shaping urban America at the time. When Jim Wheeler asked me to participate in this session, my first thought was to talk about the Project—where it came from, what it hoped to accomplish, how it succeeded and how it failed, and what it meant to me as a geographer new to the field.

This paper is organized into three parts: (1) a background section that describes the intellectual environment into which the Project was born, specifically as it relates to my academic training, (2) a review of the various components of the project itself, and (3) a discussion of how it was used and not used and what that says about geography in general and urban geography in particular.

## Background

The urban geography I first encountered as a graduate student at Ohio State University in the early 1970s embraced positivism, spatial analysis, and quantification. We looked for inspiration to the fields of economics and statistics in the search for order in the size and location of villages, cities, and metropolitan regions and in patterns of residential differentiation. Census data were mined extensively to uncover differences in urban social space and economic activity. What seemed to us then as high-speed computers enabled mass manipulation of data and comparisons of one city with another. My first urban geography course in the winter of 1971 with Larry Brown used Brian Berry and Frank Horton's new reader, *Geographic Perspectives on Urban Systems*, and, with the single-minded conviction that only graduate students possess, we thought of this as the Holy Grail of urban geography. Published in 1970, this volume took a locational analytic approach to urban geography and made extensive use of data from the U.S. Census to depict urban realms, rank-size relationships, the economic structure of cities at varying sizes in the urban hierarchy, population density gradients, social area analysis and urban land-use models, intra-urban mobility and waves of succession, and models of transportation demand (Berry and Horton, 1970). I remember learning a lot about Chicago!

At the same time, we were exposed to two lines of criticism regarding locational analysis. From the humanistic side, scholars such as David Lowenthal and Yi Fu Tuan denied objective reality and argued for a phenomenological perspective in which each individual is recognized as having a lived world of experience within which decisions are made and reflected in attitudes toward nature and tastes in landscape (Lowenthal, 1961; Tuan, 1968; Herbert and Johnston, 1978). These views laid the groundwork for work in behavioral geography that we, as students at Ohio State, saw manifest in Larry Brown's work on innovation and diffusion and migration decision-making and in Reg Golledge's pioneering work in behavioral geography and spatial cognition (Brown and Moore, 1970; Golledge and Rushton, 1976; Brown, 1981). That these processes occurred in real places with unique social histories and political environments was rarely considered. We never once walked out of our building to view the urban environment that enveloped us, and we were oblivious to the great urban policy debates of the 1960s.

We were aware, however, of David Harvey's criticism that previous work on residential patterns was too preoccupied with choice and spatial outcomes and gave insufficient attention to the political and economic system that gave rise to these patterns. Harvey's (1973) much-cited volume, *Social Justice and the City*, reinterpreted urban patterns within the context of monopoly capitalism and refocused attention from urban problems themselves to the societal structures that produced them in the first place. Although it was a provocative line of research, little was revealed about urban places themselves. Urbanism was regarded less as a "thing in itself" and more as a "vantage point from which to capture some salient features in the social processes operating in society as a whole—it becomes, as it were, a mirror in which other aspects of society can be reflected" (p. 16).

So I arrived at Arizona State University as a faculty member in 1975 with an abstract and narrow view of the field of urban geography in which cities were mere laboratories in which human behaviors occurred and reflections of the political and economic system. Much to my surprise, I discovered faculty colleagues who were actually interested in metropolitan Phoenix as a place—in its history, its climate, its natural setting, its political leadership, economic development, and social order. Moreover, they expected me to teach an urban geography course grounded in the urban setting in which we lived, and students signed up for the course expecting to learn something about Phoenix. Mel Marcus, a physical geographer who was quite active and well-known in the geographic profession, advised me to investigate the Comparative Metropolitan Analysis Project as a model for the way urban geography could be studied and taught. It came to represent my intellectual model of what urban geography should be—a mix of general and specific, quantitative and qualitative, firmly grounded in knowledge about places, and aimed at connecting academic geography to the larger society in which we work.

### Project Summary

In 1970, a number of influential urban geographers, including Brian Berry from the University of Chicago, John Adams and John Borchert from Minnesota, David Ward from Wisconsin, James Vance from Berkeley, Frank Horton from Iowa, and Warren Nystrom from the AAG Central Office asked how the discipline might use the small-area data from the 1970 U.S. Census to take stock of the nation's largest cities in a wake of rapid post-War urban expansion and to assess the effect of the Great Society programs of the 1960s. A planning conference was convened in Chicago to discuss what the 1970 Census would look like, what methods would be available, and what problems deserved attention. Another small group, this one including Ron Abler, met later that year in San Francisco; and John Adams was drafted to prepare a proposal to the National Science Foundation. At the heart of the project was a comparative atlas of the 20 largest urban areas which would be complemented by interpretive monographs focused on the metropolitan areas themselves and thematic monographs dealing with issues that cut across the metropolitan areas. Work began in 1972.

The project's three elements represented a balance between uniqueness and generality. The series of metropolitan vignettes, published in 1976 by

Ballinger, viewed urban areas from the ground up, stressing their individuality. The policy volume, also published by Ballinger, cut across metropolitan areas and sought to identify general urban problems and programs to ameliorate them. The comparative atlas portrayed the spatial distribution of measurable characteristics across the 20 urban areas and was published by the University of Minnesota Press.

## Vignettes

Although the 20 vignettes are not representative of the field of American urban geography as it came to be practiced during the 1970s, they are remarkable portraits of the geography of urban America in the early-1970s. Read as individual pieces, they are strikingly different in approach, tone, length, and frankly quality; but taken together, they reveal both the sweeping forces that shaped urban America during the post-World War II era—urban expansion, racial segregation, political fragmentation, commercial and industrial relocation, freeway construction, and increasing mobility—and the highly variegated human and natural landscapes on which they occurred. Some, like the San Francisco, New Orleans, and Baltimore pieces, tend toward the ideographic; others, such as the St. Louis, Pittsburgh, and Cleveland pieces, lean more toward the more general; and still others like the Chicago, Southern Connecticut, Seattle, and Los Angeles volumes find a balance between the two. While this diversity may have been frustrating to representatives of both the nomothetic and ideographic camps, it served a larger purpose. Only Pierce Lewis's rich and detailed account of New Orleans or James Vance's colorful story of San Francisco's neighborhoods truly convey the meaning of individual places and the importance of local forces in shaping a collective identity. But having establishing that, it is all the more remarkable to see the same underlying forces of immigration, residential segregation, urban sprawl, political fragmentation, and de-industrialization appear again and again in these texts.

It is obvious that the authors had free rein in deciding what to include and how to convey the essence of their respective places. In addition to the hard facts, they used text and experience to capture a city's collective feeling about itself. Some are intensely personal renditions. Take, for example, Sherry Olson's final comments about Baltimore neighborhoods:

> Baltimore neighborhoods are a wonderful radius of action for a child. Hopscotches are drawn like nowhere else. We eat sauerkraut with our Thanksgiving turkey and black-eyed peas for New Year's. We all go out on new roller skates on Christmas even

if it snows, and visit the "Christmas garden"—model railroad displays at the firehouse. We take in the state fair in August, the City fair in September, the Fell's Point Fair in October, and all the lesser street fairs. (Olson, 1976, p. 92)

Or Peirce Lewis' observations upon being asked to prepare a vignette about New Orleans—a place that was "terra incognito" to him previously: "I have learned that it is possible to fall in love with a city on short notice, even if one may not understand her perfectly (Lewis, 1976, p. 101). Berry et al. (1976, pp. 194–195) in "Chicago: Transformations of an Urban System" make extensive use of poetry and literature to evoke the brashness and dynamism of Chicago at the turn of the century:

She is moving upward and onward,
With victory on her lips,
And a dauntless eye and a strenuous cry,
To a world that she outstrips.

Horace Spencer Fiske, *The March of Chicago*, 1903

There is also a fair bit of boosterism in these pieces. Note Michael Conzen's and George Lewis' (1976) observations about Boston:

Boston seems to provide an atmosphere of urbanity and civility that sets it apart in the minds of its people. It has an atmosphere that must be experienced personally in order to be fully understood. (p. 53)

Boston Bay has for centuries provided the dramatic entrance for the city, and Logan Airport's coastal position opposite the downtown peninsula preserves and heightens the drama of arrival. (p. 57)

Or Ron Abler, John Adams, and John Borchert's (1976) observations about life in the Twin Cities:

...bitterness and despair are rare. The freedom from the hundreds of anxieties and squabbles characteristic of cities in which the average person expects to be gulled each time he or she turns around is perhaps the most refreshing element of Twin Cities life. People feel that life and events are manageable and under control, and they are usually right. (p. 365)

Imagine what a shock it was for me to read such experiential prose from the gods of *Spatial Organization*, another sacred text in the Ohio State School of Geography (Abler et al., 1971).

That the vignettes vary in their view of the future probably reflects individual personalities of the authors, the character of the places they studied, and also the prevailing uncertainty at the time about the fate of urban America. The University of Washington team was upbeat about Seattle's future with growing public awareness of the limits of unrestrained growth and its can-do attitude (Andrus et al., 1976). Bob Sinclair and Bryan Thompson drew a far more pessimistic portrait of Detroit, declaring that "Detroit epitomizes the ills of American society in the 1970s" and that "virtually every piece of change that has been discussed in this study has been responsible for some degree of deterioration, whether in terms of strict destruction or in terms of limbo associated with delay and uncertainty" (Sinclair and Thompson, 1976, p. 352). Regarding New Orleans, Pierce Lewis lamented the fact that "as the city becomes blacker, poorer, and more segregated, as the suburbs become whiter and more sprawling, it becomes harder and harder to put the urban Humpty Dumpty back together again. New Orleans becomes more and more like the rest of the country" (Lewis, 1976, p. 204).

The vignettes were remarkably synthetic and integrative at a time when human geography took a strong turn toward reductionist and narrow social science. I have written elsewhere about the unfortunate ideological and functional separation of human and physical geography and the challenge this separation presents for a discipline trying to respond to society's demand for more environmental research (Gober, 2000). These vignettes are the great exception to my generalization—scholarship that seeks to connect established knowledge in innovative ways rather than trying to discover something new. Each situates, to a greater and lesser extent, the process of urban development within its natural setting. Los Angeles' warm, dry climate, spectacular mountain scenery, and sandy beaches, were the basis for its image of the "good life," (Nelson and Clark, 1976); Seattle's water view influenced local housing prices and its social geography (Andrus et al., 1976); and in San Francisco, the diversity of physical conditions, presence of hazards, and small geographic area shaped the modern urban region (Vance, 1976). No wonder it was my physical geographer colleague, Mel Marcus, who first advised me to take a look at the vignettes!

## Urban Policymaking

The policy volume cut across metropolitan areas and included discussions of land speculation, housing abandonment, urban renewal, environmental quality, public education, crime, health-care delivery, open-space preservation, housing and transportation related to the elderly, metropolitan governance, and legislative redistricting. Several authors began with a detailed analysis of the problem itself, for example, land speculation and abandoned housing, while others presented a particular federal policy, such as open-space preservation, and assessed its impact across cities. Brian Berry began the volume with an excellent summary of each piece and an eloquent plea for more bridge building between the world of policy and the research activities of academic geographers (Berry, 1976).

Despite the avowed goal of engaging urban geographers in national policy debates, the Project did not lead urban geography down the policy path. A review of articles in *Urban Geography*, the *Annals*, and the *Professional Geographer* during the late 1970s and early-1980s reveals a gradual shift away from the study of problem-oriented subjects like crime, poverty, residential segregation, discrimination, and neighborhood revitalization to theoretically driven topics like residential search and satisfaction, residential mobility and migration, the locational behavior of different types of economic activity, transportation modeling, and political economy.

*Comparative Atlas*

The Atlas was based on the premise that intelligent policy-making requires accurate assessments of places—whether and to what extent they are affected by a particular problem and how effective a given program might be in alleviating a particular problem. Intended consumers were politicians, public officials, business people and scholars who designed programs for American's urban areas. Maps were drawn largely from 1970 Census data. There were at least 27 highly generalized maps for each place, including maps of terrain, land use, housing, population density, age structure, race and ethnicity, household size occupation, and income and special topics relevant to individual metropolitan areas. The final 22 chapters mapped policy-relevant characteristics such as overcrowding, rent, commuting, poverty, and premature birth across the 20 areas. According to Ron Abler, long and heated discussions centered on what variables to include and how the 20 cities would be chosen.

**Discussion**

While one can quibble, some 25 years after the fact, about the unevenness of contributions, differing writing styles and philosophies about how to write a policy paper, which variables to include and which ones to exclude from the Atlas, and the nuances of cartographic presentation, these volumes are, in my mind, a remarkable achievement in American urban geography in the 1970s. From a disciplinary perspective, they reveal what the intellectual leaders of the 1960s thought was important to the field and to the nation. Some 65 geographers contributed to the written portions, and another 26 participated in the cartographic representations. A sizable portion of the community of American urban geography at the time was at work on this project. It bore the imprimatur of the Association of American Geography, and it was funded by the National Science Foundation.

Several of the Project's younger participants recall their excitement about working with established leaders in the field on a project of such magnitude. Ron Abler noted the significance of junior scholars feeling a part of something larger and more important than their individual careers. It was a time when there was still enough flexibility in the academic system that someone could spend several years away from a specialized research program to work on a project of disciplinary importance, as in the policy volume and Atlas, and of local interest, as in the vignettes.

And yet, the Project seems to have disappeared from our disciplinary consciousness. While we talk often and passionately about the contributions of geographers at the Office of Strategic Services during World War II, the spatial analysts of the 1960s, and the Marxist critique, we rarely speak of the Comparative Metropolitan Analysis Project as an activity of urban geography in the 1970s. It is largely invisible in urban geography textbooks; it was not mentioned in the urban geography's contribution to *Geography in America* (Marston et al., 1989); and it is rarely cited by the discipline's practitioners. According to the Social Science Citation Index, the edited volumes were cited a total of only 13 times between 1980 and 1990. Several of the vignettes were cited frequently, notably Pierce Lewis' New Orleans piece (11 times) and Brian Berry et al.'s Chicago piece (9 times), but the rest were cited only once or twice. The question in my mind then becomes how it is possible that a project of such magnitude and with such potential significance for the field was written off so quickly and so completely. I offer my interpretation as a starting point in this discussion.

First, influenced by the writings of David Harvey and others, urban geography turned away from its early concern with spatial analysis and came to focus on the social forces that create spatial patterns rather than the patterns themselves. We can debate the forces internal and external to the discipline that caused this trend, but it was fatal to a project whose core was an atlas featuring spatial patterns of urban characteristics. The field also turned away temporarily from the study of historical landscapes and regional geography—a trend that ran counter to the direction of the vignettes, which was to highlight the individuality of places and showcase their unique social and environmental histories.

The promise of integration offered by the vignettes never caught hold in a discipline consumed by specialization, reductionism, and the disengagement of human and physical geography. Urban geographers and others on the human side of geography came to regard themselves as belonging to the discipline of "human geography" distinct from "physical geography," and physical geographers turned away to form closer ties to allied fields. The dominant research strategy was to break apart the complicated urban system and study its constituent parts, the antithesis of integration. In an age of increasing specialization and fragmentation few geographers had the training to produce broad, reflective studies of American urbanization. Equally few had the background or the perspective to appreciate its significance.

The challenge of national urban policy, so powerfully articulated in the policy volume, went largely unanswered by a discipline consumed by internal squabbles and academic debate and separated from its practitioners in the worlds of government and the private sector. Our journals were geared toward an academic audience, our monthly jobs listings were dominated by faculty positions in colleges and universities, and our national meetings were attended almost exclusively by faculty members and students. Too few geographers moved easily between the worlds of policy and academic research, and our presence in the national policy community, with a few notable exceptions, was pitifully small. The Project did not fulfill its mandate of steering the field toward more policy-oriented research because the field chose not to move in that direction.

Several other factors also limited the Project's impact. First, it did not receive a glowing review—not at all unusual in a discipline socialized to eat its young. My experience on interdisciplinary review panels has convinced me that we are ferocious in our self-criticism, largely, I believe, so as to legitimize ourselves among science and social science peers. Maurice Yeates, writing in the June 1978 issue of the *Annals,* relegated the vignettes to "brief regional geographies of the areas concerned," criticized the shift in scale from local-level to regional-level analyses, and accused the Project of "turning back the geographic clock" to focus on the unique. He noted that the policy pieces were formulated to focus on a particular problem rather than the integrated nature of the process that led to it, and he came down very hard on the Atlas for demonstrating "very little integrative and conceptual geographic skill" (Yeates, 1978, p. 314). His bottom line can best be characterized as damning with faint praise when he congratulated the directors for getting the job done and noting that the materials met an "acceptable level of professional expertise." Somewhat more positive reviews appeared in the *Professional Geographer* where Clyde Browning (1977), although acknowledging the unevenness of the vignettes, noted their value when read "in toto." Clifford Tiedemann (1978) also was generally complimentary. He was careful, however, to clarify that the Project represented "a picture of the state of one wing of our discipline" and lamented the fact that "other aspects of the discipline have not had similar opportunities to present comparable packages" (p. 223).

A second limitation was the sheer enormity of the Project. To be sure, it had ambitious goals and, read as a package, it made sense, but I doubt many geographers had the fortitude to slog through even a portion of the four volumes of vignettes, constituting 1,525 pages, the 565-page policy volume, or the eight-pound Atlas with its 1,000 maps. Moreover, policy-makers were hardly prepared to absorb the great mass of material, especially with its academic bent. We have learned much in recent years about the way policy-makers receive and use the products of scientific research. Long treatises in academic prose, no matter what their quality, are unlikely to be understood or used.

Third, the published products of the Project, especially the vignettes, were not well marketed by Ballinger. A commercial publisher, accustomed to national promotion, Ballinger was ill-equipped to market the vignettes to local audiences.

## Final Thoughts

A review of the Comparative Metropolitan Analysis Project and the discipline's response to it reveals both the strengths and weaknesses of American urban geography in the 1970s. That disciplinary leaders were able to mount a large and comprehensive effort to chronicle the state of American urban places at that moment in time is, in itself, an impressive accomplishment. That they had difficulty mustering the discipline's talent to produce broad, reflective pieces and to incorporate them into teaching and research after an era of highly specialized graduate education that focused on spatial analysis and academic criticism is not surprising. Tepid response to the Project reflected a shift in the discipline's paradigms away from spatial analysis, widespread acceptance of reductionism as a mode of intellectual inquiry, and unwillingness of a critical mass of geographers to undertake policy-oriented research. We were not well positioned to showcase our integrative perspective and tackle society's social problems.

On a personal note, I continue to flit around the edges of urban geography, to use its ideas to more fully understand the urban environment in which I live and to wait for urban geography to give society what it really wants—meaningful information about cities and useful ideas about how to make them better places to live.

## References

Abler, R. F., 1976, *A Comparative Atlas of America's Great Cities: Twenty Metropolitan Regions*. Minneapolis, MN: Association of American Geographers and University of Minnesota Press.

Abler, R. F., Adams, J. S., and Borchert, J. R., 1976, The Twin Cities of St. Paul and Minneapolis. In. J. S. Adams, editor, *Contemporary Metropolitan America: Nineteenth Century Inland Centers and Ports*. Cambridge, MA: Ballinger, 355–423.

Abler, R. F., Adams, J. S., and Gould, P. R., 1971, *Spatial Organization: The Geographer's View of the World*. Englewood Cliffs, NJ: Prentice Hall.

Adams, J. S., editor, 1976a, *Contemporary Metropolitan America: Cities of the Nation's Historic Metropolitan Core*. Cambridge, MA: Ballinger.

Adams, J. S., editor, 1976b, *Contemporary Metropolitan America: Nineteenth Century Ports*. Cambridge, MA: Ballinger.

Adams, J. S., editor, 1976c, *Contemporary Metropolitan America: Nineteenth Century Inland Centers and Ports*. Cambridge, MA: Ballinger.

Adams, J. S., editor, 1976d, *Contemporary Metropolitan America: Twentieth Century Cities*. Cambridge, MA: Ballinger.

Adams, J. S., editor, 1976e, *Urban Policymaking and Metropolitan Dynamics: A Comparative Geographical Analysis*. Cambridge, MA: Ballinger.

Andrus, A. P., Beyers, W. B., Boyce, R. R., Eichenbaum, J. J., Mandeville, M., Morrill, R. L., Stallings, D., and Sucher, D. M., 1976, Seattle. In. J. S. Adams, editor, *Contemporary Metropolitan America: Nineteenth Century Ports*. Cambridge, MA: Ballinger, 425–498.

Berry, B. J. L., 1976, On geography and urban policy. In John S. Adams, editor, *Urban Policymaking and Metropolitan Dynamics*. Cambridge, MA: Ballinger, 3–18.

Berry, B. J. L, Cutler, I., Draine, E. H., Diang, Y., Tocalis, T. R., and de Vise, P., 1976, Chicago: Transformations of an urban system. In. J. S. Adams, editor, *Contemporary Metropolitan America: Nineteenth Century Inland Centers and Ports*. Cambridge, MA: Ballinger, 181–283.

Berry, B. J. L. and Horton, F. E., 1970, *Geographic Perspectives on Urban Systems: With Integrated Readings*. Englewood Cliffs, NJ: Prentice-Hall.

Brown, L. A., 1981, *Innovation Diffusion: A New Perspective*. New York, NY: Methuen.

Brown, L. A. and Moore, E. G., 1970, The intra-urban migration process: A perspective. *Geografiska Annaler*, Vol. 52, 1–10.

Browning, C. E., 1977, Contemporary Metropolitan America, Volumes 1 to 4 [book review]. *Professional Geographer*, Vol. 29, 416–418.

Conzen, M. P. and Lewis, G. K., 1976, Boston: A geographical portrait. In J. S. Adams, editor, *Contemporary Metropolitan America: Cities of the Nation's Historic Metropolitan Core*. Cambridge, MA: Ballinger, 51–138.

Gober. P., 2000, In search of synthesis. *Annals of the Association of American Geographers*, Vol. 90, 1–11.

Golledge, R. G. and Rushton, G., 1976, *Spatial Choice and Spatial Behavior: Geographic Essays on the Analysis of Preferences and Perceptions*. Columbus, OH: Ohio State University Press.

Harvey, D., 1973, *Social Justice and the City*. Baltimore, MD: The Johns Hopkins University Press.

Herbert, D. T. and Johnston, R. J., 1978, Geography and the urban environment. In D. T. Herbert and R. J. Johnston, editors, *Geography and the Urban Environment*. New York, NY: John Wiley, 1–33.

Lewis, P. F., 1976, New Orleans—The making of an urban landscape. In. J. S. Adams, editor, *Contemporary Metropolitan America: Nineteenth Century Ports*. Cambridge, MA: Ballinger Publishing Company, 97–216.

Lowenthal, D., 1961, Geography, experience, and imagination: Towards a geographical epistomology. *Annals of the Association of American Geographers*, Vol. 51, 241–260.

Marston, S. A., Towers, G., Cadwallader, M., and Kirby, A., 1989, The urban problematic. In G. L. Gaile and C. J. Willmott, editors, *Geography in America*. Columbus, OH: Merrill, 651–672.

Nelson, H. J. and Clark, W. A. V., 1976, The Los Angeles metropolitan experience. In. J. S. Adams, editor, *Contemporary Metropolitan America: Twentieth Century Cities*. Cambridge, MA: Ballinger, 227–295.

Olson, S., 1976, Baltimore. In. J. S. Adams, editor, *Contemporary Metropolitan America: Nineteenth Century Ports*. Cambridge, MA: Ballinger, 1–95.

Sinclair, R. and Thompson, R., 1976, Detroit. In J. S. Adams, editor, *Contemporary Metropolitan America: Nineteenth Century Inland Centers Ports*. Cambridge, MA: Ballinger, 285–354.

Tiedemann, C. E., 1978, Urban Policymaking and Metropolitan Dynamics: A Comparative Geographical Analysis [book review]. *Professional Geographer*, Vol. 30, 222–223.

Tuan, Y. F., 1968, Discrepancies between environmental attitudes and behaviour, examples from Europe and Canada. *Canadian Geographer*, Vol. 12, 176–191.

Vance, J. E., 1976, The American city: Workshop for a national culture. In. J. S. Adams, editor, *Contemporary Metropolitan America: Cities of the Historic Metropolitan Core*. Cambridge, MA: Ballinger, 1–49.

Yeates, M., 1978, Contemporary Metropolitan America, Urban Policymaking and Metropolitan Dynamics, Comparative Atlas of America's Great Cities [book review]. *Annals of the Association of American Geographers*, Vol. 68, 309–316.

# CHAPTER 14
# Emerging Political Paradigms

LARRY R. FORD[1]

## ABSTRACT

In a way, the 1970s were a golden age for urban geography. Although spatial science was still dominant, a number of new approaches, methodologies, and worldviews were added to the field. For a while, it seemed as though many of these approaches could be mixed and matched in complex ways in order to adequately address the complexity of the urban world. The era also ushered in a new political awareness, however, and urban geographers were increasingly differentiated into those who wanted to incorporate a political stance into their work and those who advocated scientific neutrality. Over the past two decades, urban geography has become fragmented as the "politicals" have moved into social theory while the "neutrals" have merged with the boom in high technology.

[1]Correspondence concerning this chapter should be addressed to Larry R. Ford, Department of Geography, San Diego State University, San Diego, CA 92182; telephone: 619-594-5486; fax: 619-594-4938; e-mail: larryf@mail.sdsu.edu

The inevitable age of specialization has been accentuated by polarized
political stances. A focus on the city has been largely replaced by specialization in social theory or technology and mainstream urban geography
may be disappearing. [Key words: flagship subdiscipline, political correctness, historic preservation, disappearing center.]

## The 1970s: A Golden Age?

It is quite common for both individuals and entire cultures to reminisce
about a golden age some time in the past when everything was better than
it is today. We do this even though most of us realize that most golden ages
have been romanticized into something a bit more golden than they actually
were. With this in mind, I suggest that the 1970s were, for urban geography, at least a potential golden age if not a real one. Perhaps I see the 1970s
as a special time simply because that was when I first started teaching
urban geography as a faculty member at San Diego State University and
everything was still new and exciting. But I think there is more to my
perception of a golden era than youthful enthusiasm. It was a time of disagreement, controversy, and confusion, but it was also a relatively tolerant
and optimistic time. There was also a feeling that urban geography was
very important, maybe even the flagship of the discipline.

I doubt if there has ever been an era when there was only one acceptable paradigm or research agenda in urban geography but for a long time
there were not many. During the first half of the 20th century, the literature grew slowly but inexorably, with the first comprehensive reader
appearing only in 1959 (Mayer and Kohn, 1959). Throughout this period,
urban geography included a variety of emphases such as studies of site and
situation, the location of economic activities in cities, city structure and
land-use models, the distribution of ethnic and social groups, comparative
urbanization, transportation studies, and landscape tastes and preferences.
For the most part, all of these subdisciplines managed to coexist in relative
peace. The field was also politically quiescent.

During the 1960s, several new and exciting research agendas came to
the fore. The one that has gotten the most attention is the spatial analytical
approach embedded in what was known as the quantitative revolution,
but there were others as well. Kevin Lynch introduced mental maps and
environmental perception in 1960 while Gilbert White and others gradually (re)introduced concern for the physical environment of cities (Lynch,
1960; White and Haas, 1975). Some maintain that intolerance and hostility began to appear at this time due in part to the arrogance of many

extreme "quantifiers" and the uncompromising positions taken by gloom and doom environmentalists (Burton, 1963). On the other hand, while there were a number of strong-willed true believers (I vaguely remember one presentation that argued that wolves defended a nested hierarchy of hexagonal territories), for young graduate students it was a time of great excitement. For most urban geographers, positions and strategies had not yet hardened to the point that the profitable exchange of ideas could not continue (Taaffe, 1974).

## Let a Thousand Flowers Bloom

For a time, it seemed that many, if not all, of the new research agendas and paradigms could be combined in a new and productive urban geography that would be the flagship of a revitalizing discipline. I, for one, saw no reason why this should not come to pass. During the 1960s I studied at Ohio State University and experienced the halcyon years of the quantitative revolution and then moved on to the more culturally oriented University of Oregon for my doctorate. Along the way, I worked in a variety of both quantitative and qualitative paradigms and saw no reason why they should be mutually exclusive, that is, first you do some statistical analysis and then you follow up with some fieldwork and interviews. When I began teaching at San Diego State in 1970, I added a new emphasis, applied urban geography, as I served on a number of city boards and planning groups. Still, everything seemed to fit together nicely. When I gave presentations to community groups and planning agencies I often used geographic literature from authors as diverse as Tuan (1976), Lowenthal (1968), Berry (1973), Harvey (1973), and Ley (1974). Specialty groups were established in the Association of American Geographers in 1979, and the urban geography group quickly emerged as one of the largest. To a very real degree, it was the study of the city that brought a wide variety of scholars together.

Urban geography in the 1970s included at least 12 distinctive yet complimentary branches dealing with both intraurban and interurban issues. At the risk of oversimplifying, I would classify the branches, along with a tiny sample of practitioners, as follows:

Historical: Conzen 1977, Ward 1971, Goheen 1974.
Cultural Landscape: Jakle and Mattson 1981, Lewis 1976, Rubin 1979.
Social and Ethnic: Cybriwsky 1978, Rose 1972, Harries 1974.
Spatial Analytical: Yeates 1974, Berry and Horton 1970.
Urban Structure/Models: Bourne 1971, Adams 1969, Johnston 1973.
Attitudes and Values: Tuan 1976, Lowenthal 1968, Entrikin 1976.
Political: Cox 1973, Ley 1974, Morrill 1981.

Comparative/International: Wheatly 1971, Vance 1977, Bonine 1979.
Economic: Manners 1974, Pred 1977.
Perception/Behavior: Golledge and Rushton 1976, Downs and Stea 1973.
Transportation: Wheeler, 1974.
Physical/Environmental: White and Haas 1975, Detwyler and Marcus 1972.

Often one or more of these approaches were combined in interesting and productive ways, that is, social-historical, planning-political, or comparative-landscape. A few urban geographers, John Borchert comes to mind here, were able to combine many of the topical approaches as they told us how cities and urban systems came to be (Borchert, 1967). More often, individual geographers specialized in one or two approaches but during the 1970s even some former advocates of one approach mellowed and agreed that understanding real places required an appropriate balance of data and research agendas (King, 1976). Although methodological battles were more than occasionally spirited if not histrionic, students usually benefited from exposure to the great diversity of opinions that were paraded before them in the journals and at professional meetings. My first course in urban geography, for example, was with Henry Hunker in 1964. I went on to have classes with geographers as diverse as S. Earl Brown, Leslie King, Kevin Cox, and Edward Taaffe, at Ohio State and Edward Price, Everett Smith, and Clyde Patton at the University of Oregon. I also had courses and seminars in statistics, art, advertising, urban and regional economics, political science, and anthropology. Although everyone tried to sell their brand of scholarship as the best, it was usually done in a friendly, low-key way. The burden was on the student to mix and match as he/she acquired the background and skills needed for particular kinds of research. At least in my memory, a healthy skepticism of all "new geographies" abounded.

The glue that held everything together was a pervasive and profound interest in cities. As long as the goal was to understand cities, there was a reason to come together. In retrospect, however, the glue did not hold and much of the work that followed tended to move our collective focus away from the city.

## Political Correctness and the Age of Specialization

Given the explosion of journals, professional organizations, academic disciplines, electronic data, and skills such as computer science, GIS, and remote sensing, it is no surprise that urban geography, like many other fields, has become increasingly fragmented and disjointed. It would be very strange if this were not the case. This is not really new. Those with a strong preference for mathematical rigor and scientific data have traditionally focused on different kinds of topics than those with a more artistic bent. Time does not permit most students to gain computer skills, learn foreign languages, understand social theory, and do intensive field work in one graduate program. A related, but much more serious problem is that many potential urban students have now lost their focus on the city and have instead drifted toward specialization in technology, social theory, or environmental science. The center is not holding and what was once a major flagship of the discipline of geography is adrift. It may eventually be possible to refocus the efforts of a significant number of geographers on cities now that more than half the world's population lives in urban places and human problems have become urban problems. The challenge before urban geographers is greater than ever. Yet another obstacle looms before us on the path to disciplinary coherence, the twin issue of political agenda and political correctness.

It has sometimes been said that geographers "eat their young" since intercine battles are often vitriolic and seemingly blown out of proportion. In no area has this been truer than in urban subfield. One way of viewing this is to say that the golden age of diversity in the 1970s failed to adequately set the stage for the inevitable age of specialization that followed. But political agendas have been important as well. The various subfields and topical interests within urban geography have become even more divergent by becoming joined at the hip with political stances. Consequently, certain types of urban geographers see "others" as not only out of sync methodologically (as they have since the late 1950s) but also politically. The latter condition often results in even more serious condemnation by colleagues than the former.

As the pieces of what in an ideal world might make up a diverse and complex urban geography have become more separated and segregated, arrogant disinterest has been replaced by political repugnance. Some social theorists, for example, see spatial analysts and GIS practitioners as not only operating in an unproductive (positivist) manner but also as "working for the man" in an oppressive capitalist system, thus serving to make matters worse. Meanwhile, the spatial/GIS contingent not only sneers when critiquing the "meaningless gibberish" of postmodern deconstructionists but also condemns them for potentially alienating agencies that might bring grants and contracts to geography departments by pointing up possible flaws in the new magic technologies. Meanwhile, the urban environment group views both of the first two as fiddling while Rome (the planet) burns instead of trying to influence important political decisions. I am, of course, exaggerating just a little bit here (although not a lot). What follows is a subtler version of the increasing importance of political worldview based on a few of my own experiences.

## Historic Preservation, Gentrification, Displacement, Public Space and the Homeless: Trying to Cover All of the (Political) Bases

I was interested in old neighborhoods and social change long before I knew about urban geography. I grew up in the "inner city" of Columbus, Ohio, and my house later sold for one dollar in an urban homesteading program. My Linden neighborhood had an interesting mixture of Italians, African Americans, Appalachian Whites and stolid Midwesterners. I learned a lot about "culture" in school and on the streets. While I was a student at Ohio State, I started hanging out in an emerging historic neighborhood known as German Village. The area was begun in the 1820s and fell into disrepute with the "anti-Hun" hysteria of World War I. My jug band (de Gaulle Bladders) played in the local beer gardens and we got to know quite a few neighborhood residents. As I learned about urban geography, I began to monitor the interesting changes that were taking place around me. My old neighborhood was getting poorer while German Village was becoming increasingly middle class.

At the University of Oregon I became interested in architecture and landscape interpretation, and, after I began teaching in 1970, I returned to both Linden and German Village in order to flesh out some of theories of urban change with studies of historic preservation policies. During the early 1970s, I served on the Historic Site Board for the City of San Diego and I began giving presentations and publishing papers on historic preservation, including some on the revival of German Village (Ford and Fusch, 1978). During the Bicentennial era, studies of architectural heritage were much in vogue and several urban geographers joined together to create special sessions at meetings and exchange information. In conceptualizing my studies, I was informed by a vast literature in urban geography ranging from works by Adams and Bourne to those by Lowenthal and Tuan. By the early 1980s, however, things had changed. For many urban geographers, it was no longer politically correct to write about architectural conservation without focusing on displacement and the homeless (Smith, 1979).

The incorporation of (often Marxist) social theory into urban geography was for the most part a good thing that added sophistication to the analysis of urban problems, but it also tended to create a cadre of "true believers" who were unwilling to accept contradictory data. It became increasingly difficult to discuss historic preservation without questions (comments) from the audience assuring me that I was aiding and abetting the oppression of the poor by saying anything good about architectural conservation. I have studied gentrification and displacement in San Diego, Seattle, San Francisco, New York City, and London and I know the problems associated with excessive preservation and gentrification reasonably well. It is just that occasionally I want to talk about other aspects of the topic (Ford, 1979, 1984). I have also studied a number of places, including my old neighborhood, where historic conservation programs have not led to displacement. In many neighborhoods, a little gentrification is a good thing as it helps to retard redlining and abandonment. I am concerned that many socially aware young urban geographers are dissuaded from studying "good" gentrification lest they be shunned as politically incorrect. The danger lies in oversimplifying the divergent trends shaping the urban mosaic. Relatively few urban geographers have challenged the dominant line (Bourne, 1993).

A similar problem has emerged in studies of urban public space. Politically correct urbanists consistently report that public space has been privatized and commodified and that we may even be experiencing the end of public space (Sorkin, 1992). These problems do exist and should not be ignored but public space is not at equal risk everywhere. However, it is hard to argue this point without seeming to be a socially insensitive chamber-of-commerce-style booster. Yet there are many examples of new, successful, heavily used urban public spaces. Why must we examine one trend and not the other? Being critical can mean differentiating the good from bad and recognizing which is which.

Meanwhile, urban geographers who are not interested in challenging the status quo in the ways discussed above are increasingly drawn to the supposed neutrality of high tech, spatial analytical approaches. This is also where much of the grant money is because advocates are more than willing to work for government agencies and the private sector without saying things that might rock the boat. But of course, saying nothing critical is a powerful political statement too. In this realm, criticizing science and technology is sometimes seen as politically incorrect. A related problem in the positivist camp is a lack of passion. In recent years, a raft of academic and popular books have appeared loudly condemning the placeless, impersonal, oppressive modern American city but few are by urban geographers (see, for example, Kunstler, 1993).

## Is There A Mainstream Urban Geography?

Urban geography has become polarized with, for example, social theorists and GIS/spatial modelers operating in completely different ballparks. To some degree, these extremes have always existed in the field, but as late as the 1970s the center was still dominant. Today, the center seems to be disappearing. Social theory and software have replaced the city as the main focus of much "urban" geographic research. In my experience on hiring committees, most of the applicants for urban positions are interested in cities only tangentially and instead present themselves as experts in aspects of either theory or technology. Some would argue, of course, that this is quite acceptable and that there is no need for mainstream urban geography as it was once constituted. Perhaps, but I would like to think otherwise. Because urban geography was the focal point for much of the enthusiasm for "new" geographies as far back as the 1950s, what happens to our subdiscipline could easily been seen as emblematic of what is happening to geography as a whole. Will the center hold? Does the discipline have a focus?

In the age of specialization, some diversity is inevitable, but enthusiasm for extreme political correctness in urban geography tends to accentuate the extremes. The fact that many exciting new concepts emerged in urban geography during the 1970s suggests that the period was a golden age of sorts. The fact that we failed to find political stances that would help to bring these concepts together to shape a coherent field makes it an *almost* golden era.

## References

Adams, J., 1969, Directional bias in intra-urban migration. *Economic Geography*, Vol. 45, 302–323.

Berry, B. J. L., 1973, *Growth Centers in the American Urban System*. Cambridge, MA: Ballinger.

Berry, B. J. L. and Horton, F., 1970, *Geographic Perspectives on Urban Systems*. Englewood Cliffs, NJ: Prentice Hall.

Borchert, J., 1967, American metropolitan evolution. *Geographical Review*, Vol. 57, 302–332.

Bonine, M., 1979, The morphogenesis of Iranian cities, *Annals of the Association of American Geographers*, Vol. 69, 208–224.

Bourne, L. S., editor, 1971, *Internal Structure of the City: Readings on Space and Environment*. New York, NY: Oxford University Press.

Bourne, L. S., 1993, The myth and reality of gentrification: A commentary on emerging urban forms. *Urban Studies*, Vol. 30, 95–107.

Burton, I., 1963, The quantitative revolution and theoretical geography. *Canadian Geographer*, Vol. 7, 151–162.

Conzen, M., 1977, The maturing urban system in the United States, 1840–1910. *Annals of the Association of American Geographers*, Vol. 67, 88–108.

Cox, K., 1973, *Conflict, Power, and Politics in the City*. New York, NY: McGraw-Hill.

Cybriwsky, R., 1978, Social aspects of neighborhood change. *Annals of the Association of American Geographers*, Vol. 68, 17–33.

Detwyler, T. and Marcus, M., 1972, *Urbanization and Environment*. Belmont, CA: Duxbury.

Downs, R. and Stea, D., editors, 1973, *Image and Environment*. Chicago, IL: Aldine.

Entrikin, J. N., 1976, Contemporary humanism in geography. *Annals of the Association of American Geographers*, Vol. 66, No. 4, 615–632.

Ford, L., 1979, Urban preservation and the geography of the city in the USA. *Progress in Human Geography*, Vol. 4, 211–238.

Ford, L., 1984, The burden of the past: Rethinking historic preservation. *Landscape*, Vol. 28, 41–48.

Ford, L. and Fusch, R., 1978, Neighbors view German village. *Historic Preservation*, July, 37–41.

Goheen, P. G., 1974, Interpreting the American city: Some historical perspectives, *Geographical Review*, Vol. 64, 362–384.

Golledge, R. and Rushton, G., 1976, *Spatial Choice and Spatial Behavior*. Columbus, OH: The Ohio State University Press.

Harvey, D., 1973, *Social Justice and the City*. London, UK: E. Arnold.

Jakle, J. A. and Mattson, R. L., 1981, The evolution of the commercial strip. *Journal of Cultural Geography*, Vol. 1, 12–25.

Johnston, R. J., 1973, *Spatial Structures*. New York, NY: St. Martin's.

King, L. J., 1976, Alternatives to a positive economic geography. *Annals of the Association of American Geographers*, Vol. 66, 293–308.

Kunstler, J. H., 1993, *The Geography of Nowhere*. New York, NY: Simon and Shuster.

Lewis, P., 1976, *New Orleans—The Making of an Urban Landscape*. Cambridge, MA: Ballinger.

Ley, D., 1974, *The Inner City Ghetto as Frontier Outpost*. Washington, DC: Association of American Geographers.

Lowenthal, D., 1968, The American scene. *Geographical Review*, Vol. 58, 61–88.

Lynch, K., 1960, *The Image of the City*. Cambridge, MA: MIT Press.

Manners, G., 1974, The office in metropolis: An opportunity for shaping metropolitan America. *Economic Geography*, Vol. 50, 93–110.

Mayer, H. and Kohn, C., editors, 1959, *Readings in Urban Geography*. Chicago, IL: University of Chicago Press.

Morrill, R. L., 1981, *Political Redistricting and Geographic Theory*. Washington, DC: Association of American Geographers.

Pred, A. R., 1977, *City Systems in Advanced Economies*. New York, NY: Wiley.

Rose, H. M., 1972, The spatial development of Black residential subsystems. *Economic Geography*, Vol. 48, 45–65.

Rubin, B., 1979, Aesthetic ideology and urban design. *Annals of the Association of American Geographers*, Vol. 69, 339–361.

Smith, N., 1979, Toward a theory of gentrification: A back to the city movement by capital, not people. *Journal of the American Planning Association*, Vol. 45, 538–547.

Sorkin, M., 1992, *Variations on a Theme Park: The New American City and the End of Public Space*. New York, NY: Noonday.

Taaffe, E. J., 1974, The spatial view in context. *Annals of the Association of American Geographers*, Vol. 54, 1–16.

Tuan, Y. F., 1976, Humanistic geography. *Annals of the Association of American Geographers*, Vol. 66, 266–276.

Vance, J., 1977, *This Scene of Man: The Role and Structure of the City in the Geography of Western Civilization*. New York, NY: Harper's College Press.

Ward, D., 1971, *Cities and Immigrants*. New York, NY: Oxford University Press.

Wheatly, P., 1971, *The Pivot of the Four Corners: A Preliminary Inquiry into the Origins and Character of the Ancient Chinese City*. Chicago, IL: Aldine.

Wheeler, J., 1974, *The Urban Circulation Noose*. Belmont, CA: Wordsworth.

White, G. F. and Haas. J. E., 1975, *Assessment of Research on Natural Hazards*. Cambridge, MA: MIT Press.

# IV
## Urban Geography in the 1980s

# CHAPTER 15

# Introduction—The Sea Change of the 1980s
## *Urban Geography*
## *As If People and Places Matter*

PAUL L. KNOX[1]

## ABSTRACT

The literature in urban geography grew rapidly in the 1980s, with a general shift of academic interest toward cities and urbanization that was prompted by a global economy that was increasingly articulated through networks of cities, and by an awareness that the world's population was rapidly becoming increasingly urbanized. The literature also grew more diverse as researchers took on the economic, social, cultural, and political

---

[1]Correspondence concerning this chapter should be addressed to Paul L. Knox, Dean, College of Architecture and Urban Studies, Virginia Tech, Blacksburg, VA 24061-0205; telephone: 540-231-6416; fax: 540-231-6332; e-mail: knox@vt.edu

changes that were occurring as a result of new spatial divisions of labor and the "new economy." Within this expanded literature, and within the social science literature generally, the theoretical roles of space and place were strongly reasserted. [Key words: urban geography, sociospatial dialectic, built environment, structuration, cultural geography.]

Looking back now at scholarship in urban geography during the 1980s, one can see the elements—somewhat disparate at the time—of what was to become a more catholic subdiscipline. During the 1980s, the sense of competition among a few rival approaches (Wheeler, 2002) was eclipsed by a breadth of scholarship in which urban geographers have subsequently been able, as Ron Johnston (1984) put it, found ways to accommodate not only the general (things that are universally applicable), but also the unique (things that are distinctive, but whose distinctiveness can be accounted for by a particular combination of general processes and individual responses) and the singular (things that are interesting and remarkable but about which no general statements can be made). The 1980s were marked by a great ferment in human geography in general, a ferment that was paralleled in other social sciences and intensified by an unprecedented degree of intellectual trafficking, both between human geography and other social sciences, and between Anglo-American and continental European scholarship.

Whereas the 1970s saw the dominance of the spatial analytical tradition in urban geography challenged by the emergence of behavioral, humanistic, phenomenological, and historical materialist approaches, the 1980s forced many urban geographers to consider afresh both the nature of their subject matter and their theoretical and methodological approaches to it. All of us were grappling with the empirical consequences and theoretical implications of a sea change in economic, political, and cultural life in the world's leading economies. The pivotal point had been the "system shock" to the international economy that had occurred in the mid-1970s. The Eurodollar market, swollen by the U.S. government's deficit budgeting and by the huge reserves of U.S. dollars resulting from the quadrupling of crude oil prices in the wake of an OPEC embargo, quickly evolved into a new, sophisticated system of international finance, with new patterns of investment and disinvestment, contributing to new and modified geographies at every scale. Toward the end of the decade, amid a phase of deindustrialization, Thatcherism and Reaganomics propagated a climate of privatization based on the idea that welfare states not only had generated unreasonably high levels of taxation, budget deficits, disincentives to work and save, and a bloated class of unproductive workers, but also that they fostered "soft" attitudes toward "problem" groups in society: disadvantaged households and individuals (whose numbers were increasing as a result of the new spatial divisions of labor associated with the onset of globalization). The U.S. president's *Commission for a National Agenda for the Eighties* accepted the inevitability of a "nearly permanent" underclass, but nevertheless there was a marked shift away from collective consumption and Keynesian economic management toward deregulated capital accumulation. New social formations were emerging as part of post-industrial societies in most OECD countries. New urban forms were beginning to emerge in response to socioeconomic polarization and to the recentralization of high-order producer services. Meanwhile, the transnational material culture associated with the new economy was accelerating the attenuation of the meaning of place in people's lives.

So, just as urban geographers had developed neat models of the sociospatial structure of "the city", so cities and urban society began to change in unexpected and unpredictable ways. Generalizations drawn from factorial ecologies and the modeling of consumer behavior in residential markets suddenly seemed much less compelling to many of us—a "normal science" that seemed abstracted from contemporary patterns and processes. Similarly, mainstream cultural geography in the United States seemed all but irrelevant to the cultural vitality and hybridity of cities: its focus on the rural and the archaic seemed hopelessly esoteric, while its super-organicism abstracted culture from the dynamics of the contemporary political economy and the material conditions of urban society (Duncan, 1980). Thus the turn to broader social scientific frameworks and, more specifically, to contemporary adaptations and interpretations of Marxian and Weberian theory. Urban geographers began to read the influential work of French structuralists and post-structuralists such as Louis Althusser, Pierre Bourdieu, Michel de Certeau, Michel Foucault, and Jean-Francois Lyotard. Topics that had hitherto been approached in empirical, descriptive terms now came into question in terms of their social construction, identity, and contestation. Questions of race and ethnicity, for example, traditionally examined in terms of segregation indexes and historical patterns of migration and residential mobility, were now excavated in terms of cultural hegemony and the interactions between political culture and economic circumstance (Jackson, 1987, 1989; Anderson, 1988; Smith, 1989). Questions of gender were approached in terms of the lived relations of women's lives as part of processes of production and reproduction under capitalism (Hayden, 1980; McDowell, 1983; Brownill, 1984; Little et al., 1988; Mackenzie, 1988). Meanwhile, with the emergence of new social formations and new urban form, some new topics were identified. Among them were homelessness (Mair, 1986; Dear and Wolch, 1987), gentrification (Smith, 1982, 1987; Rose, 1984), and the emergence of gay and lesbian spaces in cities (Castells, 1983; Lauria and Knopp, 1985; Knopp, 1987).

David Harvey's interrogations of the political economy of urbanization (1982, 1989a) were broadly influential in setting out a Marxian interpretation of the relations between capital, class, and the built environment. Harvey's writing, grounded in historical materialism, did not lend itself easily to the sociocultural dynamics of cities that were beginning to interest many young geographers. Nevertheless, Harvey recognized the importance of this dimension of contemporary urbanization in *The Condition of Postmodernity* (1989b). Harvey's work opened windows for many geographers on to the broader landscape of ideas encompassed by structuralist theory. Equally, his books did a great deal toward gaining attention and respect for human geography among the practitioners of other social science disciplines, but it was Ed Soja (1980, 1989) who reasserted most forcefully the importance of space and place in social theory. Soja put space and place in a central role in constituting, constraining, and mediating social relations. His notion of a sociospatial dialectic—a continuous two-way process in which people create and modify urban spaces while they are conditioned in various ways by the spaces in which they live and work—has proven central to understanding the processes of contemporary urbanization. It provides a framework for an urban geography as if people and places matter.

Important contributions to theorizing the social construction of urban spaces and places were made by several geographers in addition to Harvey and Soja (for example: Ley, 1983; Thrift, 1985; Sack, 1988), but it was Anthony Giddens, a sociologist, who developed what is arguably the most robust framework for understanding the operation of the sociospatial dialectic (Giddens, 1981, 1984). Giddens' structuration theory, centering around processes of time-space routinization and time-space distanciation, led us to see urban places as continuously "becoming" (Pred, 1984, 1985). Structuration theory also helped resolve what for me had been the competing attractions of Marxian and Weberian approaches, recognizing the importance of knowledgeable actors, or agents, and of institutional arrangements (which both enable and constrain action) in the production and reproduction of specific social contexts, or structures (Cullen and Knox, 1981; Knox and Cullen, 1981a, 1981b, 1982).

Moving as I did in the early 1980s from a small department of geography in the United Kingdom (where I had pursued research in regional economic geography, medical geography, and rural geography as well as urban geography) to a department of urban affairs and planning in one of the leading colleges of architecture in the United States prompted me to focus or at least to try to focus my teaching in urban social geography (Knox, 1982) and my research on the social production of the built environment (Knox, 1984, 1987). In this context I found Michael Ball's work on the structures of building provision (Ball, 1986) to be particularly helpful. Ball drew attention to the social relations, always specific to time and place, among key actors landowners, investors, financiers, developers, builders, design professionals—and the regulatory institutions of government, the professions, business organizations, and organized labor. The potential of this framework has not (yet) been realized, largely, perhaps, because of the confidentiality surrounding property and construction deals, but also because of the sheer complexity of business and professional relations surrounding the production of the built environment.

In contrast, the emergence of a new cultural geography has generated a rich vein of scholarship with important implications for urban studies. Strongly influenced by French critical social theorists especially Michel Foucault and Henri Lefebvre—two of the most eminent social philosophers of the late 20th century, who recognized space and spatial practices as central to explanations of the dynamics of contemporary society—British geographers forged a new cultural geography that drew on both North American humanistic approaches and British approaches to social geography. This new cultural geography, with its focus on place, consumption, and cultural politics, emerged in parallel with the "cultural turn" associated with postmodernity. Theorizing the slippery concept of postmodernity gave rise to some arcane, involuted, and self-referential writing (in the same way that Marxian social theory had seemed to turn in on itself in the late 1970s and early 1980s) that no doubt earned the new cultural geography a bad name in certain quarters. Meanwhile, the mainstream of the new cultural geography, represented by Dennis Cosgrove, Stephen Daniels, James Duncan, and Peter Jackson, developed around landscapes and their meanings (Cosgrove, 1984; Cosgrove and Jackson, 1987; Cosgrove and Daniels, 1988; Duncan and Duncan, 1988; Jackson, 1989). As a result, the built environment became a subject of much greater potential interest. Before the new cultural geography, urban landscape analysis tended to be, figuratively, one-dimensional, emphasizing the development of physical form—morphogenesis—at the expense of the lived social relations of spatial practices and cultural politics. But, seen in this broader context, urban landscapes become palimpsests of meaning. This encourages us to understand the built environment through peoples' ways of seeing and, in particular, to interpret the symbolic aspects of built landscapes and to analyze the social contestation of built space among various socioeconomic, ethnic, and lifestyle groups.

By the end of the 1980s, urban geography seemed to be thriving. With the world's population becoming increasingly urbanized, and a global economy increasingly articulated through networks of cities, there was a general shift of academic interest toward cities and urbanization. An immense literature had accrued through the 1980s. In addition to the work briefly described above, there were important additions to the literature in the spatial analytical tradition (see, for example, Cadwallader, 1985) and in the behavioral and humanistic approaches that had emerged in the 1970s (e.g., Eyles, 1981; Tuan, 1986, 1989). Membership in both the AAG Urban Geography Specialty Group and the IBG Urban Geography Study Group were at all-time highs, and the contributions of urban geography were increasingly recognized both within human geography and more broadly within the social sciences.

## References

Anderson, K. J., 1988, Cultural hegemony and the race definition process in Chinatown, Vancouver, 1890–1980. *Environment and Planning D: Society and Space*, Vol. 6, 127–149.

Ball, M., 1986, The built environment and the urban question. *Environment and Planning D: Society and Space*, Vol. 4, 447–464.

Brownill, S., 1984, From Critique to intervention: Socialist feminist perspectives on urbanization. *Antipode*, Vol. 16, 21–34.

Cadwallader, M., 1985, *Analytical Urban Geography*. Englewood Cliffs, NJ: Prentice Hall.

Castells, M., 1983, *The City and the Grassroots*. Berkeley, CA: University of California Press.

Cosgrove, D., 1984, *Social Formation and Symbolic Landscape*. London, UK: Croom Helm.

Cosgrove, D. and Daniels, S., 1988, *The Iconography of Landscape*. Cambridge, UK: Cambridge University Press.

Cosgrove, D. and Jackson, P., 1987, New directions in cultural geography. *Area*, Vol. 19, 95–101.

Cullen, J. D. and Knox, P. L., 1981, The triumph of the eunuch: Planners, urban managers, and the suppression of political opposition. *Urban Affairs Quarterly*, Vol. 17, 149–172.

Dear, M. and Wolch, J., 1987, *Landscapes of Despair*. Princeton, NJ: Princeton University Press.

Duncan, J., 1980, The superorganic in American cultural geography. *Annals of the Association of American Geographers*, Vol. 70, 181–198.

Duncan, J. and Duncan, N., 1988, (Re)reading the landscape. *Environment and Planning D: Society and Space*, Vol. 6, 117–126.

Eyles J., 1981, Why geography cannot be Marxist: Towards an understanding of lived experience. *Environment and Planning A*, Vol. 13, 1371–1388.

Giddens, A., 1981, *A Contemporary Critique of Historical Materialism*. Three volumes. Cambridge, UK: Polity.

Giddens, A., 1984, *The Constitution of Society: Outline of the Theory of Structuration*. Cambridge, UK: Polity.

Harvey, D. W., 1982, *The Limits to Capital*. Oxford, UK: Blackwell.

Harvey, D. W., 1989a, *The Urban Experience*. Oxford, UK: Blackwell.

Harvey, D. W., 1989b, *The Condition of Postmodernity*. Oxford, UK: Blackwell.

Hayden, D., 1980, *The Grand Domestic Revolution*. Cambridge, MA: MIT Press.

Jackson, P., editor, 1987, *Race and Racism: Essays in Social Geography*. London, UK: Allen and Unwin.

Jackson, P., 1989, *Maps of Meaning*. London, UK: Unwin Hyman.

Johnston, R. J., 1984, The world is our oyster. *Transactions, Institute of British Geographers,* Vol. 9, 443–459.

Knopp, L., 1987, Social theory, social movements, and public policy: Recent accomplishments of the gay and lesbian movements in Minneapolis, Minnesota. *International Journal of Urban and Regional Research,* Vol. 11, 243–261.

Knox, P. L., 1982, *Urban Social Geography.* London, UK: Longman.

Knox, P. L., 1984, Styles, symbolism and settings: The built environment and the imperatives of urbanised capitalism. *Architecture et Comportement,* Vol. 2, 107–122, Editions Georgi, Switzerland.

Knox, P. L., 1987, The social production of the built environment: Architects, architecture and the post modern city. *Progress in Human Geography,* Vol. 11, 354–377.

Knox, P. L. and Cullen, J. D., 1981a, Planners as urban managers: An exploration of the attitudes and self-image of senior British planners. *Environment and Planning A,* Vol. 13, 885–892.

Knox, P. L. and Cullen, J. D., 1981b, Town planning and the internal survival mechanisms of urbanised capitalism, *Area,* Vol. 13, 183–188.

Knox, P. L. and Cullen, J. D., 1982, The city, the self and urban society. *Transactions, Institute of British Geographers,* Vol. 7, 276–291.

Lauria, L. and Knopp, L., 1985, Toward an analysis of the role of gay communities in the urban renaissance. *Urban Geography,* Vol. 6, 152–169.

Ley, D., 1983, *A Social Geography of the City.* New York, NY: Harper & Row.

Little, J., Peake, L., and Richardson, P., editors., 1988, *Women in Cities: Gender and the Urban Environment.* New York, NY: New York University Press.

McDowell, L., 1983, Towards an understanding of the gender division of urban space. *Environment and Planning D: Society and Space,* Vol. 1, 59–72.

Mackenzie, S., 1988, Building women, building cities: Toward a gender sensitive theory in the environmental disciplines. In C. Andrew and B. Milroy, editors, *Life Spaces.* Vancouver, British Columbia, Canada: University of British Columbia Press.

Mair, A., 1986, The homeless and the post-industrial city. *Political Geography Quarterly,* Vol. 5, 351–368.

Rose, D., 1984, Rethinking gentrification. *Environment and Planning D: Society and Space,* Vol. 2, 47–74.

Sack, R., 1988, The consumer's world: Place as context. *Annals of the Association of American Geographers,* Vol. 78, 642–664.

Smith, N., 1982, Gentrification and uneven development. *Economic Geography,* Vol. 58, 139–155.

Smith, N., 1987, Gentrification and the rent gap. *Annals of the Association of American Geographers,* Vol. 77, 462–465.

Smith, S. J., 1989, *The Politics of "Race" and Residence.* Cambridge, UK: Polity.

Soja, E., 1980, The sociospatial dialectic. *Annals of the Association of American Geographers,* Vol. 70, 207–225.

Soja, E., 1989, *Postmodern Geographies. The Reassertion of Space in Critical Social Theory.* London, UK: Verso.

Thrift, N., 1985, Flies and germs: A geography of knowledge. In D. Gregory and J. Urry, editors, *Social Relations and Spatial Structures.* Basingstoke, UK: Macmillan.

Tuan, Y-F., 1986, Strangers and strangeness. *Geographical Review,* Vol.76, 10-19.

Tuan, Y-F., 1989, Surface phenomena and aesthetic experience. *Annals of the Association of American Geographers,* Vol. 79, 233–241.

Wheeler, J. O., 2002, Urban Geography in the 1970s. *Urban Geography,* Vol. 23, 397–402.

# CHAPTER 16

## Urban Geography in Transition
### *A Canadian Perspective on the 1980s and Beyond*[1]

LARRY S. BOURNE[2]

## ABSTRACT

The history of most academic disciplines reveals distinct periods of research activity characterized by competing intellectual paradigms,

[1]The research on which this chapter is based was supported by a grant from the Social Sciences and Humanities Research Council of Canada (SSHRCC). This continuing support is gratefully acknowledged, as is the competent and efficient research assistance provided by Kathy Mortimer, a Ph. D. student in the Department of Geography. Jim Simmons, Jim Lemon, and Dan Hiebert provided critical comments on a draft of this paper but bear no responsibility for the shortcomings of the outcome.
[2]Correspondence concerning this chapter should be addressed to Larry S. Bourne, Department of Geography and Program in Planning, University of Toronto, Toronto, Ontario M5S 3G3 Canada; telephone: 416-978-1593; fax: 416-978-6729; e-mail: Bourne@geog.utoronto.ca

varying student interest and volatile public awareness. This paper asks whether the 1980s was one of those periods for Canadian urban geography. One perspective views the 1980s as a rather dry valley of declining interest, productivity and relevance. In effect, the decade served as a transition period from the vibrancy of urban research in the 1960s and 1970s to the pluralism and renewed urban interest of the 1990s. To test these propositions five indices were developed to measure levels of activity and output in Canadian urban geography and related fields, from the 1960s to the present: government funding, the number and composition of academic journals, government publications, and membership in academic and professional organizations. The results are mixed, but revealing. There is no clear valley of decline in the 1980s, no single trajectory in the evolution of urban geography, and no apparent resurgence of interest or activity during the 1990s. Instead, research activity in the sub-field appears to have declined, to have become more fragmented and dispersed, and to have been subjected to the effects of increased privatization and professionalism. The concluding section explores the reasons for this experience, and emphasizes the fundamental importance of context and timing—changing national social conditions, the political climate and the growth of the urban system under study—in shaping the urban agenda generally and Canadian urban geography in particular. [Key words: Canada, urban geography, urban growth.]

… urban geography in Canada appears to have been driven less by intellectual whims and perturbations than in most other countries. Indigenous Canadian urban geography, instead, has remained predominantly empirical and solidly pragmatic, and relatively immune to many of the epistemological debates that have permeated British and continental European geography. The urban literature mirrors this pragmatism, as does the sociology of the discipline itself.

(Presidential Address, Canadian Association of Geographers, 1996).

## Introduction: Context, Culture and Timing

All academic disciplines go through evolutionary cycles and more-or-less distinctive phases or periods of transition. Typically, each period is characterized by diverse, if not competing, intellectual paradigms, research priorities, methodological preferences and philosophical styles. Each period conforms roughly with the appearance of a different cohort of personalities and human capital. Each period also displays fluctuations in levels of research funding and output, swings in public awareness and student interest, and shifts in policy applications and institutional legitimacy, as the context, content and culture of academic research changes.

This paper examines the longer-term evolution of urban geography in Canada in terms of changes in its context, character and viability. It asks whether the 1980s was one of those distinctive transition periods for urban geography and for urban studies more broadly. One perspective views the decade, in retrospect, as a rather dry valley (called a "bourne"[3]) of declining funding, reduced public support and diminished student interest, as well as a period of institutional retrenchment and reduced visibility and policy relevance. It can also be viewed, on the other hand, as a period of transition from one set of epistemological and methodological regimes to another and more diverse set. For urban geography, it can be seen as a decade characterized by continued intellectual divergence and fragmentation, by the growing prominence of sectoral over place-based research, and by the increasing commodification and professionalization of applied research and practice.

It can thus be argued that the decade served as an uneasy transition between the energy and vibrancy, indeed the exploratory nature of urban research during the 1950s, 1960s and 1970s, with its relative homogeneity of research styles, methods and purposes, and the apparent resurgence and intellectual plurality of geographical research on urban issues during the 1990s and the current decade.

Drawing primarily on Canadian experience and examples, the paper explores the factors shaping urban and geographical research over the post-World War II period, with an emphasis on the position of the 1980s. It stresses the importance of understanding the external context of academic research: first, setting the 1980s in a longer time frame; second, setting geography, and urban geography specifically, in the framework of other disciplines, both academic and applied; and, third, by relating research to changes in national socioeconomic conditions, rates of urban growth, and changes in political ideologies and institutional structures.

All of these contextual variables, as well as the accidents of timing, have shaped the urban research agenda and the character of the subdiscipline. Just as urban geography cannot be separated from the broader discipline of geography, it cannot also be separated from trends in cognate disciplines, notably in planning and urban studies and in the social sciences

---

[3]In middle English one definition of the word "bourne" is a small valley that is seasonally dry.

generally. Nor can the characteristics of urban geography and urban research in Canada be separated from the geography of urban Canada and the processes, direction and timing of urban growth and change in the country.

Specifically, the paper undertakes to test the assertion that the 1980s was a lost decade and that this was the combined result of factors both internal and external to the discipline. Looking back at the 1980s, however, is really a means to a broader end—understanding the present and anticipating the future. The paper concludes by reviewing the factors shaping urban geography and then extracting some broad lessons for the future of the subdiscipline and for urban research writ large.

## A Potted History: Voices From the Past

Urban geography and urban studies have surprisingly long histories in Canada, beginning with early (late 19th and early 20th centuries) studies of settlements and regions within the fields of economic history and political economy (e.g., Innis, 1950). Urban planning as applied practice also has a long history, which is not surprising since many of the country's initial settlements were intentionally planted, physically planned, and subsequently heavily subsidized by the colonial powers of the day. As a legislated series of instruments of development control, however, and as a set of professional organizations, urban planning has been around since the early 1900s (Adams, 1921; Gertler and Crawley, 1977; Hodge, 2003).[4] In the academic world, planning instruction has been on offer in Canada since the 1920s, and as specific university courses since the early 1930s.

As a formal discipline, geography arrived somewhat later. The first professor of geography in English-speaking Canada was Griffith Taylor, appointed at the University of Toronto in 1935. Although part of the initial wave of British and Commonwealth geographers, ironically, he arrived from Chicago. Among Taylor's various legacies was that he subsequently wrote the first textbook in Canada on urban geography, an appropriately eclectic volume, published in 1949 (Taylor, 1949). For a number of reasons, and unlike several studies of the urban condition, such as the often appalling living conditions of older cities (Ames, 1897; Carver, 1948), or of the social construction of urban space (Seeley et al., 1956), which emanated

---

[4]As an early example of both institutional change and the globalization of planning styles and expertise, Thomas Adams was brought to Canada from Britain in 1917 as town planning advisor to the Conservation Commission. He later became the first director of the Town Planning Institute of Canada, and then left the country in 1923 after the Commission was abolished, and moved to become director of the Regional Plan of New York and its Environs.

from other cognate disciplines during the same period, this volume appears to have left little imprint on the country's physical, social or intellectual landscapes, or on the discipline of geography itself.

The early and explosive growth of the subfield has been carefully documented in a number of government reports, bibliographic publications and university reviews during the 1950s and 1960s, such as those by Stamp (1952), the Canadian Council on Urban and Regional Research (CCURR, 1966), and by J. W. Simmons (1967). Much of the early work in geography was excellent, but often limited to a specific city or region (Spelt, 1955; Kerr and Spelt, 1965). It could be argued, on the basis of these reviews, that some of the best and most visible work in urban geography in Canada was done during the 1950s and 1960s. Nevertheless, the largest and most thorough empirical analysis of urban growth and change in Canada was undoubtedly the three volume set of benchmark reports produced by Michael Ray and his colleagues in the mid-1970s (Ray et al., 1976), and in related publications on particular regions (Yeates, 1974). This was perhaps the first series of studies drawing on the new methodologies and data sources of the day.

Timing was critical here. This was a period when urban questions were still fresh, when new census data and government funding sources became available, when high-speed computers were introduced, when scientific methodologies were in the ascendency, when universities were undergoing massive expansions, and when we actually knew relatively little about the properties and growth of cities. Urban research, in effect, was exploring new and largely uncharted ground.

There has, to my knowledge, been no similarly comprehensive analysis or subsequent overview of urban research in Canada, although the geographical literature has since expanded five-fold (Bourne and Ley, 1993;Yeates, 1999; 2001). A number of specialized volumes and edited texts have attempted to provide overviews of the Canadian urban condition, or provide summaries of particular sectors, issues or research styles, or to identify U.S.–Canadian urban contrasts, but none has covered the entire field of urban geography per se (Bourne, 1971, 1989; Bunge and Bordessa, 1975; Lemon, 1985, 1996; Goldberg and Mercer, 1986; Jones and Simmons, 1993; Kobayashi, 1994; Caulfield and Peake, 1995; Harris, 1996; Ley, 1996; Bunting and Filion, 2000; Bourne and Simmons, 2002, 2003).

Throughout this period urban geography in Canada struggled to remain distinctive while at the same time enthusiastically, and often uncritically, incorporating ideas, images and styles from British, French and especially American urban geography. All of the prominent British and American schools of the period—Berkeley, Chicago, Iowa, Michigan,

Wisconsin, Oxford, London—were well represented within Canadian geography. This infusion of talent not only led to an enriched diversity in geographical research, but also a further fragmentation of research cultures, interests and styles.

There is, moreover, little doubt from the evidence above of a steady decline in place-based research, represented by regional and urban geography, and the ascendency of sector-based and issue-driven approaches. Perhaps urban geography, which by definition requires a synthesis of different bodies of theory (e.g., positivist, Marxist, humanist, feminist and other social theories), literally all sectoral interests (e.g., economic, social, political) and expertise (e.g., quantitative, qualitative) in the study of particular places, is now simply too demanding, too complicated, compared to single-issue research. Interestingly, the recent issue of the flagship journal *The Canadian Geographer* (Kobayashi, 2001) commemorating the 50th anniversary of the formation of the Canadian Association of Geographers, does not have a section on urban, or even regional, research or policy issues.

### Context: Urban Growth, Politics and Institutions

The initial blossoming of research in urban geography, and in urban studies and planning, through the decades of the 1950s, 1960s and 1970s, should come as no surprise. Conditions demanded it; public resources permitted it; social attitudes allowed it; institutions facilitated it. The immediate post-World War II period in Canada, as in the United States, was a time of very high population growth, continued urbanization and widespread urban growth. Table 16.1 charts the trajectory of city sizes and urban growth rates in Canada for the four decades from 1961 to 2001 by city-size category for all urban places with more than 10,000 population.

In the early part of the period, notably the 1950s and 1960s, growth rates were very high. Both fertility rates and immigration levels were also high. Productive capacity and employment opportunities expanded rapidly, not only in manufacturing and the resource industries but in both services and the public sector. Real incomes, and thus consumption levels, especially in terms of the demand for housing and living space, rose dramatically. Growth was spatially ubiquitous; almost all parts of the country grew, but especially the metropolitan areas. Since almost everyone and every region shared in the benefits (and costs) of the urban explosion it was easier to identify urban issues as national political issues and to access the resources to address those issues.

The country's urban system expanded and became more strongly differentiated by size, function and location. New spaces of development

**TABLE 16.1** The Historical Trajectory of Urban Growth Rates in Canada, by City-Size Category, 1961–2001

| CMAs/CAs: Number of cities by size category[a] | 1961 | 1971 | 1976 | 1981 | 1986 | 1991 | 1996 | 2001 |
|---|---|---|---|---|---|---|---|---|
| >1,000,000 | 2 | 3 | 3 | 3 | 3 | 3 | 3 | 4 |
| 300,000–1,000,000 | 6 | 8 | 8 | 8 | 9 | 10 | 11 | 11 |
| 100,000–300,000 | 12 | 16 | 16 | 16 | 16 | 15 | 19 | 19 |
| 30,000–100,000 | 27 | 41 | 48 | 54 | 52 | 54 | 49 | 49 |
| 10,000–30,000 | 75 | 67 | 66 | 64 | 69 | 65 | 65 | 56 |
| Total, all urban places | 122 | 135 | 141 | 145 | 149 | 147 | 147 | 139 |

| CMAs/CAs: Growth rates by percentage | 1961–1966 | 1966–1971 | 1971–1976 | 1976–1981 | 1981–1986 | 1986–1991 | 1991–1996 | 1996–2001 | 1971–2001[b] |
|---|---|---|---|---|---|---|---|---|---|
| >1,000,000 | 15.5 | 10.8 | 5.6 | 3.8 | 6.0 | 11.5 | 8.2 | 7.2 | 50.1 |
| 300,000–1,000,000 | 12.6 | 11.0 | 8.4 | 6.9 | 5.5 | 9.3 | 4.8 | 6.2 | 48.5 |
| 100,000–300,000 | 11.8 | 8.8 | 5.6 | 2.4 | 4.6 | 6.4 | 3.9 | 3.2 | 29.5 |
| 30,000–100,000 | 9.2 | 6.1 | 7.3 | 7.0 | 2.9 | 8.4 | 6.5 | 1.1 | 33.4 |
| 10,000–30,000 | 10.4 | 5.6 | 3.9 | 7.0 | -0.1 | 3.6 | 2.7 | -1.2 | 30.6 |
| Total, all urban places | 13.1 | 9.6 | 6.3 | 5.0 | 4.9 | 9.3 | 6.2 | 5.2 | 43.4 |
| Nonurban places | | | | | | | | | 24.7 |
| All places | | | | | | | | | 39.1 |

[a] Number of cities at beginning of each period. The decline after 1986 is because of annexations.
[b] We use the 1971-2001 period because of inconsistencies in pre-1971 urban boundaries.
*Note:* CMA = Census metropolitan area (over 100,000). CA = Census agglomeration (10,000–100,000). Nonurban = Non-CMA/CA. *Source:* Statistics Canada. Calculations by the author.

were opened, especially in the north, and new settlements were established over a wide geographical area. During the 1950s and 1960s the national settlement space, the ecumene, expanded dramatically, as did expectations of continued future growth and prosperity. At the local scale, suburbs expanded around all cities. Those cities, in turn, struggled to cope with growth and to provide basic services. Infrastructure was inadequate; social services were overwhelmed, or often nonexistent.

In a somewhat belated, and at times haphazard, response to these pressures, new infrastructure funding was provided, highways were expanded, new subways were built and public goods were substantially improved. In a period of moderately progressive liberalism, new social institutions were created at national, provincial and local levels. Urban research institutes were established in universities across the country to document the dynamics of urban growth and the changing conditions of urban life.[5] Regional and metropolitan governments and coordinating agencies were established by the provincial governments to facilitate the funding of infrastructure and service provision (Frisken, 1994). The decades of the 1960s and 1970s were indeed the heyday of regional planning and metropolitan governance in Canada (Wolfe, 2000; Hodge and Robinson, 2001).

At the federal level there was also a proliferation of new institutions, many with spatial (or place-based) mandates and explicit policies of income redistribution. New ministries of state were initiated for regional economic expansion, and northern and rural development, and new mechanisms for interregional fiscal transfers were established. In the face of accumulated housing deficits from the pre-World War II period, documented by Carver (1948) and others, the mandate of CMHC, the federal housing authority, was broadened (Miron, 1993). That mandate also included, through the provision of social housing, a further commitment to income redistribution, decisions which—like the subsequent introduction of national health care insurance—had long-term (and largely positive) impacts on the character and viability of Canadian cities, especially the inner cities.

In 1970 a new Ministry of State for Urban Affairs (MSUA) was created as a coordinating body for federal government policies and activities in the urban realm, supported by Lithwick's (1970) now-classic report on the condition of the cities. There was a sense of an urban crisis unfolding, despite the fact that by the 1970s urban growth rates were dropping

---

[5]Among the more prominent of these institutions were the Centre for Urban and Community Studies at the University of Toronto, INRS-Urbanisation at l'Universite du Quebec in Montreal, the Institute for Urban Studies at the University of Winnipeg, and the Centre for Human Settlements at the University of British Columbia. Geographers have played prominent roles in all of these institutes.

(Table 16.1). Even the Science Council stepped in to offer advice on how to apply science to the management of urban development (Solandt, 1971). In parallel, at the provincial level a variety of other institutions, agencies and regulatory frameworks were introduced to deal with urban affairs and municipal planning, and separate housing institutions were created to stimulate both private and public sector housing production.

By the end of 1970s, however, and especially because of the recession of the early 1980s, this flood of institutional expansion, research activity and policy innovation quickly declined. In some instances initiatives effectively ceased, and policies were reversed. Public interest in urban issues appeared to shift once again, and by most measures actually declined. As the political pendulum swung back to the center, and then to the right, senior governments almost everywhere in the country retreated from their earlier initiatives in regional development and planning. Unfortunately, this was precisely the time they were needed. In parallel, the pendulum of political ideology and policy shifted from redistribution to wealth creation. The rate of reform of municipal governance structures also declined, and public infrastructure investments waned. As a small but symbolically meaningful political gesture, MSUA was closed by the federal government in 1979, and replaced with, essentially, nothing.[6]

Not until the mid-1990s, after a decade or more of benign neglect, was there an apparent revival of interest in urban issues, in reforming urban institutions, governance and planning, and in the study of cities and city regions (Frisken, 1994). By the late 1990s, the urban question appeared to be back on the national political agenda. Cities were now seen to matter in the public sphere, even in terms of macroeconomic policy. They were cited as the engines of national economic growth, the agents of international competitiveness and cultural inclusiveness, and the centers of innovation and creativity (Gertler, 2001; Bradford, 2002; TD Economics, 2002). Although it is obvious that cities have served all of these purposes for some time, the recent marshalling of arguments and evidence in their defence provides another example of the importance of context and timing, not to mention symbolism, in shaping the content and trajectory of research.

The same was true in planning. In Canada as in other countries, the appearance of now standard but equally vague keywords - smart growth, new urbanism, sustainable development—is at least suggestive of the ascendency of the urban question. By 2001, some 30 years after the initial

---

[6]Among the complex factors explaining the demise of MSUA was the very strong resistence from the provincial governments to the intrusion of the federal government into urban affairs. Under the Canadian constitution, cities (meaning municipalities) are the legal creatures of the provinces and have no standing at the national level.

attempt, another federal initiative began with the goal of creating a national urban strategy (Sgro, 2002). It remains to be seen whether this initiative succeeds; whether context, politics and timing will permit a consensus on policy to emerge, and then to permit implementation, and if so, in what form.

## The Rise, Decline and Rise of Urban Interests

Why was there an increase in interest in urban matters during the 1960s and 1970s, and an apparent decline in the 1980s? Has there been a resurgence of urban research fortunes during the 1990s? And, what import did this ebb-and-flow have for urban research in general and urban geography in particular? Among the factors that might support the assertion of a rise in the level of disinterest in the 1980s, aside from the changing balance of national and provincial politics, and political expediency, the most obvious is that the rate of national population growth dropped sharply in the late 1970s and early 1980s, because of both lower fertility rates and much lower immigration levels. Overall, the rate of economic growth also declined, and unemployment levels increased as a result, particularly during the sharp recession of the early 1980s. In parallel, average urban growth rates declined, as Table 16.1 demonstrates, and the process of metropolitan concentration ceased, at least temporarily. Ironically, as an example of the inevitable lag between perception of a trend, the collective policy response and subsequent research activity and output, the 1990s witnessed a major resurgence in urban growth and specifically in the growth of the larger metropolitan regions.

The public interest, allowing for the usual time lag, shifted from managing growth and redistribution to promoting growth and accommodating decline. Policy interest and research priorities, to the extent these can be deduced from the public record, shifted from a focus on places to a focus on sectors, from an emphasis on outcomes to one on processes, and from basic to applied research. There was a widespread and misguided feeling that many of the key urban issues inherited from the 1960s and 1970s had been solved, or at least seriously addressed, while other policy initiatives had proven to be unsuccessful, or politically unsustainable. Scarce public resources could be spent elsewhere, in other sectors or other places, and for other purposes (e.g., health care). New urban institutions, regional plans and metropolitan governments were already in place. Minimal public transit systems were also in operation, and high levels of housing production and service provision had been realized, at least in the major cities. As a result, urban issues, which always had only a marginal hold on the national stage, slipped off the political agenda, and out of the minds of many in the public sector, as well as in education circles and in the community at large.

By the mid-1990s, however, it had become more or less obvious that urban problems had not been solved, or in some cases not even confronted. Other relatively new, or simply reformatted, issues came to the forefront. The context for both research and policy had once again changed. Among those trends, and their underlying dynamics, were a now familiar list: the rate of change in the nation's demography accelerated; fertility levels declined to historic lows, with the obvious result of an aging population (Moore et al., 1997). This transition was combined with a tripling of the rate of immigration, beginning in the late 1980s, and now those immigrants were overwhelmingly drawn from nontraditional sources and telescoped on a few metropolitan gateways (Bourne and Rose, 2001; Ley and Hiebert, 2001; Li, 2003). Urban and regional growth became more uneven; the larger metropolitan areas saw rapid growth, while other smaller centers witnessed sustained decline. This combination of trends meant that immigration policy became the country's implicit population policy, as well as its urban policy. The locational concentration of urban growth, however, made it more difficult to generate a national consensus on public policy priorities or research needs.

Economic growth also returned, especially after the 1991–1993 recession, but in a very different global and continental context. That context was characterized by the unequal geographical imprint of widespread economic restructuring, stimulated by trade liberalization, accelerating global competition and tighter economic integration with the U.S. (Britton, 1996; Coffey and Shearmur, 1996; Gertler, 2001; Bourne and Simmons, 2003). After a decade of indifference other urban problems resurfaced: increased traffic congestion, higher levels of atmospheric pollution, municipal fiscal stress, the challenge of adapting to ethnocultural diversity, housing affordability and homelessness, suburban poverty, and economic competitiveness and innovation, to name only a few.

It is not surprising that national conditions—politics, the state of the economy, demographic change, and urban growth rates—have influenced the perceptions of urban problems and the governance of research. Nor is it unexpected that these images, after a time lag, would be reflected in the priorities of the various levels of government and funding agencies, and that these in turn, with another time lag, would shape the activities of the urban research community. For example, funding priorities among granting agencies shifted from basic research to emphasize current problems and more pragmatic policy solutions to those problems.

The methods employed by the institutional system—notably governments and the funding agencies—to organize and direct the research response also changed, emphasizing structural change in the governance of research as much as substantive shifts in the topics of concern. They did so not only by setting the agenda but by encouraging interdisciplinary and collaborative research, team work, the establishment of national centers of excellence and regional networks of researchers, and public-private partnerships. Examples of those networks in which geographers have played a prominent part include research on gender and work, social inclusion and diversity, housing, immigration and settlement, new technologies, and globalization and the new economy. This list will sound familiar to colleagues in most other countries. These are all valid initiatives, but they also convey a sense that a new model of research governance is unfolding in response to contextual change, a model whose effects have yet to be determined. It is also clear from this list that sectoral rather than place-based (e.g., urban and regional) issues have continued to dominate the research agenda.

## Empirical Evidence: Urban Research Before, During and After the 1980s

Is this brief history of a changing context of research reflected in the nature and vitality of urban geography and urban studies generally in Canada? What are the relationships, if any, between the contextual framework outlined above and the level and composition of activity in academic and professional geography? How has the subdiscipline changed, and why? It is not proposed here to undertake an intellectual history of the subdiscipline, an exercise that is beyond the mandate of this paper, but it would be an interesting and valuable undertaking. It would also be a useful foundation for an assessment of the present condition and future prospects of urban geography.

Instead, the following empirical analysis focuses on constructing a set of simple indices intended to describe how the level of activity and content of geography (and urban planning) have changed through the transition from the 1960s and 1970s through the 1980s to the 1990s and beyond. Five sets of indices are examined here: (1) levels of research funding; (2) the number and viability of urban journals and publications; (3) the volume, composition, and content of academic publications in geography; (4) government publications and documents in urban affairs; and (5) membership in academic geography and associated professional organizations.

These indices, although admittedly crude, can be seen as constituting the contextual environment, the inputs, intermediate outputs, final demand,

and supply-side elements of the time-series equation for research in the subdiscipline and in its cognate fields of enquiry.

The results of the empirical analysis, not unexpectedly, are mixed. There is limited evidence of a substantial dry period during the 1980s in research funding, academic publications, government documents, student interest and professional membership. There is also relatively little evidence, at least based on these indices, of a resurgence of interest in urban geography and urban affairs during the 1990s. On the contrary, and allowing for year-to-year fluctuations, the overall trend has been to a further relative decline. The following section examines each of the five indices in turn. In each case both annual data and a polynomial trend line are provided.

### Research Funding

Funding for research in geography and urban studies generally comes from a wide variety of sources, in both the public and private sectors. The principal source for academic research, however, is the Social Science and Humanities Research Council of Canada (SSHRCC). Founded in 1977, SSHRCC took over programs previously funded by the social sciences division of the Canada Council, and it continues to support both independent and collaborative group projects, as well as mission-based strategic and community partnership programs.

Examination of the record of SSHRCC funding for 29 largely discipline-based categories annually since 1979 reveals an interesting and contradictory picture for geography and for urban and regional studies (Figure 16.1). Overall funding levels, not surprisingly, have increased

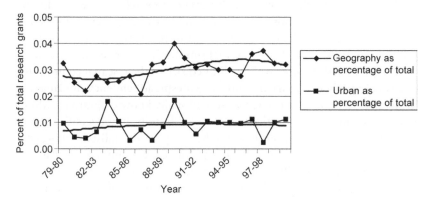

**Figure 16.1** Social Sciences and Humanities Research Council of Canada (SSHRCC) research grants to selected disciplines, 1979–2000.

modestly over time as the size of the academic and research communities, and government budgets, have grown, but with wide swings in real dollars. Geography has gradually increased its proportion of total grant monies (from 2% to 3%). The economic difficulties facing the country in the early 1980s, noted above, however, are also evident in the time series. That decade brought three significant changes in the underlying culture of research funding: an initial freeze on total funding, an emphasis on cost-sharing through public-private research partnerships, and a shift to a strictly competitive allocation of funds among applicants as a form of quality control and implicit rationing (SSHRCC, 1988). The ideology of the market seems to have become the dominant paradigm.

Although geography has remained a small player in the overall funding pool, and its level of funding as a whole actually grew, the discipline's ranking has not increased. In fact it dropped from 10th to 12th out of 29 academic divisions between 1979 and 2000 (Figure 16.2 and Table 16.2). Surprisingly perhaps, given that funding for interdisciplinary research rose from 27th to 13th place from 1989–1990 to 2000, funding for urban and regional studies, in contrast, showed almost no change. While the latter could be a classification problem, it is likely a response, at least in part, to the increasing dominance of a discipline-based peer-review process operating throughout the social sciences in a climate of competitive funding. It may also mirror the relative weakness of interest in urban issues and research.

## Academic Journals and Urban Publications

A second set of indicators encompasses one of the standard measures used in the academic world to measure the level of research interest, productivity, and intellectual fashions: the number and focus of papers published in academic journals. The usual cautions and reservations on such crude summaries apply here. These reservations, however, are even more obvious in the Canadian case given the limited size of the research community, the small number of journals, and the widespread tendency of Canadian academics to publish in foreign (internationally known?) journals. Moreover, the country's most active urban research units, in universities and in government and the nonprofit sector, have captured a substantial proportion of the research funding and output.[7] There is no summary available of the output or the contributions of all of these research institutes, but their impact has been considerable. With these reservations in mind, the initial

---

[7]For example, the University of Toronto's Centre for Urban and Community Studies has published more than 200 urban research papers and 35 major reports since the two series were established in 1969, as well as a number of books and conference proceedings.

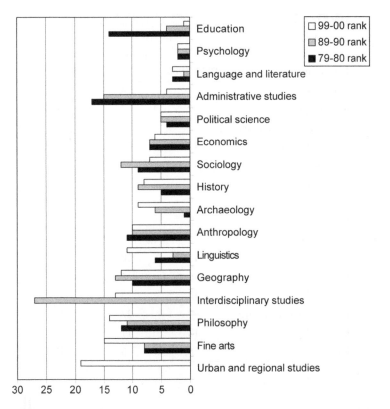

**Figure 16.2** Ranking of Social Sciences and Humanities Research Council of Canada (SSHRCC) research grants by discipline, 1979–2000. *Note:* Urban and Regional Studies became a distinct category in the 1990s.

global measure examines the number of journals published in Canada in geography and related disciplines, including urban studies, planning and regional science, and in other fields closely associated with urban geography, such as urban history and environmental studies. Here the time series data available cover a much longer period.

Figure 16.3 charts the number of urban journals and magazines available in Canada from 1920 to the present, while Figure 16.4 plots the life span of the individual publications. What is immediately evident in these profiles, aside from the small numbers and short half-lives of the journals involved, is the explosion of outlets for urban research and policy from the late 1960s until the late 1970s. This was followed by a period of relative stability during the 1980s, lasting through to the mid-1990s, and then decline through to the present. The 1990s was a period when government cutbacks in funding, institutional competition, the weak economics of

**TABLE 16.2** SSHRCC Funding for Discipline-Based Categories since 1979, Listed by Rank of Funding Received in 1999–2000

| Discipline | 1979–1980 | 1989–1990 | 1999–2000 |
|---|---|---|---|
| Education | 14 | 4 | 1 |
| Psychology | 2 | 2 | 2 |
| Language and literature | 3 | 1 | 3 |
| Administrative studies | 17 | 15 | 4 |
| Political science | 4 | 5 | 5 |
| Economics | 7 | 7 | 6 |
| Sociology | 9 | 12 | 7 |
| History | 5 | 9 | 8 |
| Archaeology | 1 | 6 | 9 |
| Anthropology | 11 | 10 | 10 |
| Linguistics | 6 | 3 | 11 |
| Geography | 10 | 13 | 12 |
| Interdisciplinary studies | — | 27 | 13 |
| Philosophy | 12 | 11 | 14 |
| Fine arts | 8 | 8 | 15 |

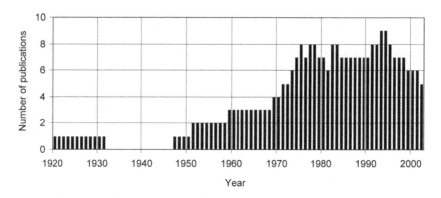

**Figure 16.3** Total number of urban publications in Canada, annually, 1920–2002.

academic publication, and, most recently, substitution through the Internet, began to constrict the proliferation of formal publications. Set in a longer time frame, the 1980s, although not a deep valley, seems to have been the initial period of retrenchment in the publication realm, as initially hypothesized, but it is less dramatic than the retreat of the late 1990s.

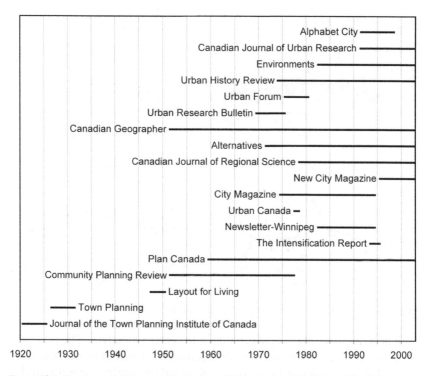

**Figure 16.4** Number and life spans of Canadian publications on urban topics, 1920–2002.

What of the number and composition of articles within these journals and magazines? What proportions are urban articles, broadly defined? Figure 16.5 provides a crude content analysis of the subject areas represented in articles published in four of the country's principal journals from the 1970s to the present: *The Canadian Geographer (TCG)*, *Cahiers de Geographie du Quebec (CdG)*, *Canadian Journal of Regional Science (CJRS)*, and the popular magazine, *Canadian Geographic (CG)*. The latter, although not considered a scholarly journal, is by far the largest seller and national distributor of geographical images and information. As such, it offers a good index of interest among the literate general public in geography and urban affairs. Other relevant journals, such as *Plan Canada* (the principal publication of the Canadian Institute of Planners) and its various predecessors, and the *Canadian Journal of Urban Research*, have more sporadic histories. They are also assumed to be largely or entirely urban in focus, and thus charting changes in their contents would not be very useful as measures of a shift in interest to or from urban issues and a geographical perspective on those issues.

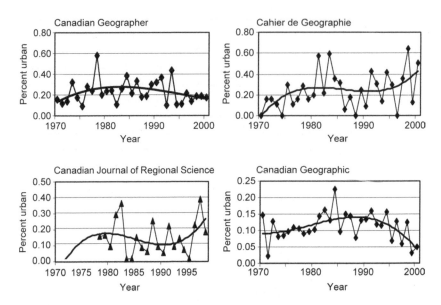

**Figure 16.5** Proportion of urban articles in four journals, 1980–2000.

Figure 16.5 summarizes the results of the content analysis. The resulting graphs for the four journals are distinctively different. Overall, the data suggest a declining output of research on urban affairs generally, and a shift in topical emphasis within the subfield. The proportion of articles in *The Canadian Geographer* that focus on urban topics shows a peak in the late 1970s, stability in the 1980s, and then a decline through to the year 2000. The trend line for the *Journal of Regional Science* shows almost exactly the opposite trajectory. Although there is a danger of reading too much into these simple trends, it is possible that there has been a switch-over in the publication of urban articles between the two journals. The urban component of the *Cahiers,* in contrast, shows a decline in the 1980s and a major resurgence in interest among the French-language community during the 1990s. The popular magazine, *Canadian Geographic,* shows the reverse trend to that of the *Cahiers,* but exhibits a trajectory of recent decline that is not unlike that of *The Canadian Geographer.*

What this diversity of experience suggests, aside from the obvious limits of such indices, is that there is no single trajectory in urban research output that applies to all outlets. Each publication has its own story to tell; each is responding to a different audience. Nevertheless, a reasonable conclusion is that interest in urban questions within the academic geographical community has declined, at least into the late 1990s and within the

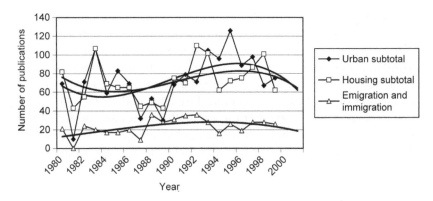

**Figure 16.6** Trends in government publications in Canada on urban and related issues, 1980–2000.

English-speaking community, but has declined less rapidly it seems than in some cognate fields.

## Government Documents and Publications

Measuring the degree of interest in and level of support of urban research by government, beyond the supply of direct funding from federal agencies other than the granting councils, is not as straightforward a task as it might seem. As one simple index we examine the volume and composition of government publications in those topical fields closely associated with geography (geography itself is not a separate category)[8] through a content analysis of the online Microlog index service. Microlog is a monthly indexing, abstracting and document delivery service that covers publications from all three levels of government in Canada, as well as from public agencies (other than universities) that receive government research grants. The number of publications serves as an index, albeit a crude and indirect measure, of the collective interests and priorities of the state, and directly as an indicator of both public sector research funding and output.

Figure 16.6 shows the temporal variations in the number of government publications in three prominent categories—urban, housing and immigration[9]—from 1980 when the Microlog records begin, through the

---

[8]Geography is not a formally recognized category in the employment/occupational classifications used by the federal government or most of its agencies in Canada.
[9]With the exception of the immigration category, the Microlog search undertaken here focused on English-language publications only. A subsequent analysis of French-language publications is clearly warranted.

year 2000. The general trend line, allowing for wide fluctuations in annual output, is upward, with a close parallel in the profile for publications in the urban and housing fields, and surprisingly much less support for government publications on immigration and settlement issues than one might have expected.[10] Despite the overall upward trend, however, there is a noticeable dip in the publication rate during the mid-1980s, followed by a return to pre-1982 levels after the 1991–1993 recession. The dry-valley hypothesis seems valid, but only in this limited instance. More significant, in looking to the future, is the apparent decline in output during the late 1990s.

*Membership and Human Capital*

It might be expected that changes in human capital would follow the ebb-and-flow of research funding described above, as well as swings in the level of social and political interest in geography and urban studies generally. It might also be expected that the output measures—the number of government publications and journal articles—would be mirrored in the size of membership rolls in relevant associations, both professional and academic. On the other hand, a "dispersion effect" might also be anticipated here. The proliferation of funding sources and outlets for academic research, and the increasing diversity (or fragmentation) of research in urban geography, would likely result in a spreading of that research over more targeted journals, including journals in cognate fields outside geography (e.g., in urban studies and planning) and outside the country. There might also be a redistribution of membership commitments from geography-based associations to other associations, notably professional associations.

Figure 16.7 examines these propositions. It offers a summary of membership change over time in the principal association for academic geographers in Canada, the Canadian Association of Geographers (CAG), combined with the much smaller and younger Canadian Regional Science Association (CRSA). For comparison, membership trends are included for the Canadian Institute of Planners (CIP),[11] one of the more prominent professional associations that consume both considerable geographical talent and a significant proportion of the state's resources for applied urban

---

[10]This modest level of internal support may in part reflect the substantial commitment of several federal government departments, agencies, and granting councils to the funding of research on immigration through a network of regionally-based centers of excellence located in key universities across the country. These are all associated with the government's contribution to the international Metropolis Project on immigration and settlement issues.
[11]CIP is a federation of provincial planning associations, in which membership in a provincial division also provides membership in the national organization.

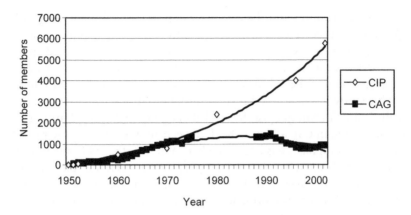

**Figure 16.7** Voting with their feet: membership trends in the Canadian Institute of Planners (CIP) and the Canadian Association of Geographers (CAG), 1960–2002.

research. The results, not unexpectedly, indicate a steady decline in total CAG membership from a peak in the mid-1980s.[12]

The reasons for this decline in membership are many, but likely reflect four principal factors: (1) the declining pool of academic geographers during the late 1980s and early 1990s; (2) a shift in program emphasis from geographical/spatial analysis to applied policy analysis; (3) the institutional reorganization and downsizing of many geography departments in universities; and (4) a loss of membership due to a simple lack of interest among academics in belonging to any association, or to increased membership in other organizations with a perceived higher external visibility and utility. Membership in CIP, in contrast, has risen fourfold over the same period, confirming the importance of the strong movement to formal certification of research expertise (and thus to closed shops), and to the further professionalization of that expertise in terms of both public awareness and the application of human capital in the applied urban field.[13]

## Conclusions and Lessons Learned

What have we learned from this analysis? Urban geography in Canada, like the urban geography of Canada itself, is contingent; it cannot and should not be detached from national conditions and from broader currents

---

[12]The CAG, unlike the AAG, does not have a specific study group focused on urban geography. A survey of the Canadian-based membership of the AAG Urban Geography study group is not available, nor would it provide a representative sample.
[13]Similar movements toward formal certification have also affected other branches of geography (e.g., geoscience).

affecting the country and the discipline at home and abroad. Geography, however it is theorized, is ultimately about real spaces and places, about people, institutions and environments, about landscapes and spatial processes. As such, it is invariably rooted in those environments, those landscapes, and in the cultures that define them and give them meaning.

This study has examined several indices that describe aspects of the evolution of urban geography, and its cognate fields of planning and urban studies, in Canada over the decades from the 1960s to the present. The specific emphasis has been on where and how the 1980s fits into this evolutionary history, but the broader purpose has been to explore the trajectory of change over four decades or more. The particular hypothesis asks whether the 1980s represented a distinct transition—a dry-valley (or bourne)—between two peaks: from the energetic, certainly exploratory, and perhaps relatively homogeneous research agenda of the 1960s and 1970s, to the more diverse, pluralistic and dispersed agenda of the 1990s. The parallel proposition stresses the importance of context and timing; that the level, composition and vitality of research activity, publication and civic participation, in urban geography and related fields, mirror national trends, urban conditions and political attitudes.

The empirical evidence on both propositions is revealing, but decidedly mixed. There is no valley in the 1980s, nor a resurgence in the 1990s, at least based on the indices used here. There is a modest association between urban growth rates and patterns and levels of urban interest over time. Rapid metropolitan growth tends to increase awareness of urban issues and researchers respond accordingly; slower metropolitan growth, or in the Canadian case a shift in growth to newer cities and regions, or decline in the periphery, reduces interest in urban affairs. Not unexpectedly, urban geographical research has remained most vibrant in universities in the larger metropolitan areas, and has declined in relative terms elsewhere.

In parallel, there is little doubt that the changing context of academic research is crucial. This context has changed frequently in response to shifts in the funding situation and political ideologies, as well as urban growth rates, and to changing perceptions of urban growth and the importance of urban problems. The institutional context, at least within the government sector and the universities, has also undergone a significant transformation. There is parallel evidence over the period of several shifts in the culture of public policy-formulation and regulation and in our understanding of the practical difficulties of managing urban areas. More broadly, there is evidence of a major shift in the design, goals and governance of urban research and information provision.

While academic research in the urban realm in geography has remained modest in scale, it has clearly become increasingly diverse and pluralistic in its approaches, composition and style of argument. The ongoing institutional reorganization of geography seems to be associated, as both cause and effect, with this proliferation of interests (or fragmentation) within the discipline. Is there an increasing danger of a "silo effect" here in the application of research expertise to urban matters? On the supply side, funding for research has become more diversified, and much more market oriented. Membership in academic associations has declined, relative to the size of the research community, and the privatization and professionalization of urban research and practice, not to mention of intellectual capital, have increased.

The specific indices employed to examine these propositions reveal a diverse series of evolutionary paths in the subdiscipline from the 1960s and 1970s through the 1980s to the present. In some instances the 1980s appear as a period of retreat and retrenchment, for other indices there is no such evidence. And, there is even less evidence of a resurgence in academic interest in urban geography or urban issues generally during the 1990s. Instead, the primary evidence points to a gradual evolution over time, a product life cycle or trajectory reflecting the changing size, diversity and situation of the subdiscipline and the path of the urban process that it seeks to understand. For the 1990s the evidence suggests a relative decline in terms of the position, the supply of human capital and research output, of urban geography and urban studies generally at the national level, despite widespread assertions of a recent resurgence.

What is not clear here is why this decline has happened, although the paper has hinted at several explanations. Is it that the kind of urban geography inherited from the 1960s and 1970s is considered by more recent cohorts of researchers (a demographic effect) to be rather old fashioned, out-of-date; or perhaps, it is too strongly associated with positivism and quantitative methodologies? The latter view in itself is strange given the increasing demand for analytical geography, spatial modelling and GIS within the private sector. Perhaps the present analysis ends at a low point in the cycle of public interest and that a subsequent analysis (post-2002) would show renewed interest and activity. Of course, it is also the case that the indices employed here do not measure either the quality of the research undertaken or its effectiveness in the larger public realm. Or is it simply that the challenges facing urban researchers—of integrating a wide range of sectoral concerns within a diversity of urban settings and environments—and to engage in both theorization and practice, are simply too difficult, and too complex? Finally, although it may be true that in a highly urbanized society all (or almost all) issues of public and academic concern are now urban issues, this view misses the important point that place matters, and now more than ever.

At the same time it is possible to argue that urban geography in Canada has not declined but has instead spread its wings. It has certainly diffused internationally in terms of the publication outlets used, and has expanded outside of academia into government, the private sector and nonprofit organizations. It is difficult, indeed impossible, to confirm or refute this hypothesis with the present limited data set. Nevertheless, there is impressionistic evidence that the contribution of geography to urban research and policy analysis generally, allowing for a smaller pool of funding overall, has actually increased in comparison to cognate disciplines such as economics and sociology. In this sense the expansion of geographical research outside of discipline boundaries could be considered a sign of success despite the decline of output within the discipline.

Three other parallel trends are worthy of emphasis and careful monitoring. One is privatization, the apparent dispersion of research activity from the academic world into the private (albeit often nonprofit) sector of research institutes and networks, and the increasing incorporation of a market paradigm for allocating research funding and establishing priorities. The second is increased professionalization, the tendency for applied geographical research to be carried out through participation in professional organizations, such as planning. Third, and of particular importance given the emphasis here on long-term trends, is the obvious centrality of timing and the time lag between urban growth trends, the availability of data to describe those trends, the funding of research and the publication of the results of that research. Indeed, the time lags from incentive (demand) to supply (funding) to final research output (publication) seems to be between five and ten years.

Perhaps the single important lesson learned here is the lesson that history itself teaches. There is, at least in the Canadian context, no single trajectory or model of change, no single period of transition, no dominant paradigm. Most academic fields, and certainly urban geography and urban studies, tend to witness evolutionary rather than revolutionary changes. In part this can be attributed to the movement of demographic cohorts of researchers—cohorts of different size and trained in different environments—through their career paths. It may also reflect the fact that the actors last longer than many of the short-term fashions that characterize research, but not as long as the longer-term evolutionary shifts in research styles. Some of these shifts may even lead to a rediscovery, indeed a reinvigoration, of long-established directions in urban research within the academic community. Or maybe not.

The apparent decline in the output and human capital associated with urban geography, at least in the academic world and in government, if accurate, poses serious questions for the future of the discipline as much as the subdiscipline itself. Addressing that decline will require a thoughtful reassessment of the effects on the viability and relevance of the discipline of continued fragmentation, competition from other disciplines, the dominance of sector-based research, and the widespread trend to further privatization and professionalization of both research and practice.

## References

Adams, T., 1921, Modern city planning: Its meanings and methods. *National Municipal Review,* Vol. 11, No. 6, 157–176.

Ames, H., 1897, *The City Below the Hill.* Montreal, Canada: Bishop.

Bourne, L. S., 1989, *Internal Structure of the City.* New York, NY, and Toronto, Canada: Oxford University Press.

Bourne, L. S., 1996, Normative Urban geographies: Recent trends, competing visions and new cultures of regulation. *The Canadian Geographer,* Vol. 40, 2–16.

Bourne, L. S. and Ley, D., editors, 1993, *The Changing Social Geography of Canadian Cities.* Montreal, Canada: McGill-Queen's University Press.

Bourne, L. S. and Rose, D., 2001, The changing face of Canada: The uneven geographies of population and social change. *The Canadian Geographer,* Vol. 45, No. 1, 105–119.

Bourne, L. S. and Simmons, J., 2002, The Dynamics of the Canadian Urban System. In M. Geyer, editor, International *Handbook of Urban Systems: Studies of Urbanization and Migration in Developed and Developing Countries.* Northampton, MA: E. Elgar, 391–418.

Bourne, L. S. and Simmons, J., 2003, New fault lines?: Recent trends in the Canadian urban system and their implications for planning and public policy. *Canadian Journal of Urban Research/Plan Canada,* Special issue, Vol. 12, 1–27.

Bradford, N., 2002, *Why Cities Matter: Policy Research Perspectives for Canada.* Ottawa, Canada: Canadian Policy Research Network.

Britton, J. N. H., editor, 1996, *Canada and The Global Economy.* Montreal, Canada: McGill-Queen's University Press.

Bunge, W. and Bordessa, R., 1975, *The Canadian Alternative: Survival, Expedition and Urban Change.* Toronto, Canada: York University. Geographical monograph #2.

Bunting, T. and Filion, P., editors, 1991, *Canadian Cities in Transition* (2nd ed.). Toronto, Canada: Oxford University Press.

Bunting, T. and Filion, P., editors, 2000, *Canadian Cities in Transition.* Toronto, Canada: Oxford University Press.

Canadian Council on Urban and Regional Research, 1966, *Urban and Regional References, 1945–64.* Ottawa, Canada: CCURR.

Carver, H., 1948, *Houses for Canadians.* Toronto, Canada: University of Toronto Press.

Caulfield, J. and Peake, L., editors, 1996, *City Lives and City Forms: Critical Research and Canadian Urbanism.* Toronto, Canada: University of Toronto Press.

Coffey, W. and Shearmur, R., 1996, *Employment Growth and Change in the Canadian Urban System.* Ottawa, Canada: Canadian Policy Research Network.

Frisken, F., editor, 1994, *The Changing Canadian Metropolis.* Berkeley, CA: Institute of Government Studies.

Germain, A. and Rose, D., 2000, *Montreal: The Quest for a Metropolis.* Chichester, UK: Wiley.

Gertler, L. and Crawley, D., 1977, *Changing Canada's Cities: The Next 25 Years.* Toronto, Canada: McClelland and Stewart.

Gertler, M., 2001, Urban economy and society: Flows of people, capital and ideas. *ISUMA—Canadian Journal of Policy Research,* Autumn, 119–130.

Goldberg, M. and Mercer, J., 1986, *The Myth of the North American City.* Vancouver, Canada: UBC Press.

Harris, R., 1996, *Unplanned Suburbs: Toronto's American Tragedy.* Baltimore, MD: The Johns Hopkins University Press.

Hodge, G., 2003, *Planning Canadian Communities.* (4th ed.). Toronto, Canada: Nelson.

Hodge, G. and Robinson, I., 2001, *Planning Canadian Regions.* Vancouver, Canada: UBC Press.

Innis, H., 1950, *Empire and Communications.* Toronto, Canada: University of Toronto Press.

Jones, K. and Simmons, J., 1993, *Location, Location, Location.* Toronto, Canada: Nelson.

Kerr, D. and Spelt, J., 1965, *The Changing Face of Toronto.* Ottawa, Canada: Geographical Branch.

Kobayashi, A., editor, 1994, *Women, Work and Place.* Montreal/Kingston, Canada: McGill-Queen's University Press.

Kobayashi, A. editor, 2001, Anniversary Issue: 50 Years After: Geographical Interpretations of Canada. *The Canadian Geographer,* Vol. 45, No. 1, Spring, 1–207.

Lemon, J., 1985, *Toronto Since 1918.* Toronto, Canada: Lorimer.

Lemon, J., 1996, *Liberal Dreams and Natures Limits: Great North American Cities Since 1600.* Toronto, Canada: Oxford University Press.

Ley, D., 1996, *The New Middle Class and the Remaking of the Central City.* Oxford, Canada: Oxford University Press.

Ley, D. and Hiebert, D., 2001, Immigration Policy as Population Policy. *The Canadian Geographer,* Vol. 45, 120-126.

Li, P., 2003, *Destination Canada.* Toronto, Canada: Oxford University Press.

Lithwick, H., 1970, *Urban Canada: Problems and Prospects.* Ottawa, Canada: CMHC.

Miron, J. editor, 1993, *Housing Progress in Canada Since 1945.* Montreal, Canada: McGill-Queen's University Press.

Moore, E., Rosenberg, M., and McGuinness, D., 1997, *Growing Old in Canada: Demographic and Geographic Perspectives.* Toronto, Canada: ITP Nelson.

Ray, D. M. et al., 1976, *Canadian Urban Trends.* 3 volumes. Toronto, Canada: Copp Clark, with Ministry of State for Urban Affairs.

Seeley, J. Jr., 1956, *Crestwood Heights.* Toronto, Canada: University of Toronto Press.

Sgro, J., 2002, *Canada's Urban Strategy: A Blueprint for Action.* Prime Minister's Caucus Task Force on Urban Issues. Ottawa, Canada: Prime Minister's Office.

Simmons, J., 1967, Urban geography in Canada. *The Canadian Geographer,* Vol. 11, 341–356.

Simmons, J. and Kamikihara, S., 2001, *Commercial Activity in Canada.* Toronto, Canada: Centre for the Study of Commercial Activity, Ryerson University.

Social Science and Humanities Research Council of Canada (SSHRCC), 1988, *Focus on Strategies.* Ottawa, Canada: Social Science and Humanities Research Council of Canada.

Solant, O. M., 1971, *Cities for Tomorrow: Some Applications of Science and Technology to Urban Development.* Ottawa, Canada: Science Council of Canada.

Spelt, J., 1955, *The Urban Development in South-Central Ontario.* Assen, Netherlands: Van Gorcum.

Stamp, L. D., 1952, *Geography in Canadian Universities, 1951: A Survey.* Ottawa, Canada: Social Science Research Council.

Stelter, G. and Artibise, A., editors, 1977, *The Canadian City: Essays in Urban History.* Toronto, Canada: McLelland and Stewart.

Taylor, G., 1949, *Urban Geography.* London, UK: Methuen.

TD Bank, 2002, *A Choice between Investing in Canada's Cities or Disinvesting in Canada's Future.* Toronto, Canada: TD Economics.

Wolfe, J., 2000, Reinventing planning: Canada. *Progress in Planning,* Vol. 57, 207–235.

Yeates, M., 1975, *Main Street.* Toronto, Canada: MacMillan.

Yeates, M., 1999, *The North American City.* (5th ed.). New York, NY: Prentice Hall.

Yeates, M., 2001, Yesterday as tomorrow's song: The contribution of the 1960s Chicago School to urban geography. *Urban Geography,* Vol. 22, 514–529.

# CHAPTER 17

# The Challenges of Research
# on the Global Network of Cities

DAVID R. MEYER[1]

## ABSTRACT

The significant rise in global commodity trade, foreign direct investment, and global financial exchange during the 1980s was a key impetus to the contemporaneous sharp jump in attention of urban geographers and other social scientists to conceptualizing and empirically examining the global network of cities. Scholars accepted an assumption that the new international division of labor recast global relations, and this has not been questioned subsequently. Research during the 1980s focused on global corporations, corporate services, financial institutions, telecommunications, and transportation as actors or modes of linking global cities.

[1]Correspondence concerning this chapter should be addressed to David R. Meyer, Department of Sociology, Box 1916, Brown University, Providence, RI 02912; telephone: 401-863-2524; fax: 401-863-3213; e-mail: David_Meyer@Brown.edu

This broad coverage framed subsequent research to the present, but the theory of the global network of cities did not advance much beyond the initial formulations. After 1990 some embellishments were made to the theory, and these were widely accepted. Although other theoretical proposals were made, none gained wide acceptance; nevertheless, scholars made some progress in clarifying the behavior of business actors. Researchers accumulated rich empirical evidence about the global network of cities, including extensive evidence on network relations. If scholars are to make greater progress in understanding the global network of cities, the theory needs deepening. A reexamination of the assumption that the new international division of labor has recast global relations might encourage greater emphasis on the principles of the intermediary behavior of the actors most responsible for global network relations. [Key words: global city, multinational corporation, networks, new international division of labor, world system.]

The global (or multinational/transnational) corporation increasingly became an object of study during the 1970s; yet this was but a prelude to the tidal wave of research which poured out the following decade. Urban geographers, along with economic and industrial geographers, joined with sociologists, political scientists, economists, and urban planners in a multidisciplinary analysis of the relation of these corporations to global cities. The activities of global corporations enhanced linkages among global cities to such an extent that researchers would claim "The world of modern capitalism is both a worldwide net of corporations and a global network of cities" (Feagin and Smith, 1986, p. 3).

In a seminal article, Cohen (1981) set out a conceptualization of the logic of the global network of cities, and his thinking was influenced by Hymer's (1971) path-breaking articulation of the relation between the activities of multinational corporations and development. Following Hymer, Cohen argued that in the old international division of labor corporations often obtained nondomestic raw materials from less-developed countries for use in manufacturing in developed countries. However, in the new international division of labor based on an organization of production and markets, corporations purchase inputs globally to supply their production facilities, which also are spread globally. These production units supply markets wherever they are located in the world, although the largest markets remain in the developed countries. With this new international division of labor, corporate-related services in commercial and investment banking, law, accounting, management consulting, advertising, and so on spread globally to service the corporations. One result of this globalization of business is an integrated worldwide financial system.

Cohen's conceptualization set out in 1981 has continued to the present as the explicit or implicit framework for examining the global network of cities. The thesis of this paper is that researchers studying global cities during the 1980s focused on clarifying components of this conceptualization through theoretical speculations and empirical studies, but they did not deepen the theoretical framework. After 1990, theoretical development made only modest progress even as empirical analyses exhibited explosive growth. Thus, the theory of the global network of cities remains undeveloped, and much empirical work, while adding rich new evidence, continues to replicate the approaches of studies carried out in the 1980s.

## The Failure to Seriously Consider the Past

Most contemporary researchers recognize that a global network of cities has existed, in at least minimal form, since the 19th century, and they also accept that merchants linked major trading centers around the world for many centuries before that. Nonetheless, few scholars see any relevance to considering the theoretical bases of a global network of cities for the period before the mid-20th century, because, it is argued, the new international division of labor has recast global relations. Yet, even if the form and content of relations among global cities has changed, this does not necessarily imply that the behavioral principles guiding the actions of individuals and firms are completely different. This neglect of the past hinders the formulation of a dynamic theory of the global network of cities.

The 1927 publication of Roderick McKenzie's "The Concept of Dominance and World-Organization" in the *American Journal of Sociology,* one of the leading sociology journals, suggests that contemporary researchers may have been too quick to dismiss the theoretical salience of the global network of cities prior to the mid-20th century. McKenzie was a human ecologist of the Chicago school of urban sociology which many urban geographers, as well as urban specialists in other disciplines, formerly subscribed to as a theoretical framework. This article is rarely cited, much less thoughtfully considered, by contemporary researchers. Regardless of its theoretical base in human ecology, the article brims with insights relevant to theorizing about the current global network of cities. McKenzie's 1927 article was published following five decades of extensive global trade whose significance was recognized by many of that era's leading social theorists, including Marx (1967), Spencer (1897-1906), Tonnies (1957), and Weber (1978).

Prior to the mid-19th century information moved physically with the transportation of people and commodities. However, following the introduction of the telegraph, telephone, and wireless communication, McKenzie (1927, p. 34) argued, sophisticated information (he called it "intelligence") moved so rapidly—almost instantaneously—that this "produced revolutionary results in spatial reorganization.… Industries and other business enterprises may now be located at great distances from the source of their management and control." He went on to cite the examples of European and United States firms which established plants in India, China, and elsewhere in the world. In McKenzie's view, modern communication produced a centralization of intelligence and control which became the basis for the dominance of some cities such as London and New York at the highest level as world financial centers. And, other, lesser world cities dominated at a lower level based on their activities as centers of specialized commodity futures such as Chicago and Liverpool in wheat or New Orleans and Liverpool in cotton. He claimed that modern communication stimulated a world regional specialization of production, and this was territorially integrated.

Although McKenzie did not elaborate his theoretical scheme, it exhibits a remarkable correspondence with Cohen's conceptualization more than 50 years later. It is not sufficient to simply dismiss McKenzie's formulation as a reflection of the old international division of labor. Certainly, the multinationals operating around McKenzie's time and several decades earlier mostly were involved with raw material production; yet some of them also operated factories in other countries. Furthermore, multinational firms such as trading companies and financial firms provided sophisticated services to the multinationals engaged in raw material extraction and processing and in manufacturing, and these multinational corporate service firms even operated in less-developed regions of the world. All of these multinationals, for example, were significantly active in Asia from the mid-19th to the early 20th centuries. The offices of multinational firms in Asian business centers headed by Hong Kong, secondarily in Singapore and Shanghai, and at a lower level in Manila, Saigon, Bangkok, and Rangoon, provided tight integration with the top global cities of London and secondarily, New York, and with lower-level cities such as Paris and Amsterdam (Meyer, 2000).

Around 1900 firms from the United States, United Kingdom, Germany, and Russia owned cotton mills, flour and oil mills, and machinery works in China's treaty ports such as Shanghai, Foochow, and Harbin. Rockefeller's Standard Oil Company had plants in China's treaty ports and across Southeast Asia, including Hong Kong, Manila, Saigon, and Bangkok. British tin smelters in Singapore had backward linkages to ore mines in Malaya and forward linkages to the tin traders in the United Kingdom who sold tin domestically and internationally. Both the Jardine, Matheson and the Butterfield and Swire trading companies which had London, Hong Kong, and Shanghai offices operated sugar refineries in Hong Kong in the late 19th century. The refineries drew on sugar cane supplies in Southeast Asia, and the sugar was marketed in Asia. These trading companies also owned steamship companies and provided shipping services for most of the important ports in Asia. Shewan, Tomes and Company had offices in Hong Kong, Shanghai, Tientsin, Kobe, London, and New York, and they managed steamship companies, manufacturing plants of multinationals, and provided services for insurance multinationals headquartered in the United States and United Kingdom (Wright, 1908; Feldwick, 1917; Lieu, 1936; Allen and Donnithorne, 1954; Marriner and Hyde, 1967; Hao, 1970; Latham, 1978; Jones, 1986; Huff, 1994). Multinational banks headquartered in London, New York, and Hong Kong offered financial services from regional headquarters and regional branch offices within Asia. The Hongkong and Shanghai Banking Corporation which was founded in Hong Kong in 1864 had major world regional offices in Shanghai and London, and branch offices throughout Asia and in Europe and North America by the late 19th and early 20th centuries (Feldwick, 1917; Jones, 1993; King, 1987, 1988a, 1988b; Wright, 1908).

The business relations of these multinationals engaged in raw material extraction and processing, manufacturing, and corporate services suggests that the global network of cities identified by McKenzie in 1927 possessed more than passing resemblance to the modern network. Thus, the claim that the new international division of labor recast global relations, compared to the old division of labor, provided a flawed basis for conceptualizing the global network of cities. During the 1980s urban geographers and other urban specialists did not critically reassess their assumption that the late 20th century was a new era in global relations, and little effort was made to deepen the theory of the modern global network of cities.

### Research during the 1980s

Friedmann and Wolff (1982) and Friedmann (1986), building on Cohen's (1981) conceptualization, set forth the most elaborate specification of a framework for studying the global network of cities, and this became, arguably, the leading agenda for researchers during the 1980s and subsequently. The focus on a research agenda, not on a theory, was clear from the start. Friedmann and Wolff (1982, pp. 309 and 329) stated, "Our paper concerns the spatial articulation of the emerging world system of production and markets through a global network of cities.... The world city perspective is a framework for the study of urbanization; it is not a theory." And, "The world city hypothesis ... is primarily intended as a framework for research. It is neither a theory nor a universal generalization about cities, but a starting point for political enquiry" (Friedmann, 1986, p. 69). This research agenda emphasized that transnational (multinational/global) corporations controlled and organized production and markets from their bases in world cities, and the central hypothesis was that *"the mode of world system integration (form and strength of integration; spatial dominance) will affect in determinant ways the economic, social, spatial and political structure of world cities and the urbanizing processes to which they are subject* (original italics)" (Friedmann and Wolff, 1982, p. 313). From the perspective of the global network of cities, this agenda was heavily weighted toward understanding the structure and functions of world cities as reflected in the activities of transnational corporations. It did not call for a theory of the behavior of these firms.

Most of the numerous studies of multinational corporations which included some aspect of themes relevant to explaining the global network of cities either focused on the geography of world-regional production or on the world city as a headquarters or production site (Peet, 1983; Dixon et al., 1986; Taylor and Thrift, 1986; Henderson and Castells, 1987; Smith and Feagin, 1987; Knight and Gappert, 1989). They paid little attention directly to the global network of cities, and network relations typically were subsumed under the rubric that specialization of global cities implied their integration. However, the nature of that integration remained vaguely specified. Researchers examined the determinants of the location choices of different levels of administrative offices of multinationals, but the network implications and the consequences for explaining the global network of cities was not developed (Dunning and Norman, 1983; Dunning and Norman, 1987). International intercorporate linkages among headquarters of firms as expressed through corporate directorate ties constitute a direct means to examine the global network of cities, and yet this avenue received little scrutiny (Green, 1983).

The failure to elaborate a theory of the global network of cities based on the organization of multinationals remains puzzling. Researchers were reminded about the importance of conceptualizing and analyzing linkages within and among units of multinationals, and the network implications were outlined, including the consequences for urban, regional, and national development (Ettlinger, 1983). The solution to the puzzle seems to rest in the limited interest in theorizing about how transnationals behaved in controlling and coordinating information and production within and between their organizations and deriving the consequences for the global network of cities. Pred (1977) had developed such a theory for the national scale which could have been readily adapted to the global scale. His model included information components, linkages through product and service flows, markets, innovations, and multiplier effects, and it could have served as a stimulus for a larger theory of the global network of cities, even if all of his arguments were not followed in detail. Nevertheless, Pred's rich theory was either given cursory attention or ignored.

Most of the modest efforts in the 1980s to elaborate a theory of, and implement empirical analyses of, the global network of cities focused on telecommunications and business services, broadly defined. The discussion of the consequences of improved global telecommunications for centralization of high-level business decision-making implicitly followed in the spirit of McKenzie's (1927) earlier formulation. Langdale (1989) formulated one of the more elaborate synthesis of generalizations about transnational production, consequences of improved international telecommunications (for example, leased networks) for centralization of decision-making, and concepts of corporate organizational structure. He made a coherent argument for linkages among leading global cities and for the existence of regional hubs of telecommunications and decision-making; his empirical evidence, though limited, provided support for the theoretical arguments. McKenzie's (1927) insight that telecommunications innovations broke the bond between the movement of information and the movement of commodities and people, thus permitting a revolution in spatial organization, was followed up implicitly with the recognition that international trade services cluster in decision-making centers, especially those with high-level financial services (O'Connor, 1987). However, the full implications for the global network of cities of the disjunction between the physical movement of commodities and movement of information, which is a more direct indicator of the control of capital exchange, remained undeveloped for the global scale, although a theoretical framework, with empirical implications, had been proposed for the national scale (Meyer, 1980).

The nature of the global network of cities was partially illuminated in research on links between headquarters and branch offices of business service firms. Distributions of the branch offices of leading global firms in banking, advertising, law, and accounting implied a network structure of global cities, even when the precise structure of the network was not specified (Moss, 1987). Detailed case studies of an individual business sector such as international property services and real estate investment, even without theoretical context, provided rich evidence for future speculations about the global network of cities (Thrift, 1986). When the explicit links between the headquarters of transnational service firms and their branches were examined, such as in international banking, the structure and degree of control exerted by firms in a global network of cities were evident, but few of these studies were carried out (Meyer, 1986). Theorizing about the behavior of financial services firms within and among global cities offered hints about the global network (Thrift, 1987); however, this approach made little headway.

The most significant theorizing about the global network of cities came from scholars working in world-system analysis. As a framework for examining the global network of cities, its roots can be traced back to Walton's (1976) call for the study of vertically integrated processes which linked the urban hinterland, urban/regional, national, and international levels. Wallerstein's (1974, 1980, 1989) seminal books on the modern world-system provided important intellectual grounding for applying this type of analysis to the global network of cities. From the world-system perspective the global economy is organized as an interstate system of production and consumption in which states operate at different levels of the global division of labor; they may be core, semiperipheral, or peripheral. These hierarchical levels from top-down reflect the differential power of states and firms to control the flow of capital and the capacity to accumulate capital. The sociologist Chase-Dunn (1984, 1985) carried out the most significant research on the systemic qualities of global cities, but much of his attention focused on the changing structural characteristics of the hierarchy (or primacy) of world cities, and this approach was followed in geography (Ettlinger and Archer, 1987). The dynamic, historical approach of world-system analysis kept its practitioners focused on long-term change, thus these researchers were less likely to be enamored of distinctions such as the "old" or "new" division of labor which the mid-20th century benchmark presumably separated. However, even with the richness of world system analysis, researchers did not expand much on the theory of the global network of cities beyond initial conceptualizations, and empirical work on networks remained modest.

Consequently, the flurry of research on the global network of cities in the 1980s left a legacy of important empirical studies which documented some key characteristics of global networks. However, the theorizing about these networks did not advance much, and, except for world-system analysis, researchers kept intact their assumption that the late 20th century was a new era in global relations. This left subsequent research in a quandary. Empirical work could flourish, but the theoretical implications of the rich empirical evidence could remain undeveloped other than providing verification of the theory or modifying it slightly. In fact, this is what happened after 1990, and only modest progress was made on deepening the theory of the global network of cities.

## The Predicament of Current Research

The 1990s opened with Sassen's (1991) extension of Cohen's (1981) and McKenzie's (1927) interpretations of global business. She argued that the territorial dispersal of economic activity, not only called for centralization of management and control, but also that this centralization was increasing as transnational firms expanded their global activities. This, in turn, stimulated a demand for higher levels of corporate services such as in finance, accounting, law, and management consulting, thus enhancing a small number of leading global cities (in Sassen's view, New York, London, and Tokyo) as centers of the production of, innovation in, and specialization of these corporate services. From this perspective, then, these global cities were hubs in the networks, because the corporate administration and the branching of the corporate services reached to other cities throughout the world.

Sassen's reinvigoration of these ideas was widely accepted, and researchers also drew heavily on Friedmann's (1982, 1986) world city hypothesis for inspiration as they embarked on extensive empirical analyses of global cities. The studies covered a panoply of variables including corporate headquarters and various types of corporate services. Much of this research constituted either a case study of one city and its attributes (Warf and Erickson, 1996; Lo and Yeung, 1998; Ciccolella and Mignaqui, 2002) or included multiple cities with a focus on one or more variables such as the number of corporate headquarters (Lyons and Salmon, 1995; Godfrey and Zhou, 1999) or of corporate services (Daniels, 1993; Leslie, 1995) for a set of cities. The rich empirical evidence of these studies constituted important building blocks for those who would construct deeper formal theory of the global network of cities. Yet, these studies did not provide formal analyses of the structure and character of the network relations among global cities, and these studies are necessary for constructing better understandings of the city networks (Smith and Timberlake, 1995a; Beaverstock, Smith, and Taylor, 2000b). Direct analyses of global networks remained less frequent because the data are difficult to assemble. Little government data are available, and data sets often have to be constructed from directories or private sources, which frequently are reluctant to make the data available or place constraints on its use.

Nonetheless, researchers added substantially to the studies of actual network links among cities which had been completed in the 1980s

(Meyer, 1986; Thrift, 1986; Moss, 1987; Langdale, 1989). The studies varied in the degree to which they examined linkages from one node to other nodes or among multiple nodes, and they included financial services (ÓhUallacháin, 1994; Agnes, 2000), corporate services (Beaverstock, Smith, and Taylor, 2000a, 2000b; Taylor, Walker, and Beaverstock, 2002), information media (Mitchelson and Wheeler, 1994), and air travel (Cattan, 1995; Smith and Timberlake, 1995b, 1998, 2002; Rimmer, 1998; Shin and Timberlake, 2000; Bowen, 2002). These network analyses, as well as the case studies of individual cities or of multiple cities with a focus on one or more variables (corporate headquarters, corporate services), provided rich evidence and broad confirmation of the conceptualization of Cohen (1981), the world city hypothesis of Friedmann (1982, 1986), and the reformulation of Sassen (1991). At the same time, they pointed to more complex, nuanced ways to understand the global network of cities, such as the variation in linkages arising from different strategic approaches of corporate headquarters and corporate services firms, the differences in physical movement through air travel versus the control linkages of corporate services firms, and world regional distinctions emerging from rapid economic growth (southeast and east Asia).

The wealth of empirical evidence about the global network of cities accumulated since 1990, however, was not accompanied commensurately by the incorporation of more sophisticated theory into the empirical analyses. Friedmann (1995) restated his hypothesis, and Sassen (1995, 2001, 2002) embellished her framework. Nevertheless, most of the empirical studies only made reference to Friedmann's world city hypothesis and/or Sassen's headquarters/corporate services framework, and some referred to world system analysis, but these studies did not deepen the theory of the global network of cities. Castells' (1996) network theory as applied to global cities did not add much beyond the accepted approaches in research. One attempt to propose a theory of the global network of cities took a behavioral approach which argued that the reactions of international business intermediaries to competition could explain important aspects of the growth and change of global cities as centers for controlling and coordinating the exchange of capital (Meyer, 1991, 1998). Researchers elaborated generalizations about the behavior of financial intermediaries (Lee and Schmidt-Marwede, 1993; ÓhUallacháin, 1994; Pryke and Lee, 1995; Agnes, 2000), corporate services' firms (Beaverstock, Smith, and Taylor, 2000a; Leslie, 1995), information firms (Mitchelson and Wheeler, 1994), and the nexus of telecommunications and financial intermediation (Langdale, 2000, 2001a, 2001b; Meyer, 2002) within and among business centers. However, the implications of these ideas for expanding the theory of the global network of cities were not articulated much. A beginning was made in the incorporation of social network theory as a means to explain the global network of cities (Agnes, 2000; Meyer, 2000). This theory was also employed in analyses of transnational corporations under the rubric of "business networks" (e.g., Yeung, 1998), but the precise relation to global cities was not the focus of that research. The extent to which social network theory will provide a way to deepen the theory of the global network of cities remains uncertain.

## The Legacy

The failure to question the assumption that the global network of cities under the new international division of labor recast global relations sets an intellectual tone which encourages researchers to focus either on static analyses or on short-term change. And, even if analyses cover several decades, no theoretical framework exists for interpreting the results except as confirmation of existing theory or as a slight modification of it. The decade of the 1980s set the research agenda for subsequent studies of the global network of cities. The theory was framed in broad strokes which constituted an outline, rather than an elaborate theory. Empirical analyses offered verification of the theory, and little effort was devoted to deepening the theory. Additions to the theory after 1990 embellished a few points, and no elaborate theory gained currency; however, researchers made some progress in clarifying the behavior of business actors. The key advance in research was the accumulation of a rich set of empirical evidence about the global network of cities, including substantial evidence on network relations. The theoretical lacuna needs addressing if scholars are to make greater progress in understanding the global network of cities. Simply adding more empirical studies will not be sufficient. And, a reconsideration of the assumption that the new international division of labor has recast global relations could open the door to greater focus on the principles of intermediary behavior of the actors (corporate headquarters, financial services, corporate services) most responsible for structuring and implementing global network relations.

## References

Agnes, P., 2000, The end of geography in financial services? Local embeddedness and territorialization in the interest rate swaps industry. *Economic Geography,* Vol. 76, 347–366.
Allen, G. C. and Donnithorne, A. G., 1954, *Western Enterprise in Far Eastern Economic Development: China and Japan.* New York, NY: Macmillan.
Beaverstock, J. V., Smith, R. G., and Taylor, P. J., 2000a, Geographies of globalization: United States law firms in world cities. *Urban Geography,* Vol. 21, 95–120.

Beaverstock, J. V., Smith, R. G., and Taylor, P. J., 2000b, World-city network: A new metageography. *Annals of the Association of American Geographers,* Vol. 90, 123–134.

Bowen, J., 2002, Network change, deregulation, and access in the global airline industry. *Economic Geography,* Vol. 78, 425–439.

Castells, M., 1996, *The Information Age: Economy, Society and Culture. Vol. 1, The Rise of the Network Society.* Cambridge, MA: Blackwell.

Cattan, N., 1995, Attractivity and internalization of major European cities: The example of air traffic. *Urban Studies,* Vol. 32, 303–312.

Chase-Dunn, C., 1984, Urbanization in the world-system: New directions for research. In M. P. Smith, editor, *Cities in Transformation: Class, Capital, and the State.* Urban Affairs Annual Reviews, Vol. 26. Beverly Hills, CA: Sage, 111–120.

Chase-Dunn, C., 1985, The system of world cities, A.D. 800–1975. In M. Timberlake, editor, *Urbanization in the World-Economy.* Orlando, FL: Academic Press, 269–292.

Ciccolella, P. and Mignaqui, I., 2002, Buenos Aires: Sociospatial impacts of the development of global city functions. In S. Sassen, editor, *Global Networks, Linked Cities.* New York, NY: Routledge, 309–325.

Cohen, R. B., 1981, The new international division of labor, multinational corporations and urban hierarchy. In M. Dear and A. J. Scott, editors, *Urbanization and Urban Planning in Capitalist Society.* London, UK.: Methuen, 287–315.

Daniels, P. W., 1993, *Service Industries and the World Economy.* Oxford, UK: Basil Blackwell.

Dixon, C. J., Drakakis-Smith, D., and Watts, H. D., editors, 1986, *Multinational Corporations and the Third World.* Boulder, CO: Westview.

Dunning, J. H. and Norman, G., 1983, The theory of the multinational enterprise: An application to multinational office location. *Environment and Planning A,* Vol. 15, 675–692.

Dunning, J. H. and Norman, G., 1987, The location choice of offices of international companies. *Environment and Planning A,* Vol. 19, 613–631.

Ettlinger, N., 1984, Comments on the concept of linkages from the perspective of corporate organization in the modern capitalist system. *Tijdschrift voor Economische en Sociale Geografie,* Vol. 75, 285–291.

Ettlinger, N. and Archer, J. C., 1987, City-size distributions and the world urban system in the 20th century. *Environment and Planning A,* Vol. 19, 1161–1174.

Feagin, J. R. and Smith, M. P., 1987, Cities and the new international division of labor: An overview. In M. P. Smith and J. R. Feagin, editors, *The Capitalist City: Global Restructuring and Community Politics.* Oxford, UK: Basil Blackwell, 3–34.

Feldwick, W., editor, 1917, *Present Day Impressions of the Far East and Prominent & Progressive Chinese at Home and Abroad.* London, UK: Globe Encyclopedia Company.

Friedmann, J., 1986, The world city hypothesis. *Development and Change,* Vol. 17, 69–83.

Friedmann, J., 1995, Where we stand: A decade of world city research. In P. L. Knox and P. J. Taylor, editors, *World Cities in a World-System.* Cambridge, UK: Cambridge University Press, 21–47.

Friedmann, J. and Wolff, G., 1982, World city formation: An agenda for research and action. *International Journal of Urban and Regional Research,* Vol. 6, 309–344.

Godfrey, B. J. and Zhou, Y., 1999, Ranking world cities: Multinational corporations and the global urban hierarchy. *Urban Geography,* Vol. 20, 268–281.

Green, M. B., 1983, The interurban corporate interlocking directorate network of Canada and the United States: A spatial perspective. *Urban Geography,* Vol. 4, 338–354.

Hao, Y. P., 1970, *The Comprador in 19th Century China: Bridge between East and West.* Cambridge, MA: Harvard University Press.

Henderson, J. and Castells, M., editors, 1987, *Global Restructuring and Territorial Development.* London, UK: Sage.

Huff, W. G., 1994, *The Economic Growth of Singapore: Trade and Development in the 20th Century.* Cambridge, UK: Cambridge University Press.

Hymer, S., 1971, The multinational corporation and the law of uneven development. In J. N. Bhagwati, editor, *Economics and World Order: From the 1970s to the 1990s.* New York, NY: Macmillan, 113–140.

Jones, S., 1986, *Two Centuries of Overseas Trading: The Origins and Growth of the Inchcape Group.* London, UK: Macmillan.

Jones, G., 1993, *British Multinational Banking, 1830–1990.* Oxford, UK: Clarendon Press.

King, F. H. H., 1987, *The Hongkong Bank in Late Imperial China, 1864–1902: On an Even Keel,* Vol. I, *The History of the Hongkong and Shanghai Banking Corporation.* Cambridge, UK, Cambridge University Press.

King, F. H. H., 1988a, *The Hongkong Bank in the Period of Imperialism and War, 1895–1918: Wayfoong, the Focus of Wealth,* Vol. II, *The History of the Hongkong and Shanghai Banking Corporation.* Cambridge, UK, Cambridge University Press.

King, F. H. H., 1988b, *The Hongkong Bank Between the Wars and the Bank Interned, 1919–1945: Return from Grandeur,* Vol. III, *The History of the Hongkong and Shanghai Banking Corporation.* Cambridge, UK, Cambridge University Press.

Knight, R. V. and Gappert, G., editors, 1989, *Cities in a Global Society.* Urban Affairs Annual Reviews, Vol. 35. Newbury Park, CA: Sage.

Langdale, J. V., 1989, The geography of international business telecommunications: The role of leased networks. *Annals of the Association of American Geographers,* Vol. 79, 501–522.

Langdale, J. V., 2000, Telecommunications and 24-hour trading in the international securities industry. In M. I. Wilson and K. E. Corey, editors, *Information Tectonics: Space, Place and Technology in an Electronic Age.* Chichester, UK: Wiley, 89–99.

Langdale, J. V., 2001a, Electronic commerce: Global-local relationships in financial services. In D. Felsenstein and M. Taylor, editors, *Promoting Local Growth: Process, Practice and Policy.* Aldershot, UK: Ashgate, 209–226.

Langdale, J. V., 2001b, Global electronic spaces: Singapore's role in the foreign exchange market in the Asia-Pacific region. In T. R. Leinbach and S. D. Brunn, editors, *Worlds of E-Commerce: Economic, Geographical and Social Dimensions.* Chichester, UK: Wiley, 203–219.

Latham, A. J. H., 1978, *The International Economy and Undeveloped World, 1865–1914.* London, UK, Croom Helm.

Lee, R. and Schmidt-Marwede, U., 1993, Interurban competition? Financial centres and the geography of financial production. *International Journal of Urban and Regional Research,* Vol. 17, 492–515.

Leslie, D. A., 1995, Global scan: The globalization of advertising agencies, concepts, and campaigns. *Economic Geography,* Vol. 71, 402–426.

Lieu, D. K., 1936, *The Growth and Industrialization of Shanghai.* Shanghai, China: China Institute of Pacific Relations.

Lo, F. and Yeung, Y., editors, 1998, *Globalization and the World of Large Cities.* Tokyo, Japan: United Nations University Press.

Lyons, D. and Salmon, S., 1995, World cities, multinational corporations, and urban hierarchy: The case of the United States. In P. L. Knox and P. J. Taylor, editors, *World Cities in a World-System.* Cambridge, UK: Cambridge University Press, 98–114.

Marriner, S. and Hyde, F. E., 1967, *The Senior John Samuel Swire, 1825–98: Management in Far Eastern Shipping Trades.* Liverpool, UK, Liverpool University Press.

McKenzie, R. D., 1927, The concept of dominance and world-organization. *American Journal of Sociology,* Vol. 33, 28–42.

Marx, K., 1967, *Capital.* 3 vols. New York, NY: International Publishers.

Meyer, D. R., 1980, A dynamic model of the integration of frontier urban places into the United States system of cities. *Economic Geography,* Vol. 56, 120–140.

Meyer, D. R., 1986, The world system of cities: Relations between international financial metropolises and South American cities. *Social Forces,* Vol. 64, 553–581.

Meyer, D. R., 1991, Change in the world system of metropolises: The role of business intermediaries. *Urban Geography,* Vol. 12, 393–416.

Meyer, D. R., 1998, World cities as financial centers. In F. Lo and Y. Yeung, editors, *Globalization and the World of Large Cities.* Tokyo, Japan: United Nations University Press, 410–432.

Meyer, D. R., 2000, *Hong Kong as a Global Metropolis.* Cambridge, UK: Cambridge University Press.

Meyer, D. R., 2002, Hong Kong: Global capital exchange. In S. Sassen, editor, *Global Networks, Linked Cities.* New York, NY: Routledge, 249–271.

Mitchelson, R. L. and Wheeler, J. O., 1994, The flow of information in a global economy: The role of the American urban system in 1990. *Annals of the Association of American Geographers,* Vol. 84, 87–107.

Moss, M. L., 1987, Telecommunications, world cities, and urban policy. *Urban Studies,* Vol. 24, 534–546.

O'Connor, K., 1987, The location of services involved with international trade. *Environment and Planning A*, Vol. 19, 687–700.

ÓhUallacháin, B., 1994, Foreign banking in the American urban system of financial organization. *Economic Geography*, Vol. 70, 206–228.

Peet, R., editor, 1983, Restructuring in the age of global capital. *Economic Geography*, Vol. 59, 105–230. Special issue.

Pred, A., 1977, *City-Systems in Advanced Economies*. New York, NY: Wiley.

Pryke, M. and Lee, R., 1995, Place your bets: Toward an understanding of globalisation, socio-financial engineering and competition within a financial centre. *Urban Studies*, Vol. 32, 329–344.

Rimmer, P. J., 1998, Transport and telecommunications among world cities. In F. Lo and Y. Yeung, editors, *Globalization and the World of Large Cities*. Tokyo, Japan: United Nations University Press, 433–470.

Sassen, S., 1991, *The Global City: New York, London, Tokyo*. Princeton, NJ: Princeton University Press.

Sassen, S., 1995, On concentration and centrality in the global city. In P. L. Knox and P. J. Taylor, editors, *World Cities in a World-System*. Cambridge, UK: Cambridge University Press, 63–75.

Sassen, S., 2001, Global cities and global city-regions: A comparison. In A. J. Scott, editor, *Global City-Regions: Trends, Theory, Policy*. Oxford, UK: Oxford University Press, 78–95.

Sassen, S., 2002, Locating cities on global circuits. In S. Sassen, editor, *Global Networks, Linked Cities*. New York, NY: Routledge, 1–36.

Shin, K. and Timberlake, M., 2000, World cities in Asia: Cliques, centrality and connectedness. *Urban Studies*, Vol. 37, 2257–2285.

Smith, D. A. and Timberlake, M., 1995a, Cities in global matrices: Toward mapping the world-system's city system. In P. L. Knox and P. J. Taylor, editors, *World Cities in a World-System*. Cambridge, UK: Cambridge University Press, 79–97.

Smith, D. A. and Timberlake, M., 1995b, Conceptualizing and mapping the structure of the world system's city system. *Urban Studies*, Vol. 32, 287–302.

Smith, D. A. and Timberlake, M., 1998, Cities and the spatial articulation of the world economy through air travel. In P. S. Ciccantell and S. G. Bunker, editors, *Space and Transport in the World-System*. Westport, CT: Greenwood Press, 213–240.

Smith, D. A. and Timberlake, M., 2002, Hierarchies of dominance among world cities: A network approach. In S. Sassen, editor, *Global Networks, Linked Cities*. New York, NY: Routledge, 117–141.

Smith, M. P. and Feagin, J. R., editors, 1987, *The Capitalist City: Global Restructuring and Community Politics*. Oxford, UK: Basil Blackwell.

Spencer, H., 1897–1906, *The Principles of Sociology*. 3 vols. London, UK: Williams and Norgate.

Taylor, M. and Thrift, N., editors, 1986, *Multinationals and the Restructuring of the World Economy*. London, UK: Croom Helm.

Taylor, P. J., Walker, D. R. F., and Beaverstock, J. V., 2002, Firms and their global service networks. In S. Sassen, editor, *Global Networks, Linked Cities*. New York, NY: Routledge, 93–115.

Thrift, N., 1986, The internationalisation of producer services and the integration of the Pacific Basin property market. In M. Taylor and N. Thrift, editors, *Multinationals and the Restructuring of the World Economy*. London, UK: Croom Helm, 142–192.

Thrift, N., 1987, The fixers: The urban geography of international commercial capital. In J. Henderson and M. Castells, editors, *Global Restructuring and Territorial Development*. London, UK, Sage, 203–233.

Tonnies, F., 1957, *Community and Society*. Translated and edited by C. P. Loomis. East Lansing, MI: Michigan State University Press.

Wallerstein, I., 1974, *The Modern World-System: Capitalist Agriculture and the Origins of the European World-Economy in the Sixteenth Century*. New York, NY: Academic Press.

Wallerstein, I., 1980, *The Modern World-System II: Mercantilism and the Consolidation of the European World-Economy, 1600–1750*. New York, NY: Academic Press.

Wallerstein, I., 1989, *The Modern World-System III: The Second Era of Great Expansion of the Capitalist World-Economy, 1730–1840s*. San Diego, CA: Academic Press.

Walton, J., 1976, Political economy of world urban systems: Directions for comparative research. In J. Walton and L. H. Masotti, editors, *The City in Comparative Perspective.* New York, NY: Wiley, 301–313.

Warf, B. and Erickson, R., editors, 1996, Special Issue: Globalization and the U.S. city system. *Urban Geography,* Vol. 17, 1–117.

Weber, M., 1978, *Economy and Society.* 2 vols. G. Roth and C. Wittich, editors. Berkeley, CA: University of California Press.

Wright, A., editor, 1908, *20th Century Impressions of Hongkong, Shanghai, and Other Treaty Ports of China.* London, UK: Lloyd's Greater Britain Publishing Company.

Yeung, H. W., 1998, *Transnational Corporations and Business Networks.* London, UK: Routledge.

# CHAPTER 18

# Quality-of-Life Research in Urban Geography

MICHAEL PACIONE[1]

## ABSTRACT

The 1980s represent a significant period for the development of urban geography—a decade that advanced the reformation of urban geography as a conceptually sound, analytically powerful, integrative discipline capable of making a distinctive contribution to mainstream social science research on the city. The decade also witnessed the emergence of new theories, concepts and research themes that were to have an enduring influence on the nature of the subject. Prominent in this new research portfolio were issues related to quality of life in the city. This paper outlines the main developments in urban geography in the 1980s with particular reference to themes relating to quality-of-life research. The

[1]Correspondence concerning this chapter should be addressed to Michael Pacione, Department of Geography, University of Glasgow, Graham Hills Building, 50 Richmond Street, G1 1XH, Scotland, United Kingdom; telephone: 0044 (0) 141 548 3793; e-mail: michaelpacione@aol.com

discussion introduces the concept of "useful knowledge" within the context of "applied urban geography"; examines the key dimensions of quality-of-life research in urban geography; and concludes by adopting a prospective viewpoint to identify a number of quality-of-life issues of significance for the urban geography of the 21st century. [Key words: cities, useful knowledge, applied urban geography, quality of life.]

The "1980s" is a convenient classificatory cipher, but the period cannot be regarded as a "closed system." Leakage of ideas, debates, theory, concepts, methods, and research topics occur at both ends of the time frame. The character of urban geography in the 1980s was influenced significantly by developments in the preceding decade including rejection of positivism in human geography (Bowen, 1979), development of critical social science (Held, 1980), and the "relevance" debate within an emerging "radical geography" (Peet, 1977). At the close of the study period the dominance of the political economy perspective was challenged by postmodern theory (Jencks, 1987). Nevertheless, there is little doubt that the 1980s represent a significant period in the development of urban geography—a decade that advanced the reformation of urban geography as an analytically powerful integrative social science, and that witnessed the emergence of new theories, concepts and research questions that were to exert an enduring influence on the nature of the subject.

In this retrospective view of urban geography in the 1980s I outline the major changes that impacted upon urban geography during the decade. I focus particular attention on developments related to quality-of-life research in urban geography. I then introduce the concept of "useful knowledge" and the approach of "applied urban geography," before reviewing the key dimensions of quality-of-life research. Finally, in the concluding section I adopt a prospective viewpoint to consider the legacy of the 1980s for urban geography and identify a number of issues of significance for urban geography in the future.

## The Changing Context and Content of Urban Geography in the 1980s

Reviewing the course of American urban geography from the vantage point of 1989 Marston et al. (1989, p. 653) observed that, "during the past twenty five years, urban geography has been characterized by the use of abstract modes of thought, especially those associated with statistics and mathematics. Particular emphasis has been given to model building, which represents a pattern-seeking viewpoint that stresses recurrent connections and interrelationships." This account accords well with the content and practice of urban geography during the early part of the authors' review period when research was dominated by the philosophy of positivism. It is much less appropriate to the nature and content of urban geography in the 1970s and 1980s, a period in which a major "paradigm shift" occurred as the deficiencies of positivism for urban social research were exposed. Critique of the "spatial science" approach to urban investigation was led by the new philosophical perspectives of behavioralism, humanism and, in particular, political economy. The introduction of a neo-Marxist political economy perspective to urban geography rocked the subject to its foundations and, in the process, forced a fundamental reappraisal of the nature, scope and academic value of urban geography (and of urban geographers). Not until the "cultural turn" in social geography and the advent of postmodernism in the 1990s was the subject obliged to reexamine its academic credentials to the same degree.

The review by Marston et al. (1989) of urban geography also illustrates the process of change that is on-going within any subject. Urban geography is a dynamic subdiscipline that comprises a combination of past ideas and approaches, current concepts, and issues that are still being worked out. As Haggett (1994, p. 223) observed, it may be likened to "a city with districts of different ages and vitalities. There are some long-established districts dating back to a century ago and sometimes in need of repair; and there are areas which were once formidable but are so no longer, while others are being rehabilitated. Other districts have expanded recently and rapidly; some are well built, others rather gimcrack." As urban geography entered the 1980s it was characterized by a mixture of traditional and emerging research themes.

At the inter-urban scale investigations of "systems of cities" (Berry, 1964) continued to be undertaken based on models such as the rank-size rule, but attempts to explain city-size distributions with reference to factors such as level of economic development, political structure and history of urbanization were largely unsuccessful. Another area where traditional research perspectives lacked explanatory power concerned the processes underlying urban growth and decline. Size-ratchet explanations of urban growth, whereby large cities were hypothesized to have a built-in advantage for growth (due to their greater employment mix, political power, capital investments and economies of scale) were being undermined by the phenomenon of counterurbanization that began to affect metropolitan areas during the 1970s. Overall, as Marston et al. (1989, p. 654) concluded, "there is obviously a need to re-conceptualize the whole process of urban change." At the intra-urban scale traditional approaches continued to be employed in attempts to explain the land-use structure of cities using the methodology of bid-rent curves (Muth, 1985). Others attempted to adapt the standard monocentric model to fit the emerging polycentric structure of metropolitan areas (Erickson, 1986). But most analyses failed to afford sufficient attention to the effects of external forces such as private sector agents and government policy, or to behavioral influences in the construction of built environments.

One area of urban geography in which traditional approaches (based on central place theory) continued to demonstrate analytical clarity was the geography of urban retailing, with studies of the urban retail hierarchy maintaining an important niche in the subject (Morrill, 1987). Even here, however, new approaches based on behavioral-humanistic perspectives, were emerging with the aim of constructing models of consumer spatial behavior to explain underlying decision processes (Golledge and Stimson, 1987). A decision-making approach was also applied to investigations of intra-urban mobility and migration. New models of the residential search process incorporated the effects of information sources, length of search activity, and the spatial pattern of search (Clark, 1982). Useful though these developments were in increasing understanding of the intra-urban migration process, their analytical power was limited by the focus on the behavioural dimension. The fact that individual and household behavior was taking place within an environment of constrained choice was not incorporated fully into urban analyses until theories relating to the nature of capitalism and capitalist urban development entered the field.

Another established area of urban geographical research that came under scrutiny in the 1980s was the use of multivariate statistical techniques to identify residential areas within cities. Much of the criticism of "factorial ecologies," however, stemmed from a misplaced association of the technique with positivist science. A balanced appraisal of the advantages and disadvantages of the factorial ecology approach would acknowledge that the outcome of any study depends on the number and types of variables in the analysis, the units of observation employed and the specific factor procedure used. Properly applied factorial ecology can be a powerful and versatile research tool for analyzing urban socio-spatial structure (Erwin, 1984; White, 1987).

The conceptual and analytical deficiencies exposed in urban geography in the early 1980s opened the way for a new philosophy for a reconstructed urban geography. The favored perspective stemmed from neo-Marxist theories of uneven development that had been introduced to geography in the 1970s (Smith, 1984; Bradbury, 1985). The political economy approach did not merely knock on the door of urban geography, it lifted it from its hinges, and ensured that the subject would be changed irreversibly. The fundamental premise of political economy is that uneven development is an inherent characteristic of capitalism that stems from the propensity of capital to flow to locations that offer the greatest potential return. The differential use of space by capital in search of profit creates a mosaic of inequality at all geographic levels from global to local. Consequently, at any one time certain countries, regions, cities and localities will be in the throes of decline as a result of the retreat of capital investment, while others will be experiencing the impact of capital inflows. At the metropolitan scale the outcome of this uneven development process is manifested in sociospatial variations in life quality and, in particular, in the poverty, powerlessness, and polarization of disadvantaged residents (Harvey, 1976; Smith and Feagin, 1987; Robson, 1988). Despite subsequent critiques of the political economy approach, particularly since the introduction of postmodernism to urban research, there can be no doubt that the application of the political economy perspective to urban geography represented one of the most influential developments in the history of the subject. The viewpoint of political economy introduced the community of urban geographers to new literatures in cognate social sciences and opened the urban geographer's eyes to "the bigger picture" by identifying the fundamental structural factors and processes that condition revealed patterns of urban life.

Application of a political economy perspective illuminated many dark corners, including the differential power of the various actors involved in urban land-use change (Harvey, 1985; Palm, 1985; Logan and Molotch, 1987; Squires and Velez, 1987), the suburbanization process (Checkoway, 1980; Walker, 1981), and gentrification (Rose, 1984; Smith and Williams, 1986). It also identified a host of new research questions relating to, for example, segregation (Western, 1981; Clark, 1984), social exclusion (Hughes; 1987), ghettoization (Deskins, 1981), citizenship and public participation (Johnson, 1983; Dreier, 1984; Pacione, 1988), sociospatial inequalities in service provision (Dear and Taylor, 1982), local economic development (Soja et al., 1983; Levine, 1987), and critiques of urban policy and planning (Krumholz, 1986; Lawless, 1988; Pacione, 1990a)—all of which have remained in the mainstream of research in urban geography.

### The Relevance Wave

Urban geography entered the decade of the 1980s on the crest of a "relevance wave" energized by structurally informed policy-oriented critical research of the 1970s. During the 1980s, of the many questions and issues brought to prominence by the political economy perspective, a principal research focus was on sociospatial variations in life quality within (and among) cities. The trend towards "relevant" research in urban geography also moved the subdiscipline closer to the kind of work undertaken by "applied" geographers. Many, (though not all), socially concerned urban geographers were able to identify with the ethos and approach of applied geography. The coming together of these two research strands led to an approach that I have referred to elsewhere as "applied urban geography" (Pacione, 1999). As Figure 18.1 indicates, not all socially concerned urban geographers were prepared to sign up to the approach that characterized applied urban geography. An unbridgeable gap developed between radical Marxist researchers and their liberal counterparts over the most appropriate means to achieve the common goal of a more equitable distribution of society's resources. While the latter preferred action within the existing structure of society, the former advocated an approach aimed at a fundamental restructuring of the prevailing social order. In essence, the liberal approach represented a continuation of the philosophy that underlay much of the applied geography of the inter-war and immediate post-war periods, with the significant difference that the topics for investigation had changed from land-use issues to questions of social welfare. Work on the new social issues in the 1970s included mapping of spatial variations in quality of life as a means of monitoring the distributional effects of social policies (Knox, 1975). By the 1980s simple mapping had been augmented by research employing the analytical insights of political economy. Revealed patterns were no longer seen as end products of research but as a basis for critical examination of the structural underpinnings of social inequality. The liberal approach remained an anathema to "radical" urban geographers. The essence of the Marxist critique of applied urban geography (and of applied social science more generally) is that it leads to ameliorative policies that merely patch up the present system, aid the legitimation of the state and bolster the forces of capitalism, with their inherent tendencies to create inequality. For radical geographers, participation in policy evaluation and formulation is ineffective, since it hinders the achievement of the greater goal of revolutionary social change. In terms of praxis, the outcome of this perspective is to do nothing short of a radical reconstruction of the dominant political economy.

While the analytical and emancipatory value of the Marxist critique of capitalism is acknowledged, its political agenda and in particular opposition to any action not directed at revolutionary social change has found limited favor among applied urban geographers. To ignore the opportunity to improve the quality of life of some people in the short term in the hope of achieving possibly greater benefit in the longer term is not

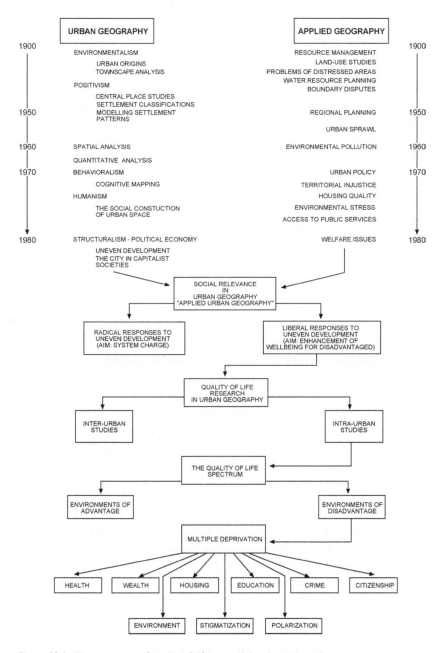

**Figure 18.1**  The provenance of quality-of-life research in urban geography.

commensurate with the ethical position implicit in the problem-oriented approach of applied urban geography. Further, in terms of political reality, the Marxist critique of capitalism and radical proposals for social redistribution had little if any chance of popular acceptance in the context of Western society in the last quarter of the 20th century. "Relevant" urban geography based on the Marxist political agenda failed to make the step from conceptual analysis to praxis, from desirability to feasibility.

Based on the conceptual foundations of the 1970s and 1980s there is now a significant research community within urban geography whose goal is to develop critical understanding of sociospatial differences in life quality, and to explore strategies to enhance the well being of disadvantaged people and places. In the second part of this paper I focus attention on this important area of "socially concerned, policy-oriented" research. I first define what I mean by "applied urban geography" and discuss the key principles underlying the approach. I then illustrate the ways in which "quality-of-life" research areas with strong conceptual roots in the urban geography of the 1980s have developed in the intervening period. Finally, I identify several research questions of relevance for urban geography in the early decades of the 21st century.

## Urban Geography and the Concept of Useful Knowledge

Applied urban geography may be defined as the application of geographic knowledge and skills to the resolution of urban social, economic and environmental problems. Central to the applied approach to urban geography is the concept of useful knowledge. The concept of useful knowledge will no doubt upset some geographers. Those who do not regard themselves as "applied geographers" may interpret the phrase as indicating a corollary in the shape of geographical research that is less useful or even useless. This would be a misinterpretation. The concept of useful knowledge expresses the fundamental ethos of applied geography. It makes explicit the belief that some kinds of research in urban geography are more useful than other kinds. This is not the same as saying that some geographical research is better than other work—all knowledge is useful—but some kinds of research and knowledge are more useful than other kinds in terms of their ability to interpret and offer solutions to problems in contemporary urban environments.

## Postmodernism and Applied Urban Geography

For some the idea of applied urban geography and the notion of useful knowledge is a chaotic concept that does not fit with the post-1980s "cultural turn" in social geography or the postmodern theorizing of recent years. The critique posed by postmodernism offers an opportunity to clarify the rationale underlying research in applied urban geography. This is achieved most clearly by comparing the applied geographical approach with the alternative postmodern perspective.

One of the major achievements of postmodern discourse has been illumination of the importance of difference in society as part of the post-structural theoretical shift from an emphasis on economically rooted structures of dominance to cultural "otherness" focused on the social construction of group identities. However, there is a danger that the reification of difference may preclude communal efforts in pursuit of goals such as social justice. A failure to address the unavoidable real-life question of "whose is the more important difference among differences" when strategic choices have to be made represents a serious threat to constructing a *practical politics* of difference. Furthermore, if all viewpoints and expressions of identity are equally valid, how do we evaluate social policy or, for that matter, right from wrong? How do we avoid the segregation, discrimination and marginalization that the postmodern appeal for recognition of difference seeks to counteract. The failure to address real issues would seem to suggest that the advent of postmodernism in radical scholarship has done little to advance the cause of social justice. Discussion of relevant issues is abstracted into consideration of how particular discourses of power are constructed and reproduced. Responsibility for bringing theory to bear on real-world circumstances is largely abdicated in favor of the intellectually sound but morally bankrupt premise that there is no such thing as reality. As Merrifield and Swyngedouw (1996, p. 11) expressed it "intriguing though this stuff may be for critical scholars, it is also intrinsically dangerous in its prospective definition of political action. De-coupling social critique from its political-economic basis is not helpful for dealing with the shifting realities of urban life at the threshold of the new millennium." In terms of real world problems postmodern thought would condemn us to inaction while we reflect on the nature of the issue. As we have already noted, a similar rebuttal may be directed at the Marxist critique of applied geography.

The views expressed in the above discussion do not represent an attempt to be prescriptive of all geographical research but are intended to indicate clearly the principles and areas of concern for applied urban geography. It is a matter of individual conscience whether geographers study topics such as the iconography of landscapes or the optimum location for health centers, but the principle underlying the kind of useful geography espoused by most applied geographers is a commitment to improving existing social, economic and environmental conditions. There can be no compromise—no academic fudge—some geographical research *is* more useful than other work; this is the focus of applied urban geography.

Of course there will continue to be divergent views on the content and value of geographical research. The concept of "useful research" poses the basic questions of useful for whom?, who decides what is useful?, and based on what criteria? All of these issues formed a central part of the "relevance debate" of the early 1970s. The related questions of values in research, the goals of different types of science, and the nature of the relationship between pure and applied research are also issues of central importance for applied urban geography that are addressed elsewhere (Pacione, 1999). In the present discussion I focus on the nature of urban geographical research relating to the key theme of quality of life.

## Urban Geography and Research into Urban Quality-of-Life

Growing concern for the future of cities and for the well being of city dwellers, stimulated by trends in world urbanization, the increasing number and size of cities, and the deterioration of many urban environments, has focused attention on the problems of living in the city. Central to this concern is the relationship between people and their everyday living environments or life spaces. Understanding the nature of the person-environment relationship is a quintessential geographical problem. In the context of the built environment this can be interpreted as a concern with the degree of congruence or dissonance between city dwellers and their urban surroundings, or the degree to which a city satisfies the physical and psychological needs and wants of its citizens (Pacione, 1990b).

Figure 18.2 presents a five-dimensional model for quality-of-life research. In this framework level of specificity refers to which domains of life quality are the subject of investigation. These can range from a whole life view of well being to individual domains (such as housing satisfaction) to subdomains (for example, the size and number of rooms in a dwelling unit). The main geographical contribution to the framework is in the introduction of a spatial dimension to augment previous two-dimensional considerations of social conditions against time. Second, just as the quality of individual life can be assessed at various domain levels, so society can be assessed at different geographic scales from the individual through the group or local scale to the city, regional, national, and international (Pacione, 1982). The third dimension refers to the type of quality-of-life indicator employed. Since any definition of life quality must include two fundamental elements—an internal psychological-physiological mechanism which produces the sense of gratification and external phenomena that engage that

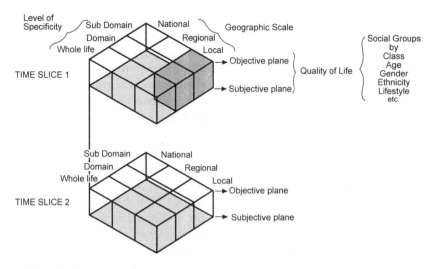

**Figure 18.2** A five-dimensional framework for quality-of-life research.

mechanism—two distinctive types of social indicators are appropriate for measuring societal and individual well being. The first type comprises objective indicators describing the environments in which people live and work. The second are subjective indicators intended to describe the ways people perceive and evaluate conditions around them. The fourth dimension is employed to measure quality of life at different points in time and to monitor the effects of policies designed to enhance quality of life for particular people and places. The fifth dimension of social groups reflects the sociospatial structure of the city and indicates the need to gauge quality of life for individual social groups differentiated along a number of dimensions including class, lifestyle, ethnicity, gender, and age within the city.

Geographical investigations of urban quality-of-life have been undertaken from two main perspectives. The first studies cities as "points in space," while the second focuses on living conditions within cities. At the inter-urban scale, a seminal study by Liu (1975) employed 132 variables related to economic, political, environmental, health, education and social conditions to rank 243 U.S. metropolitan areas. A similar procedure was employed by Boyer and Savageau (1981) to produce the popular *Places Rated Almanac.* Notwithstanding the difficulty of comparability between indices, particular interest has been drawn to temporal changes in evaluations of the best places to live. Each year since 1987 *Money Magazine* has employed a set of nine broad indicators, (related to economy, health, income, housing, education, weather, transit, leisure, and arts), to rank the 300 largest places in the United States in terms of livability (Fried, 1997).

The results of such national surveys are of more than popular interest. Being rated as a good or bad place to live can have a significant impact on a city's ability to compete in the national and global marketplace for inward investment, industry, tourism and new residents. A positive rating aids local place boosterism strategies and enhances civic pride. Conversely a poor rating can affect the local economy, offend residents and enrage civic leaders, some of whom have taken legal action to claim damages against the authors of unfavorable reports.

Other researchers working at the inter-urban scale have focused on identifying "problem cites." Rusk (1994) employed three indicators to identify 34 cities "beyond the point of no return" that typically have lost 20 percent or more of their population since 1950, have an increasingly isolated non-White minority population in their central area, and have seen a dramatic decline in the purchasing power of central city residents (Table 18.1). To these Waste (1998) added the three cities of Los Angeles, San Francisco and New York City which, along with Chicago and Philadelphia, form a group of "urban reservation" cities characterized by concentrations of extreme poverty neighborhoods. Waste (1998) also identified 14 cities with the highest rates for violent crime ("shooting gallery cities"), thereby adding Atlanta, Dallas, Houston, New Orleans, Phoenix, San Antonio and Washington DC, to produce a final list of 44 "adrenaline cities" containing 91.3 million people or 37 percent of the U.S. population and experiencing prolonged and chronic stress.

The majority of geographical research into quality of life has focused on the intra-urban scale, and on conditions at the disadvantaged end of the quality-of-life spectrum. The contemporary social significance of this research is underlined by the fact that the problems of poverty and deprivation experienced by people and places marginal to the capitalist development process have intensified over recent decades. In the U.K., during the 1980s, poverty increased faster than in any other member state of the EU so that by the end of the decade one in four of all poor families in the Community lived in Britain (Lansley and Mack, 1991). Since 1979 the gap between rich and poor has widened and by the mid-1990s over 90 percent of the nation's wealth was owned by the richest half of the population. One in three of the poorest group was unemployed, 70 percent of the income of poor households came from social security payments and nearly one in five were single-parents (Barclay, 1995). One in three of the child population and 75 percent of all children in single-parent households were living in poverty (Silburn, 1998).

The population living in poverty in the United States has also increased since the late 1970s with 13 percent below the official poverty line in 1980

**TABLE 18.1** United States Cities Past the "Point of No Return"[a]

| City | Population Loss 1950–1990 | Non-White Population in 1990 (%) | City-to-Suburb Income Ratio (%) | MSA Designation[b] | 1990 Population |
|---|---|---|---|---|---|
| Holyoke, MA | 26 | 35 | 69 | Springfield, MA | 529,519 |
| Birmingham, AL | 22 | 64 | 69 | Birmingham, AL | 907,810 |
| Flint, MI | 29 | 52 | 69 | Flint, MI | 951,270 |
| Buffalo, NY | 43 | 37 | 69 | Buffalo, NY | 1,189,288 |
| St. Louis, MO | 54 | 50 | 67 | St. Louis | 2,444,099 |
| Chicago, IL | 23 | 60 | 66 | Chicago CMSA | 8,065,633 |
| Saginaw, MI | 29 | 28 | 66 | Saginaw, MI | 399,320 |
| Baltimore, MD | 23 | 60 | 64 | Baltimore, MD | 2,382,172 |
| Dayton, OH | 31 | 36 | 64 | Dayton, OH | 951,270 |
| Philadelphia, PA | 23 | 45 | 64 | Philadelphia PMSA | 4,856,881 |
| Youngstown, OH | 44 | 35 | 64 | Youngstown, OH | 492,619 |
| Kansas City, KS | 11 | 38 | 63 | Kansas City, MO KS | 1,566,280 |
| Petersburg, VA | 7 | 74 | 63 | Richmond, VA | 865,460 |
| New Haven, CT | 21 | 47 | 62 | New Haven, CT | 530,180 |
| Milwaukee, WI | 15 | 39 | 62 | Milwaukee, WI PMSA | 1,566,280 |
| Atlantic City, NJ | 43 | 69 | 61 | Atlantic City | 319,416 |
| East Chicago, IL | 41 | 81 | 60 | Chicago CMSA | |
| Gary, IN | 25 | 85 | 59 | Gary, IN PMSA | 604,526 |
| Bessemer, AL | 17 | 59 | 58 | Birmingham, AL | |
| Chicago Heights, IL | 19 | 50 | 57 | Chicago CMSA | |
| Pontiac, MI | 17 | 52 | 55 | Detroit CMSA | 4,6655,299 |
| Elizabeth, NJ | 4 | 60 | 54 | NY-NJ CMSA | 18,027,251 |
| Cleveland, OH | 45 | 50 | 54 | Cleveland, OH | 2,759,823 |
| Perth Amboy, NJ | 4 | 65 | 53 | NY-NJ CMSA | |
| Hartford, CT | 21 | 66 | 53 | Hartford PMSA | 767,841 |

**TABLE 18.1 (continued)** United States Cities Past the "Point of No Return"[a]

| City | Population Loss 1950–1990 | Non-White Population in 1990 (%) | City-to-Suburb Income Ratio (%) | MSA Designation[b] | 1990 Population |
|---|---|---|---|---|---|
| Detroit, MI | 44 | 77 | 53 | Detroit CMSA | |
| Trenton, NJ | 31 | 59 | 50 | Trenton PMSA | 325,824 |
| Paterson, NJ | 3 | 72 | 47 | NY-NJ CMSA | |
| Benton Harbor, MI | 33 | 93 | 43 | Benton Harbor | 161,378 |
| Newark, NJ | 38 | 83 | 42 | NY-NJ CMSA | |
| Bridgeport, CT | 11 | 50 | 41 | Bridgeport PMSA | 443,722 |
| North Chicago, IL | 26 | 47 | 39 | Chicago CMSA | |
| Camden, NJ | 30 | 86 | 39 | Philadelphia PMSA | |
| East St. Louis, IL | 50 | 98 | 39 | Chicago CMSA | |
| Total | | | | | 51,489,161 |

[a] The total population in the MSAs (metropolitan statistical areas) described by Rusk as "beyond the point of no return" is 11,489,161, or 20.7% of the census figure for the 1990 U.S. population, which is 248,709,873. This table shows 34 cities within 23 MSAs.

[b] CMSA=consolidated metropolitan statistical area. PMSA=primary metropolitan statistical area.

*Source:* Waste, 1998.

and 15% in 1993 (United States Bureau of the Census, 1995). While the majority of people in poverty were White, the poverty rate was higher among minority groups, with 33% of Black and 31% of Hispanic Americans living below the poverty line compared to 12% of Whites. The incidence of poverty was also reflected in family structure and gender with 35% of female-headed families living in poverty compared to 16% of male-headed families, and 6% of married couples (Midgley and Livermore, 1998). Most of the disadvantaged live in cities, large areas of which have been devastated economically and socially by the effects of global economic restructuring, deindustrialization, and ineffective urban policies.

Urban geographers have built on the conceptual and methodological developments of the 1980s to make a major contribution to identifying the incidence, causes and consequences of disadvantage within cities. In the following discussion we consider these developments in greater detail. We identify the main theories proposed to explain the causes of deprivation. We examine the multidimensional nature of deprivation and the differential social incidence of multiple deprivation on different population groups. Using the concept of territorial social indicators we highlight spatial variations in poverty and deprivation within cities. Finally, we assess the value of an area-based approach for the analysis of urban poverty and deprivation.

## Theories of Deprivation

Identifying the causal forces underlying deprivation is of more than academic importance since the theory of deprivation espoused by policymakers determines the nature of the response. Five main models have been proposed to explain the causes of deprivation, each pointing toward a different strategy. As Table 18.2 shows, theories of deprivation range from the concept of a "culture of poverty" which regards urban deprivation as the result of the internal deficiencies of the poor to those which interpret deprivation as a product of class conflict within the prevailing social formation. The notion of a culture of poverty was first advanced in the context of the Third World and was seen as a response by the poor to their marginal position in society (Lewis, 1966). According to this thesis, realization of the improbability of their achieving advancement within a capitalist system resulted in a cycle of despair and lack of aspiration characteristic of the "culture of poverty." The related idea of transmitted deprivation focuses on the processes whereby social maladjustment is transmitted from one generation to the next, undermining the ameliorative effects of welfare programs. Particular emphasis is laid upon inadequacies in the home background and in the bringing up of children as causes of continued deprivation. The other three models take a wider perspective. The concept of institutional malfunctioning lays the blame for deprivation at the door of disjointed, and therefore ineffective, administrative structures in which the uncoordinated individual approaches of separate departments are incapable of addressing the multifaceted problem of deprivation. The theory of maldistribution of resources and opportunities regards deprivation as a consequence of the failure of certain groups to influence the political decision-making process. The final model, based on structural class conflict, stems from Marxist theory in which problems of deprivation are viewed as an inevitable outcome of the prevailing capitalist economic order.

**TABLE 18.2** Principal Models of Urban Deprivation

| Theoretical Model | Explanation | Location of the problems |
|---|---|---|
| Culture of poverty | Problems arising from the internal pathology of deviant groups | Internal dynamics of deviant behavior |
| Transmitted deprivation (cycle of deprivation) | Problems arising from individual psychological handicaps and inadequacies from one generation to the next | Relationships between individuals, families, and groups |
| Institutional malfunctioning | Problems arising from failures of planning, management, or administration | Relationship between the disadvantaged and the bureaucracy |
| Maldistribution of resources and opportunities | Problems arising from an inequitable distribution of resources | Relationship between the underprivileged and the formal political machine |
| Structural class conflict | Problems arising from the divisions necessary to maintain an economic system based on private profit | Relationship between the working class and the political and economic structure |

*Source:* Community Development Project, 1975.

While they are not mutually exclusive, each of the five theories does point to a particular policy response. Those subscribing to the culture of poverty thesis reject public expenditure on housing and other welfare items in favor of a concentration of resources on social education. Advocates of the concept of transmitted deprivation, while accepting the need for a range of antipoverty programs, emphasize the importance of the provision of facilities (such as nursery schools and health visitors) to assist child-rearing, while the solution to institutional malfunctioning has been seen in corporate management. Policies of positive discrimination are favored mostly by those who view deprivation as a result of maldistribution of resources and opportunity. Such area-based policies are dismissed by those who subscribe to the structural class conflict model of deprivation on the grounds that, being a product of the existing system responsible for deprivation, they are merely cosmetic serving to "gild the ghetto" without affecting the underlying causes of deprivation.

## The Nature of Deprivation

A fundamental issue in the debate over the nature and extent of deprivation is the distinction made between absolute and relative poverty. The absolutist or subsistence definition of poverty, derived from that formulated by Rowntree (1901, p. 186) contends that a family would be considered to be living in poverty if its "total earnings are insufficient to obtain the minimum necessaries for the maintenance of merely physical efficiency." This notion of a minimum level of subsistence and the related concept of a poverty line exerted a strong influence on the development of social welfare legislation in post-war Britain. Thus the system of National Assistance benefits introduced following the Beveridge Report (1942) was based on calculations of the amount required to satisfy the basic needs of food, clothing and housing plus a small amount for other expenses. The same principle underlies the official definition of poverty in the United States where the federal government identifies a range of poverty thresholds adjusted for the size of family, age of the householder, and number of children under 18 years of age. Poverty thresholds are updated annually and adjusted for inflation in an attempt to provide an objective measure of poverty (in 1993 the poverty threshold for a family of four was 14,763 USD).

If on the other hand we accept that needs are culturally determined rather than biologically fixed, then poverty is more accurately seen as a relative phenomenon. The broader definition of needs inherent in the concept of relative poverty includes job security, work satisfaction, fringe benefits (such as pension rights), plus various components of the "social wage" including the use of public property and services as well as satisfaction of higher-order needs such as status, power and self-esteem. An essential distinction between the two perspectives on poverty is that while the absolutist approach carries with it the implication that poverty can be eliminated in an economically advanced society, the relativist view accepts that the poor are always with us. As Figure 18.3 shows, poverty is a central element in the multidimensional problem of deprivation in which individual difficulties reinforce one another to produce a situation of compound disadvantage for those affected.

The root cause of deprivation is economic and stems from three sources (Thake and Staubach, 1993). The first arises due to the low wages earned by those employed in declining traditional industries or engaged, often on a part-time basis, in newer service-based activities. The second cause is the unemployment experienced by those marginal to the job market such as single-parents, the elderly, disabled and increasingly never-employed school leavers. Significantly, since the 1960s when poverty was largely

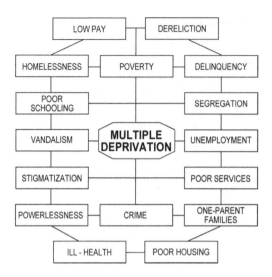

**Figure 18.3** The anatomy of multiple deprivation.

age-related, the increasing number of unemployed and growing pool of economically inactive families (such as lone parents and long-term sick) have displaced pensioners as the poorest in society. The third contributory factor is related to reductions in welfare expenditure in most Western states as a result of growing demand and an ensuing fiscal crisis.

Significantly, the complex of poverty-related problems such as crime, delinquency, poor housing, unemployment and increased mortality and morbidity has been shown to exhibit spatial concentration in cities. This patterning serves to accentuate the effects of poverty and deprivation for the residents of particular localities (Pacione, 1986). Neighborhood unemployment levels well above the national average are common in economically deprived communities. Lack of job opportunities leads to dependence on public support systems. The shift from heavy industrial employment to service-oriented activities and the consequent demand for a different kind of labor force has also served to undermine long-standing social structures built around full-time male employment and has contributed to social stress within families. Dependence upon social welfare and lack of disposable income lowers self-esteem and can lead to clinical depression. Poverty also restricts diet and accentuates poor heath. Infant mortality rates are often higher in deprived areas, and children brought up in such environments are more likely to be exposed to criminal subcultures and to suffer educational disadvantage (Pacione, 1997). The physical environment in deprived areas is typically bleak, with little landscaping, extensive areas of dereliction, and shopping and leisure facilities that reflect the poverty of the area. Residents are often the victims of stigmatization that operates as an additional obstacle to obtaining employment or credit facilities. Many deprived areas are also socially and physically isolated and those who are able to move away do so leaving behind a residual population with limited control over their quality of life. Below even this level of disadvantage some commentators have identified an underclass (Murray, 1984) inhabiting a "fourth world" (Williams, 1986, p. 21) where "all the disadvantages, inequalities and injustices of society [are] compounded among people, families and communities right at the very bottom of the social scale."

## The Geography of Deprivation

As we have seen, a strong relationship exists between poverty and deprivation, and other dimensions of urban decline. Analyses using territorial social indicators reveal that in some urban localities the intensity and sociospatial concentration of problems is severe. Figure 18.4 shows the distribution of deprivation at the District level in the U.K. This highlights the concentrations of disadvantage in inner London, the older urban industrial areas of South Wales, West Midlands, Yorkshire, the North East and North West of England, and West Central Scotland. At the metropolitan scale, as the example of London indicates (Figure 18.5), there is, in general, a gradient of deprivation that intensifies toward the "black hole" of the inner city.

Table 18.3 summarizes a number of possible explanations of inner-city decline. The relevance of each theoretical perspective for explaining inner-city decline is contingent upon the particular interaction of a variety of local and global processes; for example, the impact of market-driven suburbanization is of greater importance in the U.S. city than in either Canada or the U.K. Nevertheless, a case could be made for the salience of each of these explanations operating to some degree in most Western cities.

The structural underpinnings of the inner-city problem ensure its appearance in most of the older industrial cities of the West but the interplay of global and local forces can place a different emphasis on the components of inner-city decline in different societies. While the general problems of Britain's inner cities are mirrored in the United States, the inner-city problem in the U.S. is also conditioned by particular social factors including, for example, a greater proportion of ethnic minority residents. The correlation between ethnic minority status and residence in a U.S. inner-city problem area is demonstrated clearly by the identification of "underclass neighborhoods." These are defined as census tracts with above average rates on four variables that indicate poor integration of potential workers

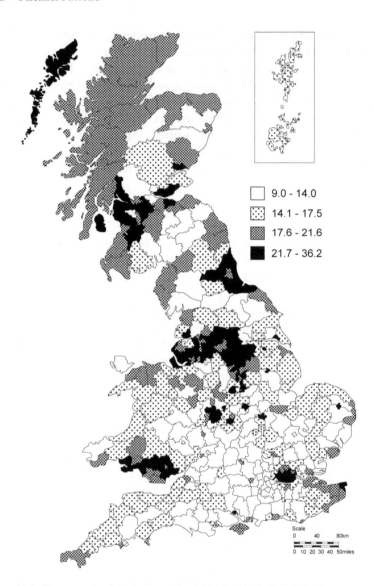

**Figure 18.4** The geography of deprivation at the district level in the United Kingdom.

into the mainstream economy (these variables measure male detachment from the labor force, percentage of households receiving public assistance, percentage of households headed by women with children, and teenage high school dropout rates). During the 1980s the incidence of such areas increased. In 1990 2.68 million people lived in such neighborhoods (an increase of 8% over 1980) with 57% Black residents, 20% Hispanic and

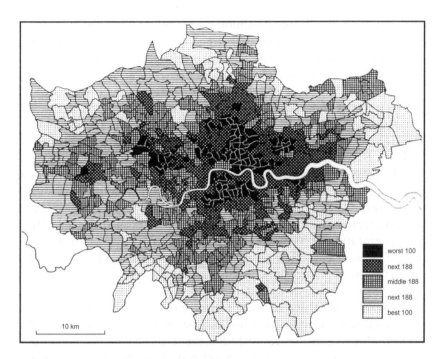

**Figure 18.5** The geography of deprivation in London.

**TABLE 18.3** A Typology of Explanations of Inner-City Decline

| Explanation | Dominant Process(es) |
|---|---|
| Natural evolution | Urban growth, ecological succession, down-filtering |
| Preference structure | Middle-class flight to the suburbs |
| Obsolescence | Aging of built environment and social infrastructure |
| Unintended effects of public policy | Suburban subsidies, including construction of freeways and aids to new single-family home ownership |
| Exploitation (1) | City manipulated by more powerful suburbs |
| Exploitation (2) | Institutional exploitation: redlining by financial institutions; tax concessions; suburbanization of factories |
| Structural change | Deindustrialization and economic decline |
| Fiscal crisis | Inequitable tax burden; high welfare, social, and infrastructure costs |
| Conflict | Racial and class polarization |

*Source:* Bourne, 1982.

20% White. If we relax the definition of disadvantage and focus solely on "extreme poverty neighborhoods" (comprising census tracts in which 40% or more of residents have money incomes below the official poverty line), many of the areas are the same as those classified as having underclass status, but in 1990 there were nearly four times as many extremely poor census tracts as underclass ones. The number of people living in extreme poverty neighborhoods rose from 5.57 million in 1980 to 10.39 million in 1990. Three-quarters lived in the central cities, with once again the majority being Black (Mincy and Weiner, 1993).

The complex of problems experienced in the inner areas of many Western cities represents a major social challenge. While many of the difficulties can be attributed to industrial decline and unemployment, others relate to personal factors such as age, infirmity or ethnicity. Still others stem from the deteriorating physical environment and affect the standard of provision of housing, education, transport, health and other social services. It is important to emphasize, however, that, although these problems are most apparent in inner-city areas, this does not mean that the underlying causes are geographical or that they are exclusive to the inner city. The "inner city" is a generic term that may usefully be seen as a metaphor for wider social problems at the heart of which is the core issue of poverty. The dispersal of the "inner city problem" to other parts of the city is particularly evident in British cities where the nature and incidence of "urban disadvantage" have been affected by public urban renewal programs and associated population movements. In Glasgow, for example, the sociospatial distribution of urban deprivation in 1991 revealed a major concentration in inner suburban areas and in the four large peripheral council housing estates. This was in marked contrast to the position in 1971 when a much higher proportion of deprived areas was located in the inner city. In general over the period the traditional inner-tenement housing areas that previously exhibited severe deprivation recorded a relative improvement in status largely as a result of a massive clearance and redevelopment program undertaken by the local authority, combined in some areas with modernization and new building aided by housing associations and private developers. This policy involved the large-scale relocation of residents in a general process of decentralization. As early as 1981 the inner-city areas contained a much reduced and ageing population living in improved accommodation. Conversely, the outer estates exhibited a younger demographic structure and, while the housing was generally well provided with the basic amenities, overcrowding was widespread. Serious social problems, such as unemployment and high proportions of single-parent households, were also prevalent. The spatial changes in the incidence of deprivation over the period were accompanied by a redistribution in terms of housing tenure with deprivation becoming concentrated increasingly in the public sector (Pacione, 1993).

A similar centrifugal spread of deprivation has been observed in North America. In U.S. cities not all of the households that participated in "the secession of the successful" to the suburbs escaped from the shadow of "inner-city" problems that have spread outward from the urban core. The suburban incidence of urban distress increased between 1980 and 1990, and by 1990 47% of the urban poor in the U.S. lived in inner-ring suburbs. Over this period the inner-ring suburbs lost 8% of their population to outer-ring suburbs as the suburbanizing wave of "White flight"continued to ripple outward.

### The Value of A Geographical Approach to Urban Quality-of-Life

For urban geographers a key question is the value of a spatial perspective on social disadvantage. Critics have dismissed area-based research on the grounds that it does not offer an explanation of the causes of revealed patterns of disadvantage. Few geographers would now deny the validity of this critique. The dialectic relationship between society and space is generally acknowledged and has informed policy-oriented analyses of multiple deprivation in which the identification of patterns is used as a basis for critique of current policy aimed at alleviating disadvantage (see, for example, Pacione, 1990a, 1995; Goodman and Webb, 1994; Smith, 1994; Gordon et al., 2000; Kearns, 2000). A second argument against spatial analysis and area-based social policy contends that as a consequence of "ecological fallacy" resources may be directed to areas in which a substantial proportion of residents do not require public assistance. It is also argued that positive discrimination and spatial targeting of resources ties people to declining neighborhoods instead of encouraging them to relocate to areas of greater opportunity.

Proponents maintain that an area-based approach is justified on several grounds. Some have argued for a degree of "area effect" since "much disadvantage is concentrated in small areas where individuals and communities suffer a multitude of disadvantages simultaneously" (Howarth et al., 1998, p. 168). As the Inner Area Studies of the 1970s concluded "there is a collective deprivation in some inner areas which affects all of the residents, even though individually the majority of people may have satisfactory homes and worthwhile jobs. It arises from a pervasive sense of decay and neglect which affects the whole area This collective deprivation amounts to more than the sum of all the individual disadvantages with which people have to contend" (Department of the Environment, 1977, p. 4). More recently, Hills (1997, p. 239) highlighted the growth of "area polarization" based on "a growth in the social and economic differences between different areas or neighborhoods," and the need for spatially targeted action to tackle spatially concentrated disadvantage. In similar vein, Massey (1996) has focused attention on "ecological mechanisms made possible by the geographic concentration of affluence and poverty" that may lead to "a deeply divided and increasingly violent social world." Policy analysts have also identified "additionality" effects from tackling problems in a defined area, including positive spillover for the area as a whole as well as for adjacent areas, and efficiency gains associated with the administration of concerted action as opposed to a "pepper pot" approach.

Areal analysis can also be used to identify spatial concentrations of particular population groups with different policy requirements. These may include the elderly and the young (with implications for provision of geriatric and pediatric services, respectively), unskilled workers (location of industry and training), and car-less households (public transport route planning). Analysis of the nature, intensity and distribution of disadvantage permits geographical and temporal comparisons within cities and regions, and facilitates monitoring of the effectiveness of remedial strategies. Further, while the long-term goal may remain a fundamental political-economic restructuring to address the roots of inequity in society, area-based policies of positive discrimination can provide more immediate benefits that enable some people to improve some aspects of their quality of life. Finally, notwithstanding criticisms of spatial targeting, area-based strategies have underlain government urban policy in the United Kingdom since the 1960s, and continue to inform initiatives aimed at alleviating the multiple disadvantages experienced by a significant number of people and places in Britain and Europe (Department of the Environment, 1995; Lawson, 1995; Parkinson, 1998; Musterd and Ostendorf, 1998). The importance of a geographical perspective on disadvantage was also highlighted in the third report of the UK Social Exclusion Unit (1998) which identified as key questions the anatomy of deprivation or what constitutes a "poor neighborhood," the number of such areas, and the location of such areas. As Rogaly (1999, p. 17) concluded, "an understanding of the spatial dimensions of poverty and social exclusion is required for the design of policies and programs aimed at tackling them."

## The Legacy of the 1980s

What did the 1980s mean for urban geography? How did the conceptual developments of the era influence the nature of the subject? What research areas were abandoned or downgraded in the course of the decade, and which new areas prospered on the basis of the intellectual debates of the 1980s? Looking back, how would we assess the contribution of the 1980s to the development of urban geography?

At the beginning of the 1980s Johnston (1982, p. 6) described a subject uncertain of itself—"at present then, urban geography is very much in a state of flux." Urban geography was "between paradigms"—conceptually only semi-clothed; having discarded the threadbare apparel of positivist science, it was trying on the new suit of political economy. By the close of the decade, however, urban geography was striding out confidently in all its finery. It had developed as a conceptually sound, analytically rigorous, academically credible, critical social science able to contribute its own particular insight on the construction and reconstruction of contemporary urban environments.

A major stimulus for the development of urban geography was the emerging concern for and research interest in quality of life (Helburn, 1982). As we have shown since the 1980s urban geographical research on quality of life has been enriched theoretically and methodologically by interaction with cognate disciplines, such as politics, sociology and environmental psychology. The definition of quality of life has been enlarged to embrace sociopsychological conditions, and new explanatory models have been developed (Grayson and Young, 1994). Use of objective social indicators has been complemented by subjective measures encapsulated in the concept of livability, and it is acknowledged that in order to obtain a proper understanding of the quality of urban life space we must consider both the city on the ground and the city in the mind. Mapping of various elements of life quality using territorial social indicators, initially viewed as an end in itself, is now more often regarded as a first stage in critical analysis of revealed patterns of urban policy or of the ethical foundations of society (Smith, 1994). These developments in quality-of-life research were part of a more general switch of focus in urban geography during the 1980s from pattern identification to seeking an understanding of underlying causal processes. In addition to identifying the sociospatial incidence of poor housing, unemployment, or ill-health urban geographers sought to understand, explain and, in some cases, prescribe alternatives to, revealed patterns.

For some observers the expansion of urban geography in the1980s was problematic. For Gregory and Walford (1989, p. 173) the growth of the subject was "so explosive that it promises to disintegrate into a myriad fragments." Carter (1995, p. 8), with perhaps a hint of regret, observed that, while it was possible in the late 1960s to give a clear shape to a nascent urban geography, today it is impossible to present a clear consensus of the content of the subject—"Urban geography is now more of an amalgam of different approaches and considerations of the city, nearer perhaps to a multidisciplinary analysis of its subject rather than a really structured systematic geographical study." Whether this is a matter of concern or of celebration is open to debate. Certainly the question of the identity and integrity of urban geography and its future as a distinctive subject are a legacy of the intellectual ferment of the 1980s that continues to generate periodic debate.

Expansion of the sphere of interest of urban geography in terms of research topics studied has meant that since the 1980s the subject has been drawn more into the mainstream of the social sciences. It is now difficult, if not impossible, to place analytical boundaries around the city. Even those, including the present writer, who maintain the value of a distinctly urban focus for their studies must integrate their inquiries into broader social processes. This can only benefit the academic health of the subject. Few urban geographers would now seek to justify their subject by claiming that explanation of urban phenomena can be restricted to the urban level. Equally, however, very few would accept claims that the concept of urban has become redundant on the grounds that, in Western societies at least, we are all urban no matter where we live. The strengths of urban geography lie in its unique spatial perspective and its integrative approach to urban analysis. The continuing vitality of the subject will be assured by employing these qualities to address the challenges of urban life in the 21st century.

## Prospective

In concluding this personal review of urban geography in the 1980s I adopt a prospective viewpoint that seeks to extrapolate some of the trends identified over the past quarter of a century in order to anticipate developments of potential significance in the subject over the next decade. Using this crystal globe I believe that the principal research topics for urban geography over this period should include the following.

First, the practice of socially relevance research established during the 1980s should continue to be developed with an aim being to increase the influence of urban geography in the spheres of politics and policy making. As we have shown, over the past 25 years the focus for urban geographers engaged in applied or socially relevant research has shifted from narrow distributional issues to concern with the political-economic factors underlying the differential distribution of society's wealth, and issues relating to the moral ideology of the state. These conceptual advances, however, have not been accompanied by a commensurate increase in the political influence of urban geographical research. While empirical research continues to record examples of "man's inhumanity to man" (Burns, 1935, p. 35) it is doubtful whether human geography is any further forward, in terms of policy relevance, than in the politically charged days of the early 1970s. A principal reason for this is that policymakers are resistant to any research that might undermine the legitimacy of the dominant ideology. The radical idealism of the socially concerned geography of the 1970s was resisted easily by the forces of political reality. Since that time social policy has remained largely impervious to geographical critique. This is not due to incompetent analysis or prognosis but reflects the failure of human geographers to place their work within a moral framework that can command the support of the population at large. As we have seen, this condition was not satisfied by the Marxist critique of capitalism and its radical prognosis for social redistribution. It should be apparent by now that any attempt to provide an effective response to the problems of sociospatial disadvantage must temper idealism with realism. In the context of advanced capitalism this requires acceptance of several premises. The first is clear cut. Market capitalism is here to stay for the foreseeable future. While the excessive inequalities of a market society is repudiated by those seeking a more just distribution of society's resources, all must acknowledge that wealth creation and wealth distribution are two sides of the same coin and that the capitalist mode of production is the most efficient engine of wealth production available. Equally, however, markets (for labor, finance, goods and services) are not created by natural or divine forces but are the product of the values, institutions, regulations and political decisions that govern them. What is required in creating an alternative moral philosophy is not rejection of the market but a shift in balance between social and economic considerations. What is needed is a social view of economics to replace the present economic view of society. This requires progress toward objectives that include socially concerned as opposed to maximum economic growth, inclusive democracy, gender and racial equality, and intergenerational equity. The volume of quality-of-life research undertaken in urban geography since the 1980s provides a solid foundation for further efforts to inject a geographical perspective into political and policymaking processes.

Second, many of the issues of interest to socially concerned geographers form an integral part of the concept of sustainable urban development which is development that meets the needs of the present without compromising the ability of future generations to meet their own needs (World Commission on Sustainable Development, 1987). This concept is founded on three principles. The first is intergenerational equity which requires that national capital assets of at least equal value to those of the present are passed on to future generations. Second, social justice requires that fair and equitable use is made of present resources in terms of meeting the basic needs of all and extending to all the opportunity to satisfy their aspirations for a better quality of life. Third, transfrontier responsibility, requires recognition and control of cross-border pollution. At the city scale the environmental costs of urban activities should not be displaced across metropolitan boundaries in order to subsidize urban growth (Haughton and Hunter, 1994). The ideal world envisaged at the Rio Earth Summit in 1992 was one in which the objectives of sustainable development would be fulfilled at all levels of spatial organization (Quarrie, 1992). Agenda 21 of the Earth Summit focused particular attention on the challenge of sustainable development at the urban scale. Concern for the sustainability of cities was expressed at two levels. The first is global and involves a range of issues concerning the long-term sustainability of the earth's environment and the implications for urban life. The world's cities cannot continue to prosper if the aggregate impact of their economic production and their inhabitants' consumption draws on global resources at unsustainable rates and deposits waste in global sinks at levels which lead to detrimental climatic change. The second is local and involves the possibility that urban life may be undermined from within because of congestion, pollution and waste generation and their accompanying social and economic consequences.

Given the diversity of cities in terms of size, population growth rates and their economic, social, political, cultural and ecological settings, it is difficult to apply the concept of sustainable development generally. In most cities there are contradictions between the goals of "sustainability" and "development." Most of the world's highly developed cities exhibit the highest per capita use of environmental capital in terms of consumption of nonrenewable resources, pressure on watersheds, forests and agricultural systems, per capita emissions of greenhouse gases and stratospheric ozone depletion gases, and excess demand on ecosystems" waste absorption capacities. By contrast, most of the world's cities making least demand on environmental capital are "underdeveloped," with high proportions of their population lacking safe and sufficient water, sanitation, adequate housing, access to health care, a secure livelihood and, often, basic civil and political rights. Under these circumstances the priorities for each city in relation to sustainability and development inevitably vary. It is unrealistic to expect poverty-stricken residents of Third World cities to attach as much importance to long-term environmental sustainability as the more comfortably placed proponents of "green politics" in advanced societies. In considering the concept of sustainable urban development, therefore, it is necessary to draw a distinction between the *"green agenda"* for long-term environmental security and the *"brown agenda"* of environmental issues associated with the immediate problems of survival and development in the cities of the Third World. As Nwaka (1996, p. 119) stated, "for us in the developing world, the 'ecological debt' to future generations is not nearly as urgent as the 'social debt' for the future if today's young people lack the standard of health, education and skills to cope with tomorrow's world." Even in the West sustainable development is not accepted universally as a key goal of urban growth, particularly if it involves constraints on personal patterns of consumption. Ideally, for richer cities with high levels of resource use, a priority should be reduction of fossil fuel use and waste generation, while maintaining a productive economy and achieving a more equitable distribution of the benefits of urban living. For poorer cities the priority is attainment of basic social, economic and political goals within a context of seeking to minimize demands on environmental capital. We do not, of course, live in an ideal world and the goals of sustainable urban development are often difficult, if not impossible, to realize. In view of the fact that the majority of the world's population lives in urban areas, the challenge of sustainable urban development would appear to be a research area in which urban geographers can and should be making a contribution.

Third, for most of the past 25 years research in urban geography has been characterized by an ethnocentric perspective focused on the cities of advanced capitalism, and in particular Europe and North America. Many of the largest and fastest growing metropolitan regions are now located in the Third World. By 2025 almost 60% of the population of less developed regions (amounting to 4 billion people) will live in towns and cities, and by 2015, 27 of the 33 world megacities, (i.e., those with more than 8 million inhabitants), will be in the Third World. These urban developments are on a scale never before witnessed and will bring new challenges for urban geography. Seeking to understand the processes of urban change in the less developed regions of the world and endeavoring to devise appropriate responses to the social, economic and environmental problems that accompany these phenomena are topics of central concern for urban geography in the 21st century.

So, finally, how do we assess the 1980s in the development of urban geography? Certainly not as a "gap decade" between the demise of positivist spatial science and rise of postmodern theory. The 1980s was a crucible of change from which an urban geography, tempered in the heat of the "relevance" and political economy debates, emerged as a conceptually sound, analytically powerful, integrative discipline capable of making a distinctive contribution to mainstream social science and with a research portfolio of direct relevance for the current and future well being of cities.

## References

Barclay, P., 1995, *Inquiry into Income and Wealth.* Volume 1. York, UK: Joseph Rowntree Foundation.

Berry, B., 1964, Cities as systems within systems of cities. *Papers of the Regional Science Association,* Vol. 13, 147–163.

Beveridge, W., 1942, *Social Insurance and Allied Services, Cmnd 6404.* London, UK: HMSO.

Bourne, L., 1982, The inner city. In C. Christian and R. Harper, editors, *Modern Metropolitan Systems.* Columbus, OH: Charles Merrill.

Bowen, M., 1979, Scientific method—after positivism. *Australian Geographical Studies,* Vol. 17, 210–216.

Boyer, R. and Savageau, D., 1981, *Places Rated Almanac.* Chicago, IL: Rand McNally.

Burns, R., 1935, Man was made to mourn. In W. Douglas, editor, *The Kilmarnock Edition of the Poetic Works of Robert Burns.* Glasgow, Scotland: Scottish Daily Express.

Checkoway, B., 1980, Large builders, federal housing programmes and post war suburbanisation. *International Journal of Urban and Regional Research,* Vol. 41, 21–45.

Clark, G., 1984, Who's to blame for racial segregation? *Urban Geography,* Vol. 5, 193–209.

Clark, W., 1982, *Modelling Housing Market Search.* London, UK: Croom Helm.

Community Development Project, 1975, *Final Report Part I: Coventry and Hillfields CDP Information and Intelligence Unit.* London, UK: HMSO.

Dear, M. and Taylor, S., 1982, *Not on Our Street: Community Attitudes Towards Mental Health Care.* London, UK: Pion.

Department of the Environment, 1977, *Inner Area Studies: Liverpool, Birmingham and Lambeth.* London, UK: HMSO.

Department of the Environment, 1995, *Our Future Homes: Opportunities, Choices and Responsibilities, Cmnd 2901.* London, UK: HMSO.

Deskins, D., 1981, Morphogenesis of a Black ghetto. *Urban Geography,* Vol. 2, 95–114.

Dreier, P., 1984, The tenants movement in the United States. *International Journal of Urban and Regional Research,* Vol. 8, 255–279.

Ericksen, R. 1986, Multinucleation in metropolitan economies. *Annals of the Association of American Geographers,* Vol. 76, 331–346.

Erwin, D., 1984, Correlates of urban residential structure. *Sociological Focus,* Vol. 17, 59–75.

Fried, C., 1997, The best places to live in America. *Money Magazine,* Vol. 20, 133–156.

Golledge, R. and Stimson, R. 1987, *Analytical Behavioural Geography.* London, UK: Croom Helm.

Goodman, A. and Webb, S., 1994, *For Richer, For Poorer: The Changing Distribution of Income in the United Kingdom, 1961–1991.* London, UK: Institute for Fiscal Studies.

Gordon, D., Adelman, L., and Ashworth, K., 2000, *Poverty and Social Exclusion in Britain.* York, UK: Joseph Rowntree Foundation.

Haggett, P., 1994, Geography. In R. Johnston, D. Gregory. and D. Smith, editors, *The Dictionary of Human Geography.* Oxford, UK: Blackwell.

Harvey, D., 1976, Class structure in a capitalist society and the theory of residential differentiation. In R. Peel, M. Chisholm, and P. Haggett, editors, *Processes in Physical and Human Geography.* London, UK: Heinemann, 354–383.

Harvey, D., 1985, *The Urbanization of Capital.* Baltimore, MD: The Johns Hopkins Press.

Haughton, G. and Hunter, C., 1994, *Sustainable Cities.* London, UK: Jessica Kingsley.

Helburn, N., 1982, Geography and the quality of life. *Annals of the Association of American Geographers,* Vol. 72, 445–456.

Held, D., 1980, *Introduction to Critical Theory: Horkheimer to Habermas.* London, UK: Hutchinson.

Hills, J., 1997, How will the scissors close? Options for UK social spending. In A. Walker and C. Walker, editors, *Britain Divided: The Growth of Social Exclusion in the 1980s and 1990s,* London, UK: Child Poverty Action Group, 231–248.

Howarth, C., Palmer, G., and Street, C., 1998, *Monitoring Poverty and Social Exclusion.* York, UK: Joseph Rowntree Foundation.

Hughes, M., 1987, Moving up and moving out: Confusing ends and means about ghetto dispersal. *Urban Studies,* Vol. 24, 503–517.

Jencks, C., 1987, *What is Postmodernism?* New York, NY: St. Martins.

Johnson, J., 1983, The role of community action in neighborhood revitalization. *Urban Geography,* Vol. 4, 16–39.

Kearns, A., Gibb, D., and Mackay, D., 2000, Area deprivation in Scotland: A new assessment. *Urban Studies,* Vol. 79, 1535–1559.

Knox, P., 1975, *Social Well-Being: A Spatial Perspective.* Oxford, UK: Oxford University Press.

Krumholz, N., 1986, City planning for greater equity. *Journal of Architecture and Planning Research,* Vol. 3, 327–337.

Lansley, S. and Mack, J., 1991, *Breadline Britain in the 1990s.* London, UK: Harper Collins.

Lawless, P., 1988, British inner urban policy. *Regional Studies,* Vol. 22, 531–542.

Lawson, R., 1995, The challenge of new poverty: Lessons from Europe and North America. In K. Funken and P. Cooper, editors, *Old and New Poverty.* London, UK: Rivers Oram, 5–28.

Levine, M. 1987, Downtown redevelopment as an urban growth strategy: A critical appraisal of the Baltimore renaissance. *Journal of Urban Affairs,* Vol. 9, 133–138.

Lewis, O., 1966, The culture of poverty. *Scientific American,* Vol. 215, 19–25.

Liu, B., 1975, *Quality of Life Indicators in the United States Metropolitan Areas, 1970.* Washington, DC: United States Government Printing Office.

Logan, J. and Molotch, H., 1987, *Urban Fortunes: The Political Economy of Place.* Berkeley, CA: University of California Press.

Marston, S., Towers, G., Cadwallader, M., and Kirby, A., 1989, The urban problematic. In G. Gaile and C. Willmott, editors, *Geography in America.* Columbus OH: Merrill.

Massey, D., 1996, The age of extremes: Concentrated affluence and poverty in the twenty-first century. *Demography,* Vol. 33, 395–412.

Merrifield, A. and Swyngedouw, E., 1996, *The Urbanisation of Injustice.* London, UK: Lawrence and Wishart.

Midgley, J. and Livermore, M., 1998, United States of America. In J. Dixon and D. Macarov, editors, *Poverty: A Persistent Global Reality,* London, UK: Routledge, 229–247.

Mincy, R. and Weiner, S., 1993, *The Underclass in the 1980s: Changing Concept, Construct, Reality.* Washington DC: Urban Institute.

Morrill, R., 1987, The structure of shopping in a metropolis. *Urban Geography,* Vol. 8, 97–128.

Murray, C., 1984, *Losing Ground.* New York, NY: Basic Books.

Musterd, S. and Ostendorf, W., 1998, Segregation, polarisation and social exclusion in metropolitan areas. In S. Musterd and W. Ostendorf, editors, *Urban Segregation and the Welfare State.* London, UK: Routledge, 1–14.

Muth, R., 1985, Models of land use, housing and rent: An evaluation. *Journal of Regional Science,* Vol. 25, 593–606.

Pacione, M., 1986, Quality of life in Glasgow: An applied geographical analysis. *Environment and Planning A,* Vol. 18, 1499–1520.

Pacione, M., 1988, Public participation in neighborhood change. *Applied Geography,* Vol. 8, 229–247.

Pacione, M., 1990a, What about people? A critical analysis of urban policy in the United Kingdom. *Geography,* Vol.75, 193–202.

Pacione, M., 1990b, Urban livability: A review. *Urban Geography,* Vol. 11, 1–30.

Pacione, M., 1993, The geography of the urban crisis: Some evidence from Glasgow. *Scottish Geographical Magazine,* Vol. 109, 87–95.

Pacione, M., 1995, The geography of deprivation in the Clydeside conurbation. *Tijdschrift voor Economische en Sociale Geografie,* Vol. 86, 407–425.

Pacione, M., 1997, The geography of educational disadvantage in Glasgow. *Applied Geography,* Vol. 17, 169–192.

Pacione, M., 1999, *Applied Geography: Principles and Practice.* London, UK: Routledge.

Pacione, M., 2001, *Urban Geography: A Global Perspective.* London, UK: Routledge.

Palm, R., 1985, Ethnic segregation of real estate agent practices in the urban housing market. *Annals of the Association of American Geographers,* Vol. 75, 58–68.

Parkinson, M., 1998, *Combating Social Exclusion: Lessons from Area-based Programmes in Europe.* Bristol, UK: Policy Press.

Peet, R., 1977, *Radical Geography: Alternative Viewpoints on Contemporary Social Issues.* London, UK: Methuen.

Quarrie, J., 1992, *Earth Summit '92: The United Nations Conference on Environment and Development.* London, UK: Regency.

Robson, B., 1988, *Those Inner Cities.* Oxford, UK: Oxford University Press

Rogaly, B., 1999, Poverty and social exclusion in Britain: where finance fits. In B. Rogaly, T. Fisher, and E. Mayo, editors. *Poverty, Social Exclusion and Microfinance in Britain.* Oxford, UK: Oxfam, 7–31.

Rose, D., 1984, Rethinking gentrification. *Environment and Planning D: Society and Space,* Vol. 2, 47–74.

Rowntree, J., 1901, *Poverty: A Study of Town Life.* London, UK: Macmillan.

Rusk, D., 1994, Bend or die: inflexible state laws and policies are dooming some of the country's central cities. *State Government News,* February, 6–10.

Silburn, R., 1998, United Kingdom. In J. Dixon and D. Macarov, editors, *Poverty: A Persistent Global Reality.* London, UK: Routledge, 204–228.

Smith, D., 1994, *Geography and Social Justice.* Oxford, UK: Blackwell.

Smith, M. and Feagin, J., 1987, *The Capitalist City.* Oxford, UK: Blackwell.

Smith, N. and Williams, P., 1986, *Gentrification of the City.* London, UK: Allen and Unwin.

Social Exclusion Unit, 1998, *Bringing Britain Together—A National Strategy for Neighborhood Renewal, Cmnd 4045.* London, UK: HMSO.

Soja, E., Morales, R., and Wolff, G., 1983, Urban restructuring: An analysis of social and spatial change in Los Angeles. Economic Geography, Vol. 59, 195–230.

Squires, G. and Velez, W., 1987, Insurance redlining and the transformation of an urban metropolis. *Urban Affairs Quarterly,* Vol. 23, 63–83.

Thake, S. and Staubach, R., 1993, *Investing in People.* York, UK: Joseph Rowntree Foundation.

United States Bureau of the Census, 1995, *Statistical Abstracts of the United States.* Washington DC: Author.

Walker, R., 1981, A theory of suburbanisation: Capitalism and the construction of urban space in the United States. In M. Dear and A. Scott, editors, *Urbanization and Urban Planning in Capitalist Society.* London, UK: Methuen, 383–429.

Waste, R., 1998, *Independent Cities.* Oxford, UK: Oxford University Press.

Western, J., 1981, *Outcast Capetown,* London, UK: Allen and Unwin.

White, M., 1987, *American Neighborhoods and Residential Differentiation.* New York, NY: Russell Sage.

Williams, S., 1986, Exclusion: The hidden face of poverty. In P. Golding, editor, *Excluding the Poor.* London, UK: Child Poverty Action Group, 21–32.

World Commission on Sustainable Development, 1987, *Our Common Future.* Oxford, UK: Oxford University Press.

# CHAPTER 19

# Coming of Age
## *Urban Geography in the 1980s*

SALLIE A. MARSTON[1] AND GERALDINE PRATT

## ABSTRACT

Urban geography in the 1980s experienced significant transformations in theory, method, and practice largely from new currents in social
theory. In this paper we describe and analyze the ways in which social
theoretical influences shaped our own work as we entered the discipline
first as graduate students and later as junior faculty. Drawn into the
social theoretical currents that were swirling both within and outside
(urban) geography, our own earliest work was an attempt to engage with
and struggle against some of these currents. In our paper we address the
theoretical, methodological and practical issues that most challenged us
as representatives of a generation of urban geographers who "came of

[1]Correspondence concerning this chapter should be addressed to Sallie Marston, Department of
Geography and Regional Development, University of Arizona, Tucson, AZ 85721; telephone: 520-
621-3903; fax: 520-621-2889; e-mail: marston@email.arizona.edu

age" in the 1980s. We specifically address our common interest in making a space for a sophisticated conceptualization of agency in a paradigm of the urban political economy that was over-determined by structural theory. We use Caroline Steedman's *Landscape for a Good Woman* as an epistemological framework for thinking through our evolving feminist work on culture and social reproduction as well as an entry point into the dramatic changes that were occurring in geographers' theorizations of capitalist urbanization in the 1980s. [Key words: social theory, spatial theory, structure, agency.]

Some of our graduate students have been hugely amused by our participation in this volume. Since hearing of it, they have been fantasizing about us performing this paper in 1980s fashions, to the background sounds of ABBA and a slide show that displays—in larger than life proportions—our 1980s permed hairdos. The imaginative pleasures of this spectacle no doubt play upon some troubling gendered stereotypes but we should also consider that clothes are objects that allow our memories to work in more specific and concrete ways. In this paper, we work with another aspect of this spectacle, this is, that our graduate students can imagine this scene in such detail precisely because so many clothing styles from the 1980s and earlier have come back into fashion.

Parallels can be drawn to intellectual fashion. Some of our graduate students are currently active and enthusiastic participants in an Althusserian reading group, a decidedly retro move. Books on the formation of American urban consumer culture and the decline of working class politics, which have lain dormant on our office shelves for decades, are being dusted off and lent to graduate students. Recent retrospectives of 1980s "classics" call attention to the opportunities that lie unrealized within them. Reconsidering Jim Duncan's 1980 paper on the superorganic in American cultural geography, for instance, Kent Mathewson argued that the failure to take up Duncan's call for a hybrid combination of humanist Marxism and American pragmatism represents "one of cultural geography's missed opportunities" that "deserves (re)visitation and elaboration" (1998, p. 571). Reflecting on the influence of Nigel Thrift's 1983 paper, published in the very first issue of *Society and Space*, John Agnew (1995) has reminded us that the paper was in large part an effort to wed Marxist insights into economic determination to a more adequate theory of subjectivity, and he regretted the voluntarist reading given to this paper through the 1980s.

A productive nostalgia for the early 1980s moves around a sense that an adequate appreciation of materiality, the economy, and of forces of determination has been lost. Certainly an antihumanist rendering of the subject, which grew in popularity in the mid-1980s and through the 1990s, placed determination at the very center of the formation of the subject. But there are now worries that the attendant focus on discourse drew attention away from the material facts of existence and structured relations of class. Another concern is that, for many, the identity politics of the 1990s replaced class politics. Moreover, the theoretical eclecticism of the 1990s, already signaled in the Duncan and Thrift papers, has led some to call for clearer theoretical definition. Renata Salecl (2000), arguing against trends in cultural theory, noted that much contemporary writing is a melange of fashionable theories. Crucial distinctions between theories are glossed over or ignored; she argued that this forges a false consensus and has the effect of dampening the capacity for critical debate.

It is these types of concerns that lead us back to the early 1980s, a time when we were both graduate students: Sallie Marston at the University of Colorado and Gerry Pratt at the University of British Columbia. If we will not be performing in exactly the way envisioned by our graduate students, we are nonetheless performing the 1980s for them in another way, by reengaging some of the central debates of the time. We do this, first, to consider how the process of interdisciplinarity at that time differed from that critiqued by Salecl; second, to remind ourselves why agency was on the theoretical agenda; and, third, to reconstruct a structure of political feeling. Dick Peet has noted that when he returned to the U.S. in 1980, after spending some time in Australia, he "found that the radical culture which had lasted at Clark University into the late 1970s was definitely gone" (1985, p. 3). If a certain type of radical culture was gone by 1980, another structure of political feeling was emerging within feminist geography something between what Carolyn Steedman characterized as "the granite-like plot around exploiter and exploited, capital and proletariat" (1986, p. 14), and the cultural relativism described by Salecl.

We turn to Carolyn Steedman's 1986 text, *Landscape for a Good Woman*, as a methodological guide for reusing our pasts in an effort to speak to the present. Asked to choose the best books for the 1980s decade, Nigel Thrift (1992, p. 721) fastened upon Steedman's (among others) as "one of the only books which manages to capture a personal struggle for understanding without unhealthy undertones of self-glorification and romanticism, and as a book which implicitly challenges quite a number of current orthodoxies." We aim to recover this tone and that challenge.

Steedman told a story about her mother's life, and especially how her mother's longing "her blind, burning envy for respectability and her corrosive desire for consumer goods" shaped her life and politics, and Steedman's own childhood. Though her mother was born into a solid, almost prototypical, working class family of weavers from rural Yorkshire, Steedman argued that the standard narratives of class and patriarchy could not render her life or account for her mother's political conservatism. Existing Marxist class narratives of the time attributed to working class people the simplest of psychologies: class consciousness and little more. There was no psychological theory to explain the complex formation of class consciousness, and Steedman's mother's political conservatism. If Marxist class analysis tended to portray working class people as passive vectors of class consciousness and little else, feminism was no more helpful. Stories of patriarchal domination did not quite capture Steedman's mother's ultimately self-defeating, but very active maneuvering of the resources at her command: these were her labor power and her capacity to produce children.

In response Steedman constructs a different sort of narrative, one in which her mother is admitted a particular and complex relationship to her historical situation. Steedman has narrated a grindingly oppressive situation but one in which systematicity is not equivalent to totality. If Steedman has directed us to the particularity of narratives, the complexity of individual lives, the openness of systems, and the agency of individuals, she did not celebrate any of this. She aimed to recuperate a structure of political feeling in which radicalism emerges from a deep sense of exclusion, and which admits people a complexity of nonheroic emotions, including the capacity "to feel desire, anger and envy—for things they did not have" (p. 23). Steedman refused to celebrate the pain of this marginalization, or claim a centrality for her mother's story. She asked for a politics "that will, watching this past say 'So what?,' and consign it to the dark" (p. 144). This is what Meaghan Morris (1998, p. xv) referred to as feminists' "enterprising attitude to the past." Rather than advancing yet another "heroic progress narrative," she argued that "feminism makes political discourse 'stammer.'" We use Steedman to signal a decisive break in the decade of the 1980s, and as our guide for situating our own stories within a larger "historical conjuncture."

## A Story of Two Academic Lives

### Coming of Age in Colorado

Of course it is impossible to go back to the 1980s and truly recover the prevailing intellectual sensibilities that constituted an urban geographic way of comprehending the world. But what *is* possible is finding a way to use a revisitation of that past to grasp the present in a more complex and comprehensive way. As a graduate student who began work at Boulder in 1980, I came to the project of urban geography with undergraduate degrees from Clark University in geography and psychology. In Worcester I was heavily influenced by the radical politics (and pedagogy) to which Peet refers. I brought that awareness to graduate school as well as a deep desire to engage in the debates about the relative importance of structure and agency that were raging through the seminar rooms of academic social science and, of course, geography. My dissertation on the development of political consciousness among Irish immigrants in 19th century Lowell, Massachusetts, was an attempt to explore how a different sort of consciousness than class informed the political identity of the Irish who were certainly caught up in the political and economic structures of early industrial urban capitalism, yet living and working on the periphery of its main streams. I think I chose this topic not because it was the burning issue of the early 1980s, but for more personal, and certainly personally confusing reasons. Having grown up in the increasingly postindustrial phase of a 19th century mill town in New England, ethnicity as social difference, was a fact of my everyday social landscape. And I wanted to understand that social difference as significant so that it resonated with my organic understandings and the stories my parents and grandparents told me about life among ethnically diverse factory workers. And when I went into the archives, like Steedman, I found the well-worn explanations that revolved around an Althusserian structural relationship between capital and labor to be instinctually unsatisfying, not because I thought they were irrelevant, but because they did not seem viscerally to account for the experiences and the meaning making that the historical subjects I was studying had produced. Moreover, I found the explanations that social historians had generated about 19th century working class life in the cities of the U.S. mostly absent of careful attention to the spaces that mercantile and industrial capital, an emerging local state, and native and newcomer groups were producing and that both constrained and enabled certain kinds of social practice and comprehensions.

My project therefore—as with so many research projects of 1980s—was about trying to complicate agency and keep it and structure in a kind of tension, a tension that would allow an explanation that was shaped not by one or the other, but by the mutually constitutive capacity of both. It was also to show how space was implicated in this productive tension. In that collective project there were at least three key figures whose work was consistently drawn upon as a way of theoretically balancing this tension and enabling agency a role in the social production of meaning and in the spaces where those meanings were made: Antonio Gramsci, Raymond Williams, and Fernand Braudel. A new appreciation for Gramsci's concept of hegemonic cultural domination (1971), based in selective systems of exclusion and inclusion, and the role of subordinated groups in reproducing and possibly resisting the hegemony of the dominant class was an important inroad in the 1980s into recognizing the power of structure without ejecting the potential of agents to comprehend and change their situation. For Gramsci, any struggle against hegemonic configurations of power and domination involves a cultural struggle and therefore requires attention to how systems of oppression and exclusion come to be constructed and represented often through the cooperation of the excluded and oppressed. Gramsci's most valuable contribution to theorizing during the 1980s was his appreciation of the operative tensions between structural reproduction and the potential for resistance.

Raymond Williams, as a long-time critic of anticultural Marxism, also became an important source of theoretical inspiration through his concept of the "structure of feeling" (the lived experience of a group over and above the structural organization of society) as well as his appreciation for language as a significant source of cultural meaning and value and the site for the construction of consequential, potentially transformative, social and historical processes (1973). And finally Fernand Braudel, whose concepts of the long and short duree and the conjunctures of human history enabled a new way of thinking about space-time that offset large-scale structural forces with the mundane but structurally constitutive (and potentially transmutable) practices of everyday life (1979, 1992a, 1992b). The long duree, which concerns the structure of society including slow, long-term, deep-level change, interacts with conjuncture, which represents shorter-term, cyclical developments in the social, political and economic realms, that have the potential to alter structure. Both of these are experienced and comprehended through the short duree which includes events that are understood as rapid occurrences brought about by the perceptions and practices of agents.

The influence of these theorists on the work of urban geographers of the 1980s was to open up a space for considering agency in new ways, especially ways that called for the evaluation of new and different empirical evidence. For instance, in my work on the Irish in Lowell I found exciting material in the popular cultural performance of the annual St. Patrick's Day parade as well as in the extensive toasting and pronouncements that characterized the Irish organization dinners that capped the day's festivities. These rituals provided insight into language and performance as sites of meaning making and value that butted up against the hegemonic practices of a dominant Yankee population and represented a structure of political feeling that derived from a deep sense of exclusion and a desire for permission to participate in the dominant society through a complex hybrid identity.

Beyond the very central theorists I have already cited, there were certainly others who were important in varying ways. Interestingly, most of the contemporary theorists whose work was especially influential in urban geography in the 1980s were continental and British sociologists/philosophers including Philip Abrams (1982), Roy Bhaskar (1975), Pierre Bordieu (1977), Anthony Giddens (1977, 1979, 1981), Michel Foucault (1980) and Henri Lefebvre (2003). The latter two, along with geographers David Harvey (1983), and Neil Smith (1984), Doreen Massey (1984), and Ed Soja (1980), were especially important for their increasingly sophisticated attempts to develop and enlarge upon a spatial theory of social life. This list of the theoretical influences on urban geographers interested in social theory is useful because it signals the 1980s as a decade of a great interdisciplinary ferment as we moved beyond the confining straits of economics and out into the wider ocean of sociology, anthropology, history, and philosophy and for a few others even further ashore into political science, psychology, literature and language, and other humanities-based disciplines.

*Growing up at UBC*

My first ever publication appeared in the premiere journal in social psychology and it was a product of the first year of my MA in the Environmental Psychology program at the University of British Columbia. In this study, we approached people at a variety of sites—from a nude beach to a rally protesting violence against women—and asked them to rate the place in terms of 105 affective adjectives, ranging from 1 (= extremely inaccurate) to 8 (= extremely accurate). We factor-analyzed these data to produce Figure 19.1 (Russell and Pratt, 1980). By this time my feelings about environmental psychology as a whole lay in the bottom left quadrant, somewhere between depressing, insignificant, and boring. The quest to reduce both a map of emotions and the meaning and eventfulness of places like a nude beach or a political rally to two-dimensional cognitive space seemed entirely misplaced, but it also seemed extremely difficult to formulate the richness of emotional life, events and places within experimental psychology.

My line of flight was the Geography Department. It is a measure of geographers' interest in the late 1970s in developing a richer psychological theory that I was admitted into the Masters program with no training in geography. In many ways the seminars offered very little insight into

**Figure 19.1** Factor analytic results showing two factors based on respondents' rating of places.

the discipline of geography; we read social theory, especially that which theorized subject formation within capitalist societies: existentialism, phenomenology, symbolic interactionism, structuration theory, critical theory and humanist Marxism. My Ph.D. dissertation bears the traces of this education. Wilbur Zelinsky certainly felt that he detected them: reacting to an article that I wrote from my dissertation that was published in the *Annals of the Association of American Geographers,* Zelinsky argued, "[e]ven applying the most generous definition of our discipline, this is not the stuff of geography... Lest this Commentary be construed as a plea for parochialism, I hasten to note that, if there has been a single consistent theme throughout my professional career, it has been the passion for extending the bounds of geographic discourse" (1987, p. 651). This type of disciplinary gate-keeping came as a complete surprise to me because it had been so absent from my experience in the discipline.

It was a surprise for another reason: my dissertation was precisely an effort to "grow up" and enter into the mainstream of urban geography. My Masters' thesis had been a study of bourgeois women's expressions of identity through their home decoration. I sat in their living rooms, did unstructured interviews, consumed large amounts of very nice coffee and heard the most extraordinary things, such as: "I don't care for French Provincial furniture—it's too frilly. People who decorate this way would not become good friends of mine. I have found this in the past" (Pratt, 1981). I perceived my dissertation to be more "serious" because it entered into debates about the political and economic implications of homeownership and, most especially, because it engaged with class.

There was a reason to be interested in the housing market in Canada in the early 1980s: there had been phenomenal house price inflation, followed by equally extraordinary mortgage rates, well over 20%. A housing crisis was proclaimed with great frequency in daily newspapers, sometimes accompanied by dramatic pictures of middle-class homeowners about to lose their house and capital investment. Housing also became a focus of theoretical debate between marxist and weberian social theorists, and it turned around the material significance of consumption-based identifications and interests. Within a Marxist analysis of class formation, homeownership was seen to be primarily of ideological significance because it drives a political and status-based wedge between working class homeowners and renters. For Weberians, capital gains through homeownership significantly altered homeowners' class positioning (Pratt, 1982). This was one corner of a set of debates about the material and political significance of consumption-based identifications which eventually led to Laclau and Mouffe's (1985) critique of Marxist class essentialism.

If there was something satisfyingly "big boyish" about jumping into debates about class, it was also a debate that focused attention on social reproduction and offered a key site to breathe agency into class analyses. Within Canadian geography, there was good intellectual company: it was an area in which a number of impressive young feminists were working (e.g., Mackenzie and Rose, 1983; Rose, 1988). In a richly textured study of Cornwall miners in the late 19th century, Damaris Rose (1988), for example, argued that homeownership was so actively sought after because it was a badge of respectability that was also meaningful to employers, and it also afforded access to land for domestic vegetable gardens, which miners could use to supplement wages and buffer the vagaries of waged labor. In other words, homeowners were more than "dupes," and there were intricate linkages between relations of production and reproduction.

Gender was beginning to come into focus for me in the early 1980s; for example, I took note in my dissertation that a good number of those I interviewed complained that owning a home in Vancouver by the 1980s required two full-time wages, and that they worried about the effects of this on their family life. But gender did not become an explicit research focus until I began teaching at Clark University, and Susan Hanson drew me into a study on gender, travel distances, and waged employment. In the summer of 1985, I was poised between projects: Dick Peet had asked me to write a chapter on reproduction, class and the spatial structure of the city for the *New Models in Geography* book that he was editing with Nigel Thrift (1989), and Susan Hanson and I were drafting our first proposal for the N.S.F. My partner likes to remember this summer as the only time in my life that I smoked cigarettes; he would regularly find me in a haze of cigarette smoke, compulsively puffing away. What I remember is a sense of disbelief and indignation about how utterly absent women were from the majority of analyses of reproduction and class in the city—a fact that I had not noticed—and certainly not *felt*—before the two projects collided. Women were automatically assumed to take on the class position of their husbands, no matter what type of work they did. And what I realized was, not only were their stories not being told, but that the city becomes a much livelier, more interesting place once women are brought into view. Many women, for example, move between class positions as they move through the city and their daily lives (Pratt and Hanson, 1988). This interest in identity, mobility, boundaries and space sustained many of us through the next decade.

### Back to the Future

In the 1985 smash movie hit, *Back to the Future,* teenager Marty McFly (played by Michael J. Fox), with the help of a zany scientist played by Christopher Lloyd, is able to travel back to the 1950s (by way of a flight-worthy DeLorean DMC-12) to meet his parents when they were in their youth and change the conditions under which they fell in love. Juxtaposed in the film are the romanticized 1950s and the more starkly rendered 1980s such that an amazon.com reviewer characterized the film as "a *Twilight Zone* episode written by Preston Sturges."

Just as Marty McFly finds his roots in his parents' romance, we have traced our roots over a comparable space of time to the 1980s; in another twist on the Twilight Zone, there is in fact much that is eerily familiar between the 1980s and now. From 1981–1989, Ronald Reagan governed as the 40th president of the U.S. Margaret Thatcher was Prime Minister of the U.K. from 1979–1990. These two individuals, as well as like-minded politicians and corporations throughout the West, were instrumental in making the world safe for globalization through some massive political and economic restructuring that left the state, particularly in its social welfare functions from the national to the local, but a shadow of its former self (Wolch, 1990). The impact on cities was dramatic, as the added effect of dwindling tax bases caused by demographic shifts to the suburbs brought big cities in the U.S., like New York and Cleveland, to bankruptcy and the Washington (state) Public Power Supply system to default on its bonds. A few key words from the 1980s sketch an oddly familiar international geopolitical context: the Iran/Contra Affair; the "evil empire," the Sandinistas, U.S. bombing raids over Libya, the *Achille Lauro*, the Falklands/Malvinas war, Strategic Defense Initiative, and the Iran-Iraq war. So too, key words such as: trickle down theory, the deserving poor, AIDS, gentrification, homelessness, cocaine, and the collapse of the savings and loan industry resonate with contemporary urban conditions.

Marty McFly goes back in time, not simply to discover his roots, but to change the facts of his existence. Not unlike the central character in *Back to the Future*, our return to the 1980s gives us an occasion to recall the decade's theoretical and political formations through our own intellectual biographies but also to tinker with what in retrospect requires changing, and to recover what we have lost or abandoned. In one way, the 1980s in urban geography signaled a decisive and widespread break from the neoclassical economic models that had been dominant for the previous two decades. And through the opening up the subfield to other ways of knowing and conceptualizing problems and questions, the 1980s stand as an important decade of intellectual growth and change. It would seem that against a background of such significant change, the old models were less satisfying in their rather mechanistic explanations for the world around us. And yet, the 1990s held its own set of problematic temptations as we moved through the seductive currents of poststructuralism and postmodernism.

Upon reflection, the strengths of the early 1980s were also weaknesses such that debate was sharply delineated along theoretical lines, and so much theory was filtered through critiques of economic reductionism and structure/agency. We can retell our biographies to accentuate this point.

> Gerry: Let's return to the smoke-filled room. Rather than scripting the chapter on class and social reproduction as a last prefeminist writing performance, it can be reimagined as a wonderfully freed moment when I could read Marxist geography in an entirely new way. At the University of British Columbia, we certainly read Marxist theory, but often with an eye to critiquing the economic determinism and reductionism that lay within it, at least as it was taken up and deployed by geographers; ironically this is a reductive way of reading Marx. Asked to write the chapter in *New Models in Geography* on social reproduction by Dick Peet at Clark

University, where Marxism was the norm against which structuration theory was critiqued, I was compelled to reengage with key Marxist urban theory (e.g., Walker, 1978; Harvey, 1982) to convey the importance of this theory within the terms set by these texts.

Sallie: As the decade wore on, it became clear to me that ethnicity as it was employed by me and other urban geographers was a "jumbo" category that inhibited the possibilities of appreciating within group differences. Yet, the truth is that I was aware of these differences right from the start of my dissertation, especially with respect to the complicating influence of religion, generation, and country of origin. At the same time, I was unable to figure out how to include Irish immigrant women in my account because their imprint never materialized at the parades, in newspaper accounts or at the evening's festivities. My rationalization for the blunt theoretical application of ethnicity as a category is that I was far too interested in arguing *against* class and *for* exploring the ways that ethnicity was enacted through and in space to see that I was perpetuating my own form of conceptual blindness. It would take me until the early 1990s to appreciate the contributions of cultural studies and its concept of difference as well as the parallel developments in feminism and the emergence of queer theory and incorporate these theoretical frames into a more complex and nuanced understanding of ethnicity.

That so much emphasis was placed on agency in the 1980s and through the 1990s is ironic, given the political and economic times. Reflecting upon Thatcherite Britain toward the end of the 1980s, Meaghan Morris (1998, p. xix) wrote of an increasingly "cramped space" in which to live and write, such that there was a loss of political optimism: "no imagining another life, no inventing different worlds, no revolutionary machine-to-come." She noted that it was the despair that came in the wake of Thatcher's famous "There is No Alternative" slogan that places great value in remembering a political structure of feeling from the early 1980s, particularly a feminist politics of survival, that of generating *just enough* conceptual space to figure out "what needed to be done in the next new space" (p. xx). This model of generating just enough conceptual space to construct a vantage point from which to proceed seems an apt description of our experiences of the 1980s: critiques of economic determinism and reductionism were flawed but created just enough space to more adequately theorize social difference.

Carolyn Steedman turned to biography in 1986 to recuperate constrained agency and psychological complexity, this is what needed to be done at that time. We turn to biography now with the desire to generate other types of conceptual space, by remembering how fully engaged urban geographers were with political economy, and a politics of survival and a determination for social change.

## References

Abrams, P., 1982, *Historical Sociology*. Ithaca, NY: Cornell University Press.
Agnew, J., 1995, Commentary 1 on Thrift, N., 1983: On the determination of social action in space and time. Environment and Planning D: Society and Space, Vol. 1, 23-57. *Progress in Human Geography*, Vol. 19, 525–526.
Bhaskar, R., 1975, *A Realist Theory of Science*. Leeds, UK: Leeds Books.
Bordieu, P., 1977, *Outline of a Theory of Practice*. Cambridge, UK: Cambridge University Press.
Braudel, F., 1979, *The Structures of Everyday Life; Civilization and Capitalism: The Limits of the Possible*, Volume 1. New York, NY: Harper & Row.
Braudel, F., 1992a, *The Wheels of Commerce*, Volume 2. New York, NY: Harper & Row.
Braudel, F., 1992b, *The Perspective of the World*, Volume 3. New York, NY: Harper & Row.
Duncan, J., 1980, The superorganic in American cultural geography. *Annals of the Association of American Geographers*, Vol. 70, 181–198.
Foucault, M., 1980, *Power/Knowledge: Selected Interviews and Other Writings 1972–1977*. New York, NY: Pantheon.
Giddens, A., 1977, *Studies in Social and Political Theory*. New York, NY: Basic Books.
Giddens, A., 1979, *Central Problems in Social Theory: Action, Structure, and Contradiction in Social Analysis*. Berkeley, CA, and Los Angeles, CA: University of California Press.
Giddens, A., 1981, *A Contemporary Critique of Historical Materialism*. Berkeley, CA, Los Angeles, CA: University of California Press.
Gramsci, A., 1971, *Selections from the Prison Notebooks*. London, UK: Lawrence and Wishart.
Harvey, D., 1983, *The Limits to Capital*. Oxford, UK: Basil Blackwell.
Laclau, E. and Mouffe, C., 1985, *Hegemony and Socialist Strategy*. London, UK: Verso.
Lefebvre, H., 2003, *La Revolution Urbaine* [The Urban Revolution]. Minneapolis, MN: University of Minnesota Press.
Mackenzie, S. and Rose, D., 1983, Industrial change, the domestic economy and home life. In J. Anderson, S. Duncan and R. Hudson, editors, *Redundant Spaces in Cities and Regions? Studies in Industrial Decline and Social Change*. London, UK: Academic Press, 155–199.
Marston, S., Towers, G., Cadwallader, M., and Kirby, A., 1989, The Urban Problematic. In G. Gaile and C. J. Wilmott, editors, *Geography in America*. Columbus, OH: Merrill, pp. 651–672.
Marston, S., 1988, Neighborhood and politics: Irish ethnicity in nineteenth century Lowell, Massachusetts. *Annals of the Association of American Geographers*, Vol. 78, 414–432.
Marston, S., 1989a, Public rituals and community power: St. Patrick's Day parades in Lowell, Massachusetts, *Political Geography Quarterly*, Vol. 8, 255–269.
Marston, S., 1989b, Adopted Citizens: Discourse and the Production of Meaning among Nineteenth Century American Urban Immigrants. *Transactions of the Institute of British Geographers*, Vol. 14, 435–445.
Massey, D., 1984, *Spatial Divisions of Labor: Social Structures and the Geography of Production*. New York, NY: Methuen.
Mathewson, K., 1998, Commentary 2 on Duncan, J. S., 1980: The superorganic in American cultural geography. *Annals of the Association of American Geographers*, Vol. 70, 181–198. *Progress in Human Geography*, Vol. 22, 569–571.
Morris, M., 1998, *Too Soon Too Late: History in Popular Culture*. Bloomington, IN, and Indianapolis, IN: University of Indiana Press.
Peet, R., 1985, Radical geography in the United States: A personal history. *Antipode*, Vol. 17, 1–6.
Peet, R. and Thrift, N., editors, 1989, *New Models in Geography*. London, UK: Unwin and Hyman.

Pratt, G., 1981, The house as an expression of social worlds. In J. S. Duncan, editor. *Housing and Identity.* London, UK: Croom Helm, 135–180.

Pratt, G., 1982, Class analysis and urban domestic property: A critical reexamination. *International Journal of Urban and Regional Research,* Vol. 6, 481–502.

Pratt, G., 1989, Reproduction, class, and the spatial structure of the city. In R. Peet and N. Thrift, editors, *New Models in Geography.* London, UK: Unwin and Hyman, 84–108.

Pratt, G. and Hanson, S., 1988, Gender, class, and space. *Environment and Planning D: Society and Space,* Vol. 6, 15–35.

Rose, D., 1988, Homeownership, subsistence and historical change: The mining district of West Cornwall in the late nineteenth century. In N. Thrift and P. Williams, editors, *Class and Space.* London, UK: Routledge and Kegan Paul, 108–153.

Russell, J. A. and Pratt, G., 1980, A description of the affective quality attributed to environments. *Journal of Personality and Social Psychology,* Vol. 38, 311–322.

Salecl, R., 2000, Disbelief in the big other in the university and beyond. *Anglistica,* Vol. 4, 49–68.

Smith, N., 1984, *Uneven Development: Nature, Capital and the Production of Space.* Oxford, UK: Basil Blackwell.

Soja, E., 1980, The socio-spatial dialectic, *Annals of the Association of American Geographers,* Vol. 70, 207–225.

Steedman, C., 1986, *Landscape for a Good Woman: A Story of Two Lives.* London, UK: Virago.

Thrift, N., 1983, On the determination of social action in space and time. *Environment and Planning D: Society and Space.* Vol. 1, 23–57.

Thrift, N., 1992, Books of the decade: an eclectic listing. *Environment and Planning D: Society and Space,* Vol. 10, 721.

Walker, R., 1978, The transformation of urban structure in the nineteenth century and the beginning of suburbanization. In M. Dear and A. J. Scott, editors, *Urbanization and Urban Planning in Capitalist Society.* London, UK: Methuen, 383–429.

Williams, R., 1973, *The Country and the City.* New York, NY: Oxford University Press.

Wolch, J., 1990, *The Shadow State: Government and the Voluntary Sector in Transition.* New York, NY: The Foundation Center.

Zelinsky, W., 1987, Commentary on "Housing tenure and social cleavages in urban Canada," *Annals of the Association of American Geographers,* Vol. 77, 651–653.

# CHAPTER 20

# Commentary
## *The Antiurban Angst of Urban Geography in the 1980s*[1]

ROBERT W. LAKE[2]

The gravitational pull of sweeping national and global changes, both within and outside the discipline, imposed a strong imprint on urban geography in the 1980s. Contextuality is a common theme in assessments of 1980s urban geography, as is the closely related theme of agency and structure. These themes, however, appear differently in each of the papers in this issue: viz., the exclusive focus on structure implicit in David Meyer's city systems perspective; Paul Knox's note of the "sea change"

[1]An earlier version of this commentary was presented at the Annual Meeting of the Association of American Geographers, New Orleans, LA, March 2003. I am greatly indebted to Bob Beauregard, Laura Liu, Kathe Newman, Mark Pendras, and Elvin Wyly for their incisive and constructive comments on earlier drafts. All opinions, statements, errors and omissions are entirely my own.
[2]Correspondence concerning this chapter should be addressed to Robert W. Lake, Center for Urban Policy Research, Rutgers University, New Brunswick, NJ, 08901; telephone: 732-932-3133; fax: 732-932-2363; e-mail: rlake@rci.rutgers.edu

introduced by global economic restructuring; the absence of agency implicit in Michael Pacione's idea of urban deprivation; the resurgence of agency as research object in Sallie Marston's and Gerry Pratt's response to the rigid structure of an earlier economism; and Larry Bourne's consideration of the agency of urban geographers to define a research agenda against the weight of surrounding trends and conditions. We made our history in the 1980s but not in conditions of our own choosing.

To appreciate the tendencies of 1980s urban geography, it is helpful to reconstruct the tenor of the times. The 1980s in the United States was the decade of acquisitiveness and polarization. Ronald Reagan occupied the White House, presiding over tax cuts, military spending, budget deficits, and deregulation. Long-term processes of White suburbanization, deindustrialization, and urban disinvestment reached their peak. It was the decade when Francis Fukuyama (1989) declared the triumph of the market and the end of history, Russell Jacoby (1987) documented the waning influence of public intellectuals, and Robert Bellah (1985) deplored blatant individualism and the death of the communitarian impulse. Charles Murray published *Losing Ground* (1984), a libertarian tract deriding government intervention for creating a permanent underclass, a condition that William Julius Wilson (1980) identified as responsible for the declining significance of race. The 1980s, then, was the decade when influential commentators proclaimed, in short order, the end of history, progress, race, community, civic responsibility, and public mindedness, and pointed to the complicity of both government and academia in fostering these trends. Today's undergraduates were born in that decade and many have never known another viewpoint, especially those for whom their parents' bumper sticker admonishing them to "Question Authority" had long ago faded into illegibility.

No more and no less than a mirror of its time, urban geography in the 1980s evinced an antiurban angst that mimicked the broader culture. That angst was revealed in two interrelated observations. First, urban geography in the 1980s abandoned the material city in a move expressive of the period's prevailing antiurbanism. Second, geography's abandonment of the city reflected, at least in part, the racial dynamics of the times.

The antiurbanism of 1980s urban geography became starkly apparent when viewed against the immediately preceding history. During the 1960s and early 1970s, both public attention and government programs focused on cities to an unprecedented extent. Urban renewal and the "war on poverty" were only the more visible of a multitude of federally funded interventions targeting cities. Accompanying these project-based programs came millions of dollars supporting urban research, in which urban geographers fully participated. As Brian Berry (2001, p. 560) wrote in his commentary on urban geography in the 1960s, "The 'urban problem' was in the air." Influenced in part by the Chicago School's themes of human ecology and neighborhood succession (Theodorson, 1961; Berry and Kasarda, 1977) and by the Marxist geographers' concern with social change, urban geographers filled the journals with studies of racial segregation, neighborhood transition, segregation indexes, and neighborhood tipping points.

In retrospect, the federal programs and the academic research they funded in the 1960s and early 1970s focused on predominantly White cities confronting racial transition. Urban renewal, slum clearance, integration management, and neighborhood redevelopment programs represented last-ditch efforts to stem the tide, to save the city as a place for White residence, employment, investment, and leisure. Much of the research of the time on what was called "race" chronicled the White city's confrontation with racial transition. Particularly in the northern industrial cities, the issue of "race" encompassed the city's transition, as perceived by urban Whites, from White to Black occupancy. But the "ghetto," a narrowly circumscribed district of non-White residence and the locus of the "urban problem," could not be contained. By the 1970s, the battle was over. Whites abandoned the city in droves. Newark, New Jersey, 91% White in 1930, was 31% White by 1980. The real war in Vietnam forced an armistice in the war on poverty. In 1973, President Nixon unilaterally declared that "the hour of [urban] crisis has passed" (quoted in Beauregard, 2003, p. 163). Targeted federal programs gave way to block grants and general revenue sharing dispersed throughout the suburbs. In the late 1970s, Jimmy Carter's Urban Development Action Grant program represented a brief Democratic flare-up of interest in urban redevelopment but, characteristically, sought to reclaim downtowns through hotel and convention-center development and explicitly ignored the neighborhoods where people of color now lived. In 1980, Ronald Reagan, in boots and cowboy hat, slammed shut the barn door of urban public policy.

Urban geographers, in company with most other urban-inflected social scientists in the 1980s, generally followed suit. The now multiracial city faded to the periphery as a site for capital investment, government programs—and academic inquiry. Places like Detroit, St. Louis, Camden, even Chicago, rarely surfaced as research sites. Few were motivated to produce a geography of Detroit's neighborhoods *after* deindustrialization and White flight. Bill Bunge's (1971) Fitzgerald expedition was rarely, if ever, repeated. Once the cities became marginalized places, their study too became marginalized, the subject no longer of the study of place but of "race," and now under the purview of Black or ethnic studies, outside the academic mainstream. The review of urban geography by Sallie Marston and her colleagues in *Geography in America* (1989, p. 658) observed that "since the early 1980s , urban geographic work paid far less attention to issues of race, ethnicity, and segregation than it did in the 1970s and 1960s," by which they meant, I think, that we paid less attention to places perceived to have completed the transition from White to Black occupancy.

Urban geographers remained busy in the 1980s but our attention had turned elsewhere: outward to the metropolitan scale, and global. We followed "growth" to the suburbs and ignored the city that remained. Our fascinations were with high tech, silicon landscapes, edge cities, financial capital, telecommunications networks, metropolitics, festival marketplaces, the new urbanism, and global cities. Not coincidentally, these fascinations addressed the few remaining vestiges of the White city or positioned the city at the metropolitan or global scales where White life now was lived. "Ghetto" became not an urban place but a teenage clothing style.

When urban geographers examined the city, we looked not at the cities as they now were but rather searched for signs of how they had been or might again be. We analyzed the desperate attempts of urban growth machines to recreate the city as a site of capital investment and middle-class White residence. To many, urban revitalization was synonymous with "re-Whitalization." We obsessed over gentrification, those tiny pockets of White revival, while ignoring the vast urban reaches comprising, we thought, only "seas of decay." When we did part those waters, it was to document the geography of deprivation, as Michael Pacione aptly recounts, but we did so through a highly ethno- and capitalocentric lens. Approached from a different perspective, the "geography of deprivation" might instead target the anomic suburb, where "diversity" is the cook and the gardener, Denny's and IHOP exhaust one's culinary options, and the nearest art museum is more than a subway ride away. To say this is neither to trivialize the reality of income inequality nor to romanticize the conditions of inner-city life (nor to express the suburban antipathy of an unreformed urbanite). Rather, it is to contrast the commonplace characterization of urban deprivation against the cultural and political complexities, largely unexamined by urban geographers, of Newark's New Community Corporation, Harlem's Abyssinian Baptist Church, the bodega and the check cashing window, Girls High School, the Department of Youth and Family Services, and the visitor's waiting room at the Cook County jail. A single Newark, New Jersey, neighborhood—15,000 residents in 120 square blocks—contains 110 churches, waiting to be recognized not as cultural sites or practices but as political moments and insurrectionary openings.

With the clarity of hindsight, it is apparent that many U.S. urban geographers in the 1980s followed the broader trend and abandoned the city. Perhaps the best we might say in this respect is that we were neither smarter nor more blinkered than anyone else. While the contextual influences spurring geography's antiurbanism originated in and applied most directly to the U.S. case, they likely exerted a strong shadow effect over what was discussed, published, cited, and academically rewarded in allied geographic communities. That urban geography followed the gravitational pull away from the city implies neither approval nor complicity, but rather that important research and conceptual opportunities remained out of reach. This is not to argue that urban geographers should have ignored their newly formed interests in regional and global topics or that these interests were in any way misguided. The point is simply that, in relentlessly pursuing new research frontiers, urban geographers were perhaps too quick to relinquish attention to the very city that had held our fascination for so long as the site of wealth generation and economic growth. This is also not to repeat the observation that urban geographers lost interest in "race," a point aptly made by Gilmore (2002), Pulido (2002), and others but rather that the discipline lost interest in cities that had become identified as racialized places.

Michael Pacione's unrestrained optimism (this issue) saw urban geography in the 1980s "striding out confidently in all its finery, a conceptually sound, analytically rigorous, academically credible, critical social science." I am somewhat less sanguine. An inescapably White discipline, we wrote (and write) about ourselves. In our defense, we might evoke a path-dependency in which the ethnographic sensibility that might have eased our antiurban angst was supported by neither the quantitative modeling nor the economistic Marxism that continued to dominate the discipline in the 1980s. For this, we needed to await the cultural turn of the 1990s. We can also see, in Sallie Marston's and Gerry Pratt's account (this issue), the hints of persistent countertrends, an emerging willingness to grapple with the intersecting dynamics of class, race, ethnicity and gender and, perhaps most importantly, the benefits of applying a critical self-awareness to the task of enlarging that "cramped space" in which we assert our agency as scholars.

## References

Beauregard, R., 2003, *Voices of Decline: The Postwar Fate of U.S. Cities* (2nd ed.). New York, NY: Routledge.

Bellah, R., editor, 1985, *Habits of the Heart: Individualism and Commitment in American Life.* Berkeley, CA: University of California Press.

Berry, B., 2001, The Chicago School in retrospect and prospect. *Urban Geography,* Vol. 22, 559–561.

Berry, B. and Kasarda, J., editors, 1977, *Contemporary Urban Ecology.* New York, NY: Macmillan.

Bunge, W., 1971, *Fitzgerald: Geography of a Revolution.* Cambridge, MA: Schenkman.

Fukuyama, F., 1989, The end of history? *The National Interest,* Vol. 16, 1–3.

Gilmore, R., 2002, Fatal couplings of power and difference: Notes on racism and geography. *Professional Geographer,* Vol. 54, 15–24.

Jacoby, R., 1987, *The Last Intellectuals: American Culture in the Age of Academe.* New York, NY: Basic Books.

Marston, S., Towers, G., Cadwallader, M., and Kirby, A., 1989, The urban problematic. In G. Gaile and C. Wilmott, editors, *Geography in America.* Columbus, OH: Merrill, 651–672.

Murray, C., 1984, *Losing Ground: American Social Policy 1950–1980.* New York, NY: Basic Books.

Pulido, L., 2002, Reflections on a White discipline. *The Professional Geographer,* Vol. 54, 42–49.

Theodorson, G., editor, 1961, *Studies in Human Ecology.* New York, NY: Harper and Row.

Wilson, W., 1980, *The Declining Significance of Race.* Chicago, IL: University of Chicago Press.

# V
## Urban Geography in the 1990s

# CHAPTER 21

# Introduction

## *The Power of Culture and the Culture of Power in Urban Geography in the 1990s*[1]

ROBERT W. LAKE[2]

So now comes culture but not, of course, without its caveats and discontents. Urban geographers would do well to heed Don Mitchell's (1996) admonition that "there is no such thing as culture" in the sense of an ontological thing on which geographers, perhaps despairing of deciphering the economy, can now focus their analytical attention. Mitchell suggested, instead, a focus on the "idea of culture" as an ideological strategy and a medium for the assertion and expression of power. This is a provocative argument that demands the difficult task of uncovering the sources of

[1]An earlier version of this commentary was presented at the Annual Meeting of the Association of American Geographers, New Orleans, Louisiana, March 2003.
[2]Correspondence concerning this chapter should be addressed to Robert W. Lake, Center for Urban Policy Research, Rutgers University, New Brunswick, NJ 08901; telephone: 732-932-3133; fax: 732-932-2363; e-mail: rlake@rci.rutgers.edu

power (sorry, the economy is still in the equation) and tracing its effects on what Susan Hanson (this issue) refers to as "people's everyday lives in cities." Each of the research areas that Trevor Barnes (this issue) marks as signs of the cultural turn in urban geography—research on public space, the culture industry, housing consumption, migration, and the urban economy—entails an explicit focus on power relations and their consequences, although the slide into anecdotal narrative exerts a powerful and dangerous attraction.

The themes of power and the political pervade these four papers on urban geography in the 1990s: power as research object and also as exerted in the continuing struggle over the direction of the discipline. Another admonition that geographers would do well to heed in this regard is Helga Leitner's and Eric Sheppard's call (this issue) for an "unbounding" of urban geography through recognition of new spatialities of the city but also through an expanded catholicity of conception and approach.

An important manifestation of both a respatialized and a conceptually more expansive urban geography is the emergence of the Los Angeles (LA) School persuasively championed by Michael Dear. But the LA School's potential for expanding the boundaries of urban geography may not yet have been fully realized. I have elsewhere urged the LA School's proponents to adhere more closely to the postmodernist impulse motivating its inauguration by accentuating the voices from below rather than categorizing them from above (Lake, 1999). Those voices—efflorescent, performative, ephemeral, cacophonous, expressing "people's everyday lives and livelihoods" (Hanson, this issue, p. 474)—resist efforts at characterization in a spatial analytic schematic, whether Burgess's concentric zones or otherwise. To unhook the city from the bedrock of analytic categories is the essence of the postmodernist challenge, as Beauregard (1991) noted more than ten years ago, and offers an exciting avenue for the further unbounding of urban geography by approaching the city as a continually unfolding performance in which any analytical categorization contains the seeds of its own contradiction and denial (Amin and Thrift, 2002).

Yet another opportunity for unbounding arises through a focus on the unlit corners that the spotlight on LA-style urbanism relegates to the shadows. As justification for foregrounding LA as the "prototype of our urban future," Dear (2002) approvingly cites Joel Garreau (1991) who observed that "Every American city that is growing, is growing in the fashion of Los Angeles." While this remains an empirically testable assertion, urban geographers should not be so blinded by the alluring ideology of "growth" that we forget or ignore its other. In the dialectic of uneven development, LA is not a substitute or a replacement for Chicago—the new city relegating

Chicago's concentric rings to the dustbin of history—but rather its dialectically necessary other. Chicago at the turn of the 21st century is not the "old" city but rather the "new" old city, the dialectical counterposition to LA's growth, and thus an equally necessary focus of research attention. To Ed Soja's (1989) rhetorical question about LA: "What better place can there be to illustrate and synthesize the dynamics of capitalist spatialization?," one answer is those benighted urban regions on whose misfortune LA's growth depends. While LA surely presents a rich and evocative urban laboratory to those fortunate enough to work within striking distance of South Central or the Hollywood hills, its lessons must surely be aligned alongside those offered by Camden, Gary, East St. Louis and, indeed, Chicago, all equally with LA emblematic of the "dynamics of capitalist spatialization." Followed to their logical conclusion, the LA School's ambitions to articulate a "revised urban theory" and a "profound realignment" of our understanding of cities will quickly transcend the spatial and conceptual confines of LA—"the city [that] posited a set of different rules for understanding urban growth" (Dear, 2002)—to also encompass cities on the departing end of the capital and labor flows to which LA's growth is inexorably connected.

What would a glimpse of "the lives and livelihoods of people" (to yet again quote Susan Hanson's felicitous phrase) in these shadowed regions reveal? Alongside the images of edge city and privatopia would arise a view of another city, one no longer fueled by capital, a view of social reproduction unhinged from capitalist production, and a view of how life is lived in the city that has been absorbed into and is sustained by the state. This is the geography of transfer payments; of life on workfare; of life in the "third sector" and the shadow state; and, in a bizarre reversal of the contemporary dominant trend toward privatization, of life lived under conditions of criminalization and incarceration by substantial segments of the urban population. Careful ethnographies responding to Susan Hanson's call for cross-scale analyses that, for example, "trace out the impacts of institutional change on the daily lives of [former] welfare recipients" could finally begin to yield the data necessary to compile a "revised urban theory" fully cognizant of the interplay of culture, power and the urban. This entails a methodological unbounding in which ethnography (once again) takes its place alongside traditional forms of urban analysis.

The plea for policy relevance constitutes yet another potential avenue for the unbounding of urban geographic research through an exploration of power and the political, for what is policymaking if not a reflection of power? If relevance means "useful for policymakers," however, it is a double-edged sword, a siren-song promising tantalizing riches in the form of financial support, disciplinary visibility, external validation, and a taste of power. Following Odysseus, we should tie ourselves to the mast. The policymaker whose attention we crave is the state, and the state occupies a contentious social position in the exercise of power. Contrary to widespread perception, the state is not the master but the slave, a caged beast trapped between its utter dependence on the whims and necessities of capital and a slavish desire to maintain its own legitimation and that of a rapacious social order, and doing so through whatever stratagems of rhetoric and policy compromise it can devise for the moment. To make this slave our master may bring riches in the short run but at a heavy price: the ceding of geography's agenda to the perspective of the policymaker. This will not empower the discipline but constrain it. Before seeking policy relevance, we may encounter the prior task of reforming or reformulating the state onto whose policies we hook our fortunes, in yet another opportunity to confront power relationships where we live as a discipline. It may help to differentiate between influencing policymakers and influencing the policy process, with the latter goal presenting a greater range of possibilities for intervention on behalf of those affected by policy outcomes. A less constraining alternative to policy relevance is policy critique, recognizing, however, that the same external perspective that fosters incisive policy assessment may leave us outside the cozy circle of policy intimates when the warm brandy is doled out around the fire.

The theme of power finally confronts the question of whose voice is heard, both within and through the discipline. Within the discipline, the internecine debates over ruling orthodoxy are surely both fruitless and counterproductive in a field that has become too large and too fragmented to be subsumed under a single conceptual or methodological signpost. Far more important is how the practice of urban geographers can best provide a space for the voices of our research subjects. Here I can only echo Helga Leitner's and Eric Sheppard's (this issue) eloquent plea for detailed activist research on the streets, documenting urban livelihoods, learning from urban residents, and helping urban residents improve their lives. Such an approach foregrounds the experiences of urban life, unfiltered through analytical categories and abstractions imposed from above. Similarly, this perspective aligns the technical skills, institutional resources, and political legitimacy of the urban geographer on behalf of the subjects of policy rather than the policymakers. It may be useful frequently to remind ourselves that people in cities are not the means to achieve our analytical purposes but rather the ends whose betterment our analytical activities should serve.

# References

Amin, A. and Thrift, N., 2002, *Cities: Reimagining the Urban*. Cambridge, UK: Polity.

Beauregard, R., 1991, Without a net: Modernist planning and the postmodern abyss. *Journal of Planning Education and Research*, Vol. 10, 189–194.

Dear, M., 2002, Los Angeles and the Chicago school: Invitation to a debate. *City and Community*, Vol. 1, 5–32.

Garreau, J., 1991, *Edge City: Life on the New Frontier*. New York, NY: Anchor.

Lake, R., 1999, Postmodern urbanism? *Urban Geography*, Vol. 20, 393–395.

Mitchell, D., 1996, There's no such thing as culture: Towards a reconceptualization of the idea of culture in geography. *Transactions of the Institute of British Geographers*, n.s., Vol. 20, 102–116.

Soja, E., 1989, *Postmodern Geographies: The Reassertion of Space in Critical Social Theory*. New York, NY: Verso.

## References

# CHAPTER 22

# The Weight of Tradition, The Springboard of Tradition
## *Let's Move Beyond the 1990s*[1]

SUSAN HANSON[2]

## ABSTRACT

My initial armchair speculation about what had transpired in urban geography in the 1990s was rudely undermined by a look at the published literature. In the first part of this paper I describe the ways in which the weight of tradition is evident in the urban geography literature of the 1990s; tradition dominates in the topics and themes that are the focus of research, the types of cities studied, the epistemological and

---

[1]My thanks to Stentor Danielson for comments on an earlier draft and for enabling me to "look at the data" by assembling an inventory of the published record in urban geography.
[2]Correspondence concerning this chapter should be addressed to Susan Hanson, School of Geography, Clark University, Worcester, MA 01610; telephone: 508-793-7323; fax: 508-793-8881; e-mail: shanson@clarku.edu

methodological approaches used, the gaps that are evident, and the disinclination to make clear the links between research and possible action. In
the second part of the paper I describe three strands of path divergence
that have emerged in the 1990s from these traditions, namely research on
globalization, research on gender, and a greater diversity of methodological approaches and data sources used. In the final section I suggest how
we might use tradition as a springboard to move urban geography
beyond the 1990s. The topics that demand a larger share of our attention
include those focused on the urban environment, sustainability, identity,
and megacities in non-OECD countries. I propose that we work to ensure
that our research on these and other topics is useful to certain audiences
and constituencies by highlighting geographic scale and cross-scale processes and by crossing the permeable boundary that delineates academe.
[Key words: geographic scale, feminist geography, path dependence, policy
relevance.]

The invitation to think about urban geography in the 1990s has turned
out to be an interesting challenge, one that has pushed me along three
intertwining paths of reflection. First, meeting this challenge has been a
sobering lesson in the deception of impressions, in the gap between what I
thought had transpired in urban geography in the 1990s and what the
written record shows was actually published in the field. Second, my
encounter with the literature led me to focus on what is new there and to
what extent these new directions match my initial impressions of urban
geography in the 1990s. Finally, because the urban geography of the 1990s
is in many respects still in progress, considering the written record of the
1990s has prompted me to reflect on what kind of future I would like to
see for urban geography and how that future might draw upon, yet depart
from, the recent past. I will take up each of these three strands in turn.

## Impressions Overruled: Meet the Published Record

After Brian Berry and Jim Wheeler asked me to participate in this session,
I had some time to ponder the challenge before writing an abstract last
August—but nowhere near enough time to undertake a thorough review
and assessment of the urban geography literature of an entire decade. So
the abstract I concocted was based on my perceptions of the literature, my
sense of what had transpired in urban geography in the 1990s, not on a
careful look at the published record. Now, perhaps the organizers of this
session actually wanted participants to write something based on their

perceptions and intuitions about what happened in urban geography during the 1990s because perceptions and intuitions can certainly be illuminating and instructive.

But I happen to love data. (In my predilection toward data, I am simply Exhibit A for the field itself, which has always been—and still remains—strongly committed to empirical research.) Please understand that I did not set out to conduct a detailed analysis of the urban geography published during the 1990s; I just don't like to make assertions without checking in with the evidence to gauge the validity of those assertions. So, with the invaluable help of a research assistant, Stentor Danielson, I set out to examine my impressions of urban geography in the 1990s in light of the data—what was published in the field. A careful look at the inventory that Stentor assembled[3] brought to mind Thomas Huxley's wry observation that "the great tragedy of science [is] the slaying of a beautiful hypothesis by an ugly fact." The data show an embarrassingly large gap between my impressions, based as they were on selective reading of the literature (and selective memory of that selective reading), and the nature of the work published in the field.

My initial hunch ("hypothesis" sounds too highfalutin in this case), based on those shadowy impressions of the 1990s literature, was this: I had thought I detected a "scale divergence" in urban geography during the 1990s, with scholars increasingly focusing on either macroscales (especially global cities) or microscales (the body). And I thought that, through this scale divergence, urban geography was becoming increasingly disconnected from the concerns of people living in cities and the concerns of people in a position to create change in cities. In my abstract, entitled "Feminists to the rescue? Scale and use-oriented research in urban geography," I suggested that, following the lead of feminist geographers, urban geographers could enhance the use value[4] of their research and teaching by emphasizing connections across scales. Although, as you can see, I have changed the title, my story is still partially about "feminism to the rescue," albeit with a different emphasis than I had originally intended.

A comprehensive look at the published record reveals above all that the enterprise of urban geography seems to be regulated by an enormous flywheel: enduring themes, research sites, and approaches in the field not only persist, they dominate. This perseverance and pervasiveness of long-standing, even classical, themes and approaches surprised me, perhaps

---

[3]This inventory includes articles published from 1990 to 2000 in the *Annals of the Association of American Geographers, Professional Geographer,* and *Urban Geography.* I take this not as a complete record, but as a reasonable indicator of what was going on in the U.S.-based, English-language branch of the field.
[4]To use Brian Berry's (1994) phrase.

because my first reaction to the request to think about urban geography in the 1990s had been to focus on what I thought had been different, not to reflect on any continuities. But continuity does have the upper hand, and it demolishes any notion of a "scale divergence" in the field. Although some urban research in the 1990s does deal with macroscale processes, such as those creating global cities, and microscale processes, such as those concerning identity, far more prevalent are studies that uphold abiding traditions in the field (many of which use census tract data, for example). In the remainder of this first section of the paper, which focuses on the published record, I briefly review the nature of the enduring themes, study sites, approaches, and gaps—yes, even the gaps in the literature are persistent—evident in a perusal of the urban geography literature of the 1990s, and then reflect on urban geography's reluctance to connect explicitly with policy concerns.

*Tradition Reigns: Enduring Themes*

Table 22.1 provides a rough breakdown of articles by topic.[5] A substantial proportion of articles take up topics with lineages that stretch back to the Chicago School of urban sociology in the 1920s: immigration and immigrants, residential segregation by race and class, urban spatial structure, the impact of neighborhood context on households and individuals. Another sizable group of papers emphasize flows (of people, goods, ideas, and above all money) both within settlements and between them and the ways in which longer-distance flows embed a place in a network of linkages and shape the urban system. Interest in such flows in urban geography also dates to the 1920s (e.g., Platt 1928). At the intra-urban scale, many articles continue to focus on a particular type of flow—the journey to work, which owes its debut as a star in urban geography to neoclassical urban economists Alonso (1964) and Muth (1969).

Segregation remains by far the dominant theme in the literature, accounting for more than a quarter of the papers reviewed. Included in this category are papers on housing, immigration, labor markets, and gentrification, all of which share the unifying theme of spatially based inequality. Despite growing interest in identity in the feminist and cultural studies literatures, urban geographers have paid scant attention to the potentially rich links between residential or commercial segregation within cities and identity issues. A few articles begin to suggest the poten-

---

[5] I say "rough" because any classification exercise like this is subjective and therefore tricky; I have tried to make the categories at least semitransparent by indicating the kinds of topics included in each.

**TABLE 22.1** Themes in Urban Geography Literature, 1990s: Proportion of Articles in Each Category[a]

| Category | %[b] |
|---|---|
| Segregation (includes housing, immigration, labor markets, gentrification) | 26 |
| Overviews and theoretical perspectives | 10 |
| Flows (e.g., airline flows, commuter flows, migration, business linkages) | 9 |
| Gender | 9 |
| Politics (e.g., NIMBYism, activism, gerrymandering) | 9 |
| Globalization (e.g., of labor, capital) | 8 |
| Location of economic activity (retail, manufacturing, services) | 8 |
| Urban spatial structure (e.g., rural-urban boundary, suburbanization) | 8 |
| Urban image (representations of specific cities) | 4 |
| Nature/urban environment (water, chemical hazards, topography) | 3 |
| Policy evaluations (e.g., banking deregulation, protection of agricultural land) | 3 |
| Culture (e.g., cultural practices, understandings of place) | 3 |
| Third World cities/megacities | 1 |

[a] Summary of articles in *Urban Geography, Annals of the Association of American Geographers, The Professional Geographer,* 1999–2000.
[b] Percentages total 101 because of rounding.

tial fruitfulness of this line of research. Laws (1993), for example, drew connections between ageist attitudes and the increasing spatial segregation of the elderly; Kirby and Hay (1997) considered sexuality-based spatial segregation/integration and impacts on gay men's identities; and in an *Urban Geography* progress report, Smith (1990) described the ways racism shapes access not only to services such as health care but also to knowledge production. These articles begin to trace out the links between segregation and identity, as does William Julius Wilson's (1987) central argument—that the disappearance of middle-class households from inner-city neighborhoods exacerbated the problems of those left behind.

*Tradition Reigns: Cities Studied*

Unlike geographers schooled in cultural ecology, urban geographers have always been parochial, and perhaps understandably so, in that they have long shown a proclivity to study the places where they live. As a result, the geographic distribution of cities studied reflects the geographic distribution of urban researchers: the U.S.-based English-language urban geography literature, even during the 1990s, is replete with studies of North

American, British, and Australian cities (in that order), all of which have no shortage of problems worthy of analysis, to be sure. But the paucity of English-language articles on urban areas located in other parts of the world is striking; most such studies reported in the journals under review were authored by scholars located in or originating from the cities they were analyzing.

A broadening of study sites in urban geography would enliven and enrich the field. North American, British, and Australian urban geographers can undertake research farther from home, either solo or in collaboration with international colleagues (and my sense is that a significant cohort among current graduate students is doing just this). In addition, as urban geography joins other social sciences in becoming more international-ized—i.e., by having a more international cadre of practitioners—the location of urban study sites should likewise become less parochial, less Anglo-centric, and more representative of the world's population distribu-tion. These shifts, which I believe are underway, in the types and locations of cities studied represent a welcome break from the past.

### Tradition Reigns: Ways of Knowing

Urban geography is an unabashedly data-based enterprise, and the data most often consulted for making sense of cities comes from the decennial census or other secondary data sources, such as flow data from private firms. Although ethnography has a venerable tradition in urban studies, dating to the Chicago School in the 1920s, only a handful (about 3%) of the articles scanned used an ethnographic approach. The reigning episte-mology within urban geography is decidedly positivist in the sense that investigators assume an agreed-upon reality and value analyses based on observation and experience of that reality. Moreover, researchers also seem implicitly to assume that the secondary data sources analyzed are a reason-able approximation of the portion of the world under study. The version of positivism that now guides urban geographic research is, however, a softer one than that practiced three or four decades ago. For example, most investigators no longer build theory via the hypothetic-deductive model or aim to discover universal laws. Nevertheless, questions of mean-ing or of chaotic categories taken off the shelf often remain submerged.

### Tradition Reigns: The Gaps Remain

The continuity of key topics, study sites, and research approaches in urban geography extends to what is missing from the record of the 1990s. Still practically nonexistent, and far too scarce in light of their importance, are studies linking segregation to questions of identity as already mentioned, studies of urban environmental issues, and studies of megacities and rapid urbanization, especially in the global south. All of these enduring gaps take on significance in the face of efforts to understand and enable urban sustainability. Although a few articles on the urban "natural" environment appeared in the 1990s, hardly any of these tackle problems that deal directly with urban sustainability (notable exceptions are Shukla and Parikh (1992), which focused on the relationship between city size and air quality and Emel (1990), which focused on water). Several of the papers that consider the urban biophysical environment do so within the framework of segregation rather than sustainability. The central problem in Meyer's (2000) study on topography, for example, Talarchek's (1990) on the urban forest, and Cutter and Solecki's (1996) on toxic releases is that of residential segregation. Another absence, related to the paucity of studies addressing urban sustainability, is the dearth of work on the links between the urban environment and health, a problem that McLafferty (1992) lamented. And with few exceptions urban geographers have not contributed to scholarship on the so-called megacities, most of which are sites of rapid urbanization in Asia, Africa, and South America; the three exceptions are Eyre (1990) and Shaw (1995), both of which focused on Bombay, and Stewart (1996), which focused on Cairo.

*Tradition Reigns: What about Policy?*

The lack of attention to identity, sustainability, environment, health, and megacities in the global south is disturbing because it connects to another enduring absence in the urban geography literature: explicit links to policy. Perhaps the term policy is a misnomer here. I use "policy relevant" in the broadest sense to refer to research that connects with policy concerns, i.e., with making change or suggesting possible changes at a variety of scales and with a variety of constituencies. It seems to me there are two concerns for urban geographers to contemplate about "policy relevance" in this broad sense of the term: one has to do with the topics or themes that are the focus of investigation; the other has to do with how we treat those themes and who we see as the potential audience for our research. With regard to the former, identity, sustainability, environment, health, megacities—these are the big urban policy issues of the day (at least insofar as they are the focus of attention for groups like the National Research Council and the United Nations), but they are not yet much in evidence on the research agenda for urban geography. With regard to the latter, ensuring that our research "makes a difference" entails crossing boundaries, most notably the boundary between the academy and the rest of the world, and doing so at multiple scales.[6] There are numerous ways in which academics can venture across that porous boundary that delineates academe and can connect with the many and diverse potential audiences for whom our research might have meaning and use value. I take up some of these ways in the third part of this paper.

Although I had thought that urban geography was increasingly distancing itself from policy concerns during the 1990s, the published record suggests that a disinclination to make policy connections is nothing new. Certainly the classical themes that have long been and remain on the agenda, such as immigration and segregation, are vitally important to policy, and yet we urban geographers remain reluctant to trace out the links between our scholarly work and the implications of that work for effecting change; we leave the making of those connections to someone else, and as a result they rarely get made.

In this regard, urban geographers are simply behaving like most U.S. social scientists. Kenneth Prewitt, a former president of the Social Science Research Council and a former director of the U.S. Bureau of the Census, puts into historical context this reluctance of social scientists to engage with policy (Prewitt, 2002). Like David Ward (1990), Prewitt argued that American social science emerged in response to the social upheavals resulting from urbanization, immigration, and industrialization toward the end of the 19th century (p. 5). The original agenda of social science was explicitly oriented toward social reform, with the goal of making "American liberal democracy work better" (Prewitt, 2002, p. 6); social scientists understood their role to be that of giving "expert advice." By the 1920s, however, social scientists were becoming concerned that dissonance within social science in the nature of that advice would jeopardize their credibility and therefore their influence with policy makers. In response to this concern, social scientists adopted a stance of scientific neutrality and objectivity and moved away from an explicitly reformist agenda (Prewitt, 2002, p. 7). As Prewitt (p. 7) saw it, the relevance and credibility of social science came then to depend on social scientists' projecting an air of apolitical objectivity: "The agenda was to remain politically relevant and socially significant, but it was to be reached, paradoxically, by developing an apolitical science. ... Social scientists will stand outside of advocacy, but with the hope that their new knowledge will be used by those whose business is advocacy. This inherent contradiction here shadows the social sciences across their entire history."

---

[6]This point was the subject of my remarks in the Presidential Plenary session at the Annual Meeting of the Association of American Geographers (AAG), New Orleans, March 2003.

It still shadows urban geography, although the question of policy, even in its most traditional garb, has not been entirely absent. Bob Lake (1994) has chastised the U.S. government for its lack of any coherent urban policy, and Brian Berry (1994) has taken urban geographers to task for their scholarly neglect of policy impacts on cities. A handful of studies has examined the impact of policies on urban areas; these include Lord (1992), who traced out impacts of banking deregulation, Law and Wolch (1991), who examine the impacts of urban policies aimed at the homeless, Reid and Yeates (1991), who consider the impacts of agricultural land protection policies, and the 1994 (No. 5) issue of *Urban Geography*, which examines the legacy of the 1954 Brown *v.* Board of Education decision on school desegregation. Still, what is most striking about "urban geography and policy" is the disinclination of authors of empirical studies, which comprise at least three-quarters of the articles in urban geography, to relate their findings to relevant policy concerns, however ecumenically construed. That these links could be made but are not seems a missed opportunity for urban geography.[7]

## Impressions Confirmed: What's New

Although path dependence—in topics and in study sites covered as well as in those avoided, in a reluctance to evaluate policies or to make policy recommendations—characterizes urban geography in the 1990s, a few strands of path divergence are also evident. Whereas our assessment of the literature disclosed that what I thought had happened in urban geography bore practically no resemblance to the vast majority of published work, it turns out that my "data-free" impressions are somewhat related to the "data-based" new elements I see emerging in the field, what I call "strands of path divergence." I identify three such strands, all of which, in one quirky way or another, emerge from and extend established traditions in the field. Two, namely, an increased interest in globalization and in gender, relate closely to what had sprung to mind in my initial "take" on urban geography in the 1990s (an increase in studies focused on macroscales and microscales and the prospects for cross-scale analyses). The third strand, an increased interest in using interviews, ethnography, and discourse analysis, has to do with epistemology and methodology; the increase in use of

---

[7]In the AAG session in which I presented (a draft of) this paper, Michael Dear perhaps quite rightly pointed out that researchers do not publish papers aimed at policymakers in academic journals because such journals are not the proper venue for reaching that audience. I, nevertheless, hold to my view that it would be useful to draw out the connections between research and possible action. Making these links explicit in academic journal articles would recognize that such articles reach multiple audiences, most importantly students, *and* leaders of NGOs, members of community groups, and sometimes even community and business leaders.

these approaches to research in urban geography also relates, but more loosely so, to my initial sense of scale-related changes within the field and opportunities for cross-scale studies.

## Globalization

In one sense studies of globalization are not fundamentally new; they simply extend to the global scale studies of the relationships between flows and nodes (bank deposits, newspaper circulation, corporate headquarters locations, telephone traffic), which date back at least to the 1920s in urban geography. But interest in economic and cultural globalization has heightened in recent years across the social sciences, and urban geography is no exception. Flow-oriented studies within urban geography that have pushed the scale of analysis to that of the globe have examined international flows of labor (Tyner, 2000), capital (Edgington, 1995; Gong, 1995), and products (Esparza and Krmenec, 1996). Urban geographers have focused on how to identify global cities (Godfrey and Zhou, 1999; Beaverstock et al., 2000) and how globalization is affecting U.S. cities (see special-issue articles in *Urban Geography*, Vol. 17, No. 1 on globalization and the U.S. city system). Geographers have also stressed the importance of geographic context in understanding globalization processes and impacts (Li, 1998; Olds, 1998).

Despite the de facto international theme of globalization, these studies within urban geography have retained a distinctly first-world orientation, in part because of their concern to identify and probe the powers of those cities (such as London, New York, Tokyo) that dominate the global urban system. Another limitation is that globalization studies in urban geography come in one dominant flavor—economistic; although few have considered culture or nature, the recent SARS epidemic suggests the need to make the links among culture, nature, and economy in a globalized world. Globalization studies to date also fall short in my view because they rarely trace out the impacts or implications of these international flows for smaller cities or for the everyday lives of people in cities of any size.

## Gender

It is the second strand of path divergence, an interest in gender, that has reinvigorated a concern with people's everyday lives in cities. I say "reinvigorated" because people's everyday lives and human welfare were an important focus of the Chicago School of urban sociology in the 1920s. Yet the Chicago School did not think even remotely about gender. Although interest in how race and ethnicity shape urban spatial structures and access to opportunity has been at the core of urban geography since its inception, the same cannot be said of gender or sexuality, which appeared on the agenda only in the 1980s (after a very few papers came out in the 1970s).

By the 1990s research on gender was well represented within urban geography, with about as many articles as those on location of economic activity, politics, flows (excluding global), and urban spatial structure. A large proportion (about two-thirds) of articles with an explicit gender focus dealt with women's relationship to paid employment, including the journey to work. This emphasis on employment reflects geographers' engagement with an idea that circulates widely throughout the social sciences, namely, that "[t]he fundamental gender question of the past two decades...has been: How equal are economic opportunities and outcomes for women and men in U.S. society" (Bianchi, 1995, p. 107). One striking aspect of gender-focused studies in urban geography, striking here because of its absence from the rest of the literature, is an emphasis on everyday life and on how geographic context affects people's everyday lives. So, for example, Wolch et al. (1993) described the everyday lives of homeless women in Los Angeles, Hays-Mitchell (1994) examined the everyday lives of street vendors in Peruvian cities, and Kwan (1999) considered the daily activity patterns of people in Columbus, Ohio. Often those studies that examine women's employment do so in the context of women's everyday lives, tracing out the links among paid work, unpaid work, housing, transportation, and child care (Peake, 1995; Gilbert, 1998; Mattingly, 1999).

A significant aspect of studies that deal with gender is that, almost by definition, they cross and integrate geographic scales. This is because gender ideologies, like those of economy or race, circulate at local, regional, national, and international scales while affecting the life chances of individuals. So, placing the lens on gender invites analyses that cross geographic scales. All of the studies mentioned above that examine people's everyday lives (Wolch et al., 1993; Hays-Mitchell, 1994; Kwan, 1999) cross geographic scales by connecting ideologies, and sometimes policies, to everyday life. Holloway's (1999) study of how neighborhood context affects the labor force participation of mothers is another example of cross-scale analysis. Through their work that highlights cross-scale processes, feminist geographers (among others) are helping urban geography to use tradition as a springboard, to move the field beyond the 1990s.

*Approaches: Interviews, Ethnographies, Discourse Analysis*

The third strand of path divergence is epistemological and methodological: an increase in the use of personal interviews, ethnographies, and discourse analysis. Although census bureaus and other government agencies remain the data sources of choice in urban geography, these other paths to understanding urban processes are also evident and have interesting implications for the combining of geographic scales. Interviews and ethnographies often emphasize microscales (individuals and bodies), whereas discourse analysis often stresses more macroscales (the metro area, the nation state, global institutions). Although this association between methodological approach and geographic scale certainly does not have to be a close one, it is an association that is frequently found in the literature. In-depth interviews and ethnographies, in particular, reflect the growing recognition that much knowledge is situated in experience and is simply not understandable when shorn of that experience.

Because different approaches yield different insights, the nascent broadening of approaches and kinds of data used in urban geography is welcome. Secondary data sources are invaluable for establishing trends and patterns, but cannot, for example, reveal meanings. As just one example, Peters and Zvonkovic (2003) undertook a detailed analysis of secondary data on overnight work-related travel, documenting the number of nights per year that members of different groups spent away from home. The authors were unable, however, to shed any light on what the results mean in terms of work-family harmony or conflict. By contrast, meaning is at the heart of Walton-Roberts's (1998) study of ethnic identity in Vancouver (in this case, the meaning of the Sikh turban) and of Vanderbeck and Johnson's (2000) study of the place and meaning of the mall in young people's understanding of their geographic context. Interviews, ethnographies, and discourse analysis, especially when combined with secondary data such as those from the Census, enable cross-scale analyses that cannot be conducted with conventional (secondary) data alone.

## The Springboard of Tradition:
## Urban Geographies in the 21st Century?

Many of my musings on urban geography in the 1990s have dwelt on the absences that seemed to shout from the published record, for I was truly surprised to find that so many contemporary problems that I see as key candidates for urban geographic scholarship are not (yet) established as sites of investigation in the field. So, in large part I have already revealed what sorts of urban geography I would like to see developing in the coming decades. To reiterate, I would welcome more urban geographers tackling urban environmental problems such as water, urban forests, or brownfields; urbanization and megacities in non-OECD countries; sustainability questions; the relationship between segregation and identity. I welcome a diversity of methodological approaches and encourage making explicit the links between research and potential change, that is, increasing the use value of research in urban geography.

Why ask urban geographers to take on these problems? Are not others already doing so? They are, but scholars from other disciplines are generally handicapped by a lack of any geographic imagination; they do not think spatially, they do not see how various elements of a problem are connected in place and over space, and the concept of scale eludes them. Why ask urban geographers to engage with these issues? Are not geographers who are steeped in the human-environment tradition better positioned than are urban geographers to take on problems such as sustainability? Not necessarily. These problems will require the theoretical and analytical insights of both the nature-society and the society-space traditions in geography (Hanson, 1999). In particular, they will require, inter alia, a spatial analytic viewpoint, which students of urban geography are likely to have.

I think that we could enhance the use value of research in urban geography by giving increased attention to connections across scales. In this proposal I am clearly drawing upon an established tradition in feminist geography (and this is why, despite the title change, the paper is still about "feminism to the rescue"), namely, emphasizing connections across scales as well as among the social, economic, political, and ecological realms. My reason for asking us urban geographers to focus more explicitly than we have on scale connections is that the urban is where multiple scales converge in particularly complex ways, and so understanding urban processes necessitates cross-scale analysis. Many instructors use simple classroom exercises to introduce students to this fundamental idea; by focusing on water, waste, immigration streams, daily commuting, or markets for local products (inter alia), students can readily identify the ways that multiple scales converge in the city and affect their own lives. Although studies of globalization, for example, do examine flows across scales, such studies would add greatly to their use value by tracing out the connections between international flows and people's everyday lives and livelihoods in places. Several feminists have done just this, e.g., Vicky Lawson's (1999) work on the impacts of structural adjustment policies on the lives of rural-urban migrants in Quito, Ecuador. Not every study has to make all the connections across all scales, but authors need to make the cross-scale linkages among their own work and that of others.

Why would illuminating cross-scale connections increase the use value of urban geographic research? First, as I have already mentioned, cross-scale processes are the essence of the urban. One cannot grasp the formation or functioning of the rapidly growing townships on the fringes of South African cities, for example, without thinking at multiple scales. Second, the concepts of geographic scale and cross-scale processes are poorly understood by potential users of urban geographic research—be they citizens, NGOs, or urban officials. As a result, their current comprehension of the problems they are grappling with is impoverished. Third, because an understanding of scale is at the heart of geography, we urban geographers are unusually well-positioned to help decision makers see the importance of understanding scale and cross-scale processes by showing them how gaining such an understanding will help them trace out the likely implications of their decisions.

In addition to shifting the topics of urban geographic research and increasing the visibility of the cross-scale analytics in our treatment of those topics, we urban geographers can, I believe, increase the use value of our studies by working more systematically than we have in the past to breach the boundary that surrounds academe. We do this every day in our teaching, simply by being engaged and engaging teachers and by encouraging our students to work in the community. We also do this every day as geographer-citizens by serving on planning boards, working with schools, leading environmental groups, helping NGOs like United Way think strategically about how to meet community needs, working with women's groups, and the like. We could doubtless do more, as Helga Leitner and Eric Sheppard suggest in their paper in this issue, to involve urban residents and community groups in our research, thereby working from the beginning of the research process with those who are potential audiences or constituencies for the research findings. At all of these "border crossings," there is the chance that the use value or policy relevance of our research will escape the bounds of academe.

## Conclusion

Urban geography remains the core of human geography. Although the GIS specialty group within the AAG has more members than the Urban Geography Specialty Group (UGSG), the UGSG is the second largest, with a substantial lead over the next largest group in human geography (Economic Geography).[8] Specialty group size is simply an indicator of what we all know: the urban continues to matter enormously. And as the world's population grows increasingly urban, the need to understand urban processes will only intensify. Cities are the places where processes from multiple

---

[8]In 1999, the GIS Specialty Group had 1,246 members, the UGSG had 739, and Economic Geography had 429.

scales and diverse people come together. The range of topics represented in the urban geography literature confirms the centrality of the urban to geography.

In shaping the urban geography of the next few decades, let us move beyond the 1990s, which, because of the weight of tradition and the strong path dependence in the field, means moving beyond the 1980s, the 1970s, and to some extent even the 1960s. I do not mean that we should toss heritage aside; we can use tradition as a springboard to tackle pressing contemporary urban concerns, to highlight the insights to be gained via cross-scale analyses, and thereby to demonstrate the value of our research and teaching for improving life in cities. Drawing upon our tradition of studying segregation and contextual effects, for instance, urban geographers can increase understandings of how geographic context shapes and sustains identities and how identities cross geographic scales; understanding the relation between identity and violence is just one example of the potential use value of such studies. Drawing upon our tradition of studying urbanization processes, urban geographers can contribute to understandings of the rapid urbanization now taking place in Asia, Africa, and Latin America. Drawing upon our tradition of examining the relationships between people and the urban environment, urban geographers can expand the concept of urban environment to include biophysical elements. Drawing upon our tradition of investigating the relationships between flows and nodes, urban geographers can contribute substantially to understanding sustainability. And drawing upon our tradition of wanting to make a difference, we urban geographers can suggest how others might make use of our ideas.

## References

Alonso, W., 1964, *Location and Land Use*. Cambridge MA: Harvard University Press.

Beaverstock, J. V., Smith, R. G., and Taylor, P. J., 2000, Geographies of globalization: United States law firms in world cities. *Urban Geography*, Vol. 21, 95–120.

Berry, B. J. L., 1994, Let's have more policy analysis. *Urban Geography*, Vol. 15, 315–317.

Bianchi, S., 1995, Changing economic roles of women and men. In R. Farley, editor, *State of the Union: America in the 1990s, v. 1, Economic Trends*. New York, NY: Russell Sage, 107–154.

Cutter, S. and Solecki, W. D., 1996, Setting environmental justice in space and place: Acute and chronic airborne toxic releases in the southeastern United States. *Urban Geography*, Vol. 17, 380–399.

Edgington, D. W., 1995, Locational preferences of Japanese real estate investors in North America. *Urban Geography*, Vol. 16, 373–396.

Emel, J., 1990, Resource instrumentalism, privatization, and commodification. *Urban Geography*, Vol. 11, 527–547.

Esparza, A. X. and Krmenec, A. J., 1996, The spatial markets of cities organized in a hierarchical system. *Professional Geographer*, Vol. 48, 367–378.

Eyre, L. A., 1990, The shanty towns of central Bombay. *Urban Geography*, Vol. 11, 130–152.

Gilbert, M. R., 1998, "Race," space, and power: The survival strategies of working poor women. *Annals of the Association of American Geographers*, Vol. 88, 595–621.

Godfrey, B. J. and Zhou, Y., 1999, Ranking world cities: Multinational corporations and the global urban hierarchy. *Urban Geography,* Vol. 20, 268–281.

Gong, H., 1995, Spatial patterns of foreign investment in China's cities, 1980–1989. *Urban Geography,* Vol. 16, 198–209.

Hanson, S., 1999, Isms and schisms: Healing the rift between the nature-society and the space-society traditions in human geography. *Annals of the Association of American Geographers,* Vol. 89, 133–143.

Hays-Mitchell, M., 1994, Streetvending in Peruvian cities: The spatio-temporal behavior of ambulantes. *Professional Geographer,* Vol. 46, 425–438.

Holloway, S. L., 1999, Mother and worker? The negotiation of motherhood and paid employment in two urban neighborhoods. *Urban Geography,* Vol. 20, 438–460.

Kirby, S. and Hay, I., 1997, (Hetero)sexing space: Gay men and "straight" space in Adelaide, South Australia. *Professional Geographer,* Vol. 49, 295–305.

Kwan, M.-P., 1999, Gender and individual access to urban opportunities: A study using time-space measures. *Professional Geographer,* Vol. 51, 210–227.

Lake, R. W., 1994, What urban policy? *Urban Geography,* Vol. 15, 205–206.

Law, R. and Wolch, J. R., 1991, Homelessness and urban restructuring. *Urban Geography,* Vol. 12, 105–136.

Laws, G., 1993, The land of old age: Society's changing attitudes toward urban built environments for elderly people. *Annals of the Association of American Geographers,* Vol. 83, 672–693.

Lawson, V. 1999, Tailoring is a profession, seamstressing is just work! Resisting work and reworking gender identities among artisnal garment workers in Quito. *Environment and Planning A,* Vol. 31, 209–227.

Li, W., 1998, Los Angeles' Chinese *ethnoburb:* From ethnic service center to global economy outpost. *Urban Geography,* Vol. 19, 502–517.

Lord, J. D., 1992, Geographic deregulation of the U.S. banking industry and spatial transfers of corporate control. *Urban Geography,* Vol. 13, 25–48.

Mattingly, D. J., 1999, Job search, social networks, and local labor-market dynamics: The case of paid household work in San Diego, California. *Urban Geography,* Vol. 20, 46–74.

McLafferty, S., 1992, Health and the urban environment. *Urban Geography,* Vol. 13, No. 6, 567–576.

Meyer, W. B., 2000, The other Burgess model. *Urban Geography,* Vol. 21, 261–270.

Muth, R., 1969, *Cities and Housing: The Spatial Pattern of Urban Residential Land Use.* Chicago, IL: University of Chicago Press.

Olds, K., 1998, Globalization and urban change: Tales from Vancouver via Hong Kong. *Urban Geography,* Vol. 19, 360–385.

Peake, L. J., 1995, Toward an understanding of the interconnectedness of women's lives: The "racial" reproduction of labor in low-income urban areas. *Urban Geography,* Vol. 16, 414–439.

Peters, C. and Zvonkovic, A., 2003, *Who are the frequent work travelers? Constructing profiles based on their work and social worlds.* Paper presented at the Business and Professional Women's Conference on How Workplace Changes Impact Families, Work, and Communities, February 28-March 1, 2003, Orlando, Florida.

Platt, R. S., 1928, A detail of regional geography: Ellison Bay community as an industrial organism. *Annals of the Association of American Geographers,* Vol. 18, 81–126.

Prewitt, K., 2002, The social science project: Then, now, and next. *Items and Issues,* Vol. 3, No. 1-2, 5–9.

Reid, E. P. and Yeates, M., 1991, Bill 90—An act to protect agricultural land: An assessment of its success in Laprairie County, Quebec. *Urban Geography,* Vol. 12, 295-309.

Shaw, A., 1995, Satellite town development in Asia: The case of New Bombay, India. *Urban Geography,* Vol. 16, 254–271.

Shukla, V. and Parikh, K., 1992, The environmental consequences of urban growth: Cross-national perspectives on economic development, air pollution, and city size. *Urban Geography,* Vol. 13, 422–449.

Smith, S. J., 1990, Race and racism: Health, welfare, and the quality of life. *Urban Geography,* Vol. 11, 606–616.

Stewart, D. J., 1996, Cities in the desert: The Egyptian new town program. *Annals of the Association of American Geographers,* Vol. 86, 459–480.

Talarchek, G. M., 1990, The urban forest of New Orleans: An exploratory analysis of relationships. *Urban Geography,* Vol. 11, 65–86.

Tyner, J. A., 2000, Global cities and circuits of global labor: The case of Manila, Philippines. *Professional Geographer*, Vol. 52, 61–74.

Vanderbeck, R. M. and Johnson, J. H., 2000, "That's the only place where you can hang out": Urban young people and the space of the mall. *Urban Geography*, Vol. 21, 5–25.

Walton-Roberts, M., 1998, Three readings of the turban: Sikh identity in greater Vancouver. *Urban Geography*, Vol. 19, 311–331.

Ward, D., 1990, Social reform, social surveys, and the discovery of the modern city. *Annals of the Association of American Geographers*, Vol. 80, 491–503.

Wilson, W. J., 1987, *The Truly Disadvantaged*. Chicago, IL: University of Chicago Press.

Wolch, J. R., Rahimian, A., and Koegel, P., 1993, Daily and periodic mobility patterns of the urban homeless. *Professional Geographer*, Vol. 45, 159–169.

# CHAPTER 23

# The 1990s Show
## *Culture Leaves the Farm and Hits the Streets*[1]

TREVOR J. BARNES[2]

## ABSTRACT

The tradition of American cultural geography was defined by studies of the rural, and in its more prosaic form focused on cataloging and mapping artifacts such as fence posts, barn types, and gravestones in order to delimit culture areas. In contrast, the city was all but ignored, treated as a cultural vacuum, and conceived only as a site of work, production, and economic relations. Hence, the importance of urban spatial

---

[1]This chapter benefitted from the comments of David Ley and Wolfgang Zierhofer. I am very grateful to both of them.
[2]Trevor J. Barnes, Department of Geography, 1984 West Mall, University of British Columbia, Vancouver, British Columbia, Canada, V6T 1Z2; telephone: 604-822-2663; fax: 604-822-6150; e-mail: tbarnes@geog.ubc.ca

science that from the late 1950s formalized that economism as central place theory, or Alonso's map of bid-rent curves, or models of retail location. Even when urban geography began eschewing formal models and theory, turning toward some kind of Marxist approach during the 1970s, the focus on things economic remained, but couched now in a different vocabulary such as rent gap, urban gatekeepers, and uneven development. The economism of spatial science and Marxism could not be sustained, however. Culture had to be let in. From 1990 pressured by outside theoretical developments such as cultural studies, and postmodernism, and changes from inside the discipline such as the rise of the new cultural geography, urban geography finally cracked, explicitly allowing culture in first as a trickle, but by the end of that decade as a flood. Culture had finally left the farm and hit the streets. [Key words: cultural geography, spatial science, Marxism.]

## Introduction

In the winter term of my first year as a graduate student at the University of Minnesota, I took a seminar in the "History of Geographical Thought" from a giant of a professor in every sense, Fred Lukermann. I barely understood anything that went on, but it was my most exhilarating intellectual experience even now 25 years later. Lukermann spoke for three hours without notes, and even his physical movements—the lumbering pacing, the gestures of his enormous hands, even the unwrapping of a toothpick—seemed as if they were scripted for the occasion. We joked as graduate students that there were probably cut out footprints pasted to his office floor so he could rehearse his movements, Arthur-Murray like. On the first evening we met, Lukermann told us what the class would be about. But just as he got to the crucial sentence, I was somehow distracted, or lost concentration, and didn't catch what he said. He never repeated that sentence, and from that point he talked only about "the problem," or "the issue" or "the central theme," or just "it."

Too shy and insecure to ask Lukermann what "it" was, the seminar became a series of weekly clues to a puzzle that I could never seem to solve. The clues were many and drawn from a dizzying range of sources: ancient Greek philosophers such as Plato and Aristotle, nineteenth-century German historians such as Dilthey and Herder, late 19th- and early 20th-century American philosophers such Pierce and Dewey, 20th-century ethnographers such as Malinowski, Radcliffe-Brown, Kroeber and Benedict, geographers from all ages from Strabo to von Humboldt to Sauer and Hartshorne, and, Lukermann's favorite, the German philosopher Immanuel Kant who began the seminar and ended it.

The exam was a take-home, and required us to summarize in a blue book the main argument of the course. The exam topic could hardly have been worse, and I was completely flummoxed. With some late nights, strong cups of tea, and muttered British cursing I managed to grind something out. A week later there was a message in my mail box asking if I would go to see Lukermann who was also Dean of Liberal Arts. I turned up at his enormous Dean's office on the East Bank of the University of Minnesota campus with much trepidation. But he was very kind, and handing back my exam said, "I think you knew better what was going on in the course than I did." Flabbergasted, I blurted "But I never knew what "it" was about." "That's culture for you," he said, "you never get to the bottom of it."

So, that was what the course was about. No wonder I didn't get it. Culture was a dirty word for me as a politicized teenager living in 1970s England: it meant class privilege, the hoi polloi, Radio 3, the opera, braying critics, and inscrutable performing and visual arts. Furthermore, at university it meant nothing intellectually. Most of my undergraduate degree at UCL was in economics. Because "culture could not be put into an equation," as my microeconomics professor professed, it was treated as a pseudo idea, as an epiphenomenon, as something that appeared real but was not. And in the geography courses I took my lecturers were hardly less scathing: "Aren't those American cultural geographers funny with their studies of barn types, fence posts, and gravestones?" But in the winter quarter of 1979 at the University of Minnesota, I had my Pauline experience. Leaving Lukermann's cavernous office, the scales fell from my eyes, and I saw the geographical world anew; I saw it in terms of culture. That perspective irrevocably stuck, informing all my subsequent work.

That work, though, has been exclusively in the field of economic geography. In this sense, I am a most unlikely contributor to this special issue, an interloper, if not an impostor. I've never written on urban geography, nor taken an urban geography class. Although in the strange-but-true world of university appointments, I was hired as an urban geographer, and have taught an urban geography course for 20 years. What most strikes me as an interested outsider about urban geography is its similar intellectual trajectory to economic geography that over the 1990s has also emphasized culture after years of neglect and even hostility (Ley, 1996a, p. 475; Thrift, 2000a, p. 692; Barnes, 2003). That is, in the last decade both disciplines appear to be moving toward Lukermann, becoming ever more cultural.

I'm not saying that Fred Lukermann engineered this result. There are forces that even as great a man as Fred Lukermann cannot control. Nonetheless, his 1979 graduate seminar brilliantly caught what was in the air, persuading me of culture's importance, and introducing a theme that is now central to urban geography, and the topic of this paper.

Specifically, my intention is to examine the processes by which culture came into urban geography, and the particular forms it has taken. The paper is divided into three short sections. First, I present brief histories of urban and cultural geography, and their relationship. Constituted in different periods, with different foci, methodologies, and internal sociological imperatives, American traditional pre-war cultural geography and post-war urban geography were never going to be a successful union. There was too much "interference" to use the sociologist of science John Law's (2000) term. Furthermore, even when urban geography began to change from its spatial science to its political economy incarnation during the 1970s, those interferences were only reconstituted in a different register and not removed. Second, I will examine the subsequent process of reconciliation between urban and cultural geography that stems from external debates in political economy and cultural studies, and an internal one around the re-conceptualization of cultural geography. Finally, I discuss some of the specific forms that the "cultural turn" in urban geography now takes. Following my own interests, I focus on especially the changing theoretical relation between culture and economy neglecting the specific issues of gender, race, and sexuality, and which are clearly just as central, if not more so, to urban geography's recent emphasis on culture.

## Two Solitudes: Cultural and Urban Geography

Historically there has been a strained relationship between American cultural and urban geography. Cultural geography, the older of the two, dominated U.S. geography during the first part of the 20th century (Cosgrove, 2000, p. 135). Its most famous exponent, Carl Sauer (1963), presented the methodological template in his 1925 essay, "The Morphology of Landscape." While later expressing regrets about its publication—Lukermann told us in his seminar that Sauer would refer to it as "That thing"—the essay fundamentally molded cultural geographical practices. Those practices turned on examining how traditional culture as an agent modified the natural environment (the medium) to produce a specific cultural landscape.

As a form of inquiry, Sauerian cultural geography was defined by a number of features. First, it focused on rural areas at the expense of urban ones (Wheeler, 1998, p. 586). A well-known urban-phobe, Sauer thought modernization had run amok in the city destroying established folk cultures and the natural environment. It was only in the countryside, on the farm, that the cultural landscape was visible. Hence, those later studies by Berkeley School apostles on barn types, fence posts, and country gravestones. This was a view of culture, according to two of those apostles, Marvin Mikesell and Philip Wagner, defined by "verifiable characteristics" that then provided the "means for classifying areas according to the character of the human groups that occupy them" (Wagner and Mikesell, 1962, p. 2). Second, to achieve such an objective, Sauer demanded detailed field inquiry involving learning foreign languages, prolonged talk with the natives, protracted periods of field study, and knowledge of both natural and human sciences. There were no quick and dirty short cuts, no ready-made protocols to follow. Researchers needed to earn their knowledge through their own talent, the dirt of their boots, and the toil of individual study. Cosgrove (1993, p. 516) labeled such a mentality "the backpack entry ticket to cultural geography." Third, Sauer's broader scheme was holistic. A cultural landscape was more than the sum of its parts involving the integration and mutual modification of culture and nature. As he said in "Morphology," "one has not fully understood the nature of an area until 'one has learned to see it as an organic unit'" (Sauer, 1963, p. 231). The corollary is skepticism of analytical separation, and which for Sauer produced in geography's past a morally and intellectually repugnant environmental determinism. Finally, and most broadly, he was an intellectual conservative, an antimodernist, appealing to history rather than to progress, to cultural tradition rather than to universal reason, and to organic complexity rather than to reductionist simplification (Mathewson, 1987; Speth, 1987). That is why his work was historical and set against the forces of a gung-ho modernity, why he emphasized local culture and was a skeptic of metropolitan power, and why he focused on the detailed holistic arrangement of cultural landscapes and eschewed analytical modeling and abstract theorization. As Sauer wrote in a letter to Campbell Pennington: "I am saddened by model builders and system builders and piddlers with formulas for imaginary universals" (letter to Campbell Pennington, February 4, 1967, quoted in Martin, 1987, p. xv).

Given the methods and goals of Sauerian cultural geography, it was no wonder that when North American urban geography systematically emerged in the mid-to-late 1950s there would be at best a frosty relation to it, if no relation at all. For urban geography defined itself from the beginning in terms of everything Sauerian cultural geography was not. It celebrated contemporary urbanism, it believed in general not individual rules of inquiry, it held up analysis, and preached the virtues of modernism including model and system building, and universals that were thought neither piddling nor imaginary, but large and real.

While there were individual Ph.Ds completed on urban geographical topics from the turn of the 20th century (Harris, 1990, p. 403, indicates that the first at Chicago was in 1907), urban geography courses such as Charles Colby's at Chicago, and even urban geographical texts such as Robert Dickinson's (1947) *City, Region and Regionalism* (Johnston, 2001), urban geography was not systematized until the mid-1950s, if not later (Yeates, 2001, p. 516). Harold Mayer (1990, pp. 419–420), Director of Research of the Chicago Plan Commission, and later appointed at the Geography Department at the University of Chicago, chaired the committee that wrote the key urban chapter in the influential volume, *American Geography: Inventory and Prospect* (Mayer, 1954). And perhaps even more influential was Mayer's edited volume with Clyde Kohn published five years later, *Readings in Urban Geography* (Mayer and Kohn, 1959). Consisting of 54 chapters divided into 18 sections, the book served as a blueprint for the new discipline (see remarks by Taaffe, 1990, p. 423; Clark, 2001, pp. 542–543; and Yeates, 2001, p. 516). Establishing the city as a legitimate objective of social scientific inquiry, contributors pressed the merits of abstract theorizing, rigorous quantitative empirical methods, and instrumental reasoning, and in so doing effortlessly folded into geography's quantitative and theoretical revolution occurring at the same time. Urban geography was spatial science (Wheeler, 2001a, 2001b).

More specifically, unlike Sauer, urban geographers were vitally concerned to record and analyze the effects of modernization in the city, not to retreat from it. The pace and changing form of urban growth in postwar America was transformative, demanding scholarly attention. Central was the urban economy, and reflected in the Mayer and Kohn reader: five chapters were on economic base, four on central place theory, six on commercial structure, six on transportation, and two on industry. The main action was not in the cultural folkways of the Louisiana Bayou, or around the peculiar fence notching on east Nebraskan ranches, but in and between cities: in the Fordist factories of the manufacturing belt, in the commercial strips and malls of the emerging automobile suburbs, and in the metropolitan downtown centers of command and control. Second, while urban geographers did fieldwork, it was not a fetish, nor did it require years of scholarly preparation. Moreover, the most important work was not done outside in the field, but inside the university: in the computer center initially patch wiring early machines, and later punching Fortran cards; in the statistics lab pressing keys on electric Friden calculating machines, and later on expensive hand-held calculators; or in the office compiling data, calculating chi-square coefficients, and drawing regression lines (Wheeler, 2001b, p. 551). Third, a key term and activity for urban geographers was analysis, the breaking down of a problem into smaller elements that could be logically related using a formal vocabulary: hence chi square coefficients, regression lines, and theories like economic-base, rank-size, or central place models. In contrast, Sauerian holistic talk was anathema: it was mystical, obfuscatory, and unscientific. Finally, in every way urban geography upheld the ideal of modernity. Modern knowledge was instrumental in achieving progress in the human condition itself. By implementing rational theories, models, and techniques cities would become better places, cumulatively enriching the everyday lives of people living within them (Adams, 2001, p. 530; Clark, 2001, p. 542; Yeates, 2001, p. 522). Hence, the Center for Urban Studies, University of Chicago, or the Bartlett School, University College London, where rational theories were translated into policy prescriptions for a new and improved urban reality.

Sometime during the 1970s, however, the waxing of urban spatial science of the 1960s turned to waning as radical political economy emerged as an increasingly important alternative. Crucial for this paper, however, is that culture continued to remain off the agenda, with the gap with conservative Sauerian cultural geography if anything widening.

So, radical political economists, like Harvey (1973) and Castells (1977), continued to insist on the centrality of the city. Indeed, urbanization was for them the pulsing heart of capitalism itself. In contrast, Marx and Engels (1967, p. 38) famously spoke about the "idiocy of rural life," putting the cultural study of pastoral folkways outside the radical urban theoretical pale.

Radical urban geographers were also not much inclined toward Sauerian fieldwork. It was not that fieldwork itself was abhorred, only its lack of connection to a wider political project. After all, Engels (1987) in *The Condition of the Working Class in England* was one of the first, if not the first, to engage in modern urban geographical fieldwork in his study of the appalling conditions of the Ur-gritty city of the British industrial revolution, Manchester. He "gained intimacy" of that city, as Steven Marcus (1974, p. 98) wrote, "by taking to the streets, at all hours of the day and night, on weekends and holidays." The point, though, was that Engel's fieldwork—his learning a new language, his talking and interacting with the natives, his meticulous field notes—had political purpose, and revealed with apocalyptic clarity four years later in *The Communist Manifesto* (1967).

While some critics charge Marxism with the holism (Elster, 1985) that also informs Sauer's work, radical urban geographers for the most part practiced a sometimes-compulsive theoretical *analysis*. Most brilliantly represented by Harvey (1982), *The Limits to Capital* especially provides a toolbox of concepts with which to take apart and scrutinize individual components of urban capitalism, and to see them in their larger logical relation. But that analysis, as is well known, provides little place for culture. Even the working class, supposedly the beneficiaries of the analysis, is often silent (Katz, 1986). The stuff of culture—meaning, symbols, signification—are treated by Harvey in his book as flotsam and jetsam, surficial and superstructural, and not part of the all-important economic base that undergirds and directs metropolitan capitalism.

Finally, modernity's ideals of progress, reason, and reductionist simplification are all represented in radical urban geography. Smith's rent-gap thesis is trumpeted as an improvement over Alonso's bid-rent model. The logic and associated diagrams of Harvey's theory of urban capital are just as hard-edged, flinty, and unyielding as Ullman's earlier logic and diagrams in his theory of location for cities. And radical urban geography was equally economically reductionist as orthodox urban geography: the former reduces urban places to the logic of capital, the latter to the logic of consumer and producer behavior. As a result, spatial science urbanism and political economic urbanism ostracized culture, leaving it back at the farm, while the economy was given pride of place within the city.

## Cultural Studies and the New Cultural Geography

The problem, of course, is that you cannot keep culture ostracized, separated, and cordoned off. Lewis Mumford wrote about *The Culture of the City* in 1938, at least a decade before urban geography formally existed. And Fred Lukermann in a series of articles from the late 1950s conducted a guerrilla campaign in a set of admittedly eccentrically chosen journals insisting that urban culture be recognized along with its economy (Barnes, 1996, chapter 9). In addition, the Chicago School of urban sociology, and on which urban geographers were to draw heavily especially during the 1960s, made issues of culture and cultural interpretation central, from William Thomas' (1918) formative early 20th-century study on the Polish peasant through to the classic urban ethnographies of the 1930s such as Frederic Thrasher's (1936) *The Gang*, or Paul Cressey's (1932) *The Taxi-Dance Hall*. Robert Park (1952, p. 15), who along with Earnest Burgess and Robert McKenzie founded the school with publication of the *The City* in 1925, was very explicit about their links with anthropology and the study of culture. "The same patient methods of observation which anthropologists like Boas and Lowie have expended on the study of life and manners of the North American Indian might be even more fruitfully employed in the investigation of the customs, beliefs, social practices, and general conceptions of life prevalent in Little Italy on the Lower North Side in Chicago, or in recording the more sophisticated folkways of the inhabitants of Greenwich Village and the neighborhood of Washington Square."

The point is that there were resources available to urban geographers either of the spatial science or radical stripe to deal with culture if they were inclined to do so. But they were not. Because of historical connotations about the meaning and practice of traditional cultural geography (and bound to the forceful personality and intellect of Sauer), and a tendency toward economism, that is, the belief that the economy is determinant and central in social life, urban culture was sidelined, or at best treated as supplemental and marginal, an extra that one could do without. However, during the 1990s things begin to change. Culture became increasingly basic and core, a theoretical necessity rather than a luxury. Propelling such a transformation are two related literatures outside of urban geography: first, cultural studies and postmodernism that retheorizes the relation between culture and economy, thereby undermining the economism of both spatial science and radical urbanism, and, second, the new cultural geography that both critiques and begins to replace the Sauerian tradition. Together they revamp urban geography. Culture leaves the farm and hits the streets.

Cultural studies emerged in postwar Britain from debates inside political economy about the relationship of culture to economy. Marx in his preface to *A Contribution to the Critique of Political Economy* (1859) asserted that "the mode of production of material life conditions the social, political and intellectual life process in general. It is not the consciousness of men that determines their being, but, on the contrary, their social being that determines their consciousness" (Marx, 1904, preface). In this view, "the mode of production," the economy, determines, and consciousness, culture is determined. From its beginning in the 1950s, cultural studies played down that economism, substituting a softer view that provided some autonomy and determining role to culture. Early statements were made by Raymond Williams (1977, p. 132–133) ("the structure of feeling"), Richard Hoggart (1957) ("the felt quality of life"), and Stuart Hall (1986) ("Marxism without guarantees"). Their importance was in showing that one could hang on to class analysis and the economy, yet also recognize values, ways of life, and emotional and political commitments that lay outside. Hence, for example, William's phrase the "structure of feeling" that connotes the "doubleness of culture ... [as both] material reality and lived experience" (Eagleton, 2000, p. 36).

The same issue of the relation between economy and culture is one of the key themes in the sometimes-allied postmodernist literature, and which becomes important from the 1970s. I say *one* of the themes because the postmodernist literature is vast and sprawling, spanning disciplines, substantive topics and foci, and philosophical positions. One corner of it, though, is concerned with moving away from a determinant relation between the economy and culture, sometimes disassociating the two altogether, and in some cases representing both as "no more than a free play of texts, representations and discourses" (Bradley and Fenton, 1999, p. 114). As within the cultural studies literature, though, there is a wide range of positions. In geography it varies from Harvey's (1989) "postmodernism," that remains within shouting distance of his earlier classical Marxism of *The Limits*, to Michael Dear's "postmodernism" of Keno Capitalism, cyburbia and privatopias (Dear and Flusty, 1998), to J. K. Gibson-Graham's (1996) "postmodernism" in which the economy is conceived as a discursive construction that will disappear if enough of us believe in that possibility and act accordingly in our everyday lives.

It is easy to get lost in the details of these positions, and the sometimes antagonistic debates among people who you think should be friendlier. The broad point is that cultural studies and postmodernism undermined economism, which was found in urban geography from its inception. The infusion of cultural studies and postmodernism into urban geography, then, began to dislodge the centrality accorded to the economy, and in doing so put the focus on what Gibson-Graham (2000) call economy's "other," culture.

The other literature making the difference is the new cultural geography. There already exist a number of reviews, assessments, dictionary entries, and even retrospectives (Philo, 2000). For the purpose of this paper, the new cultural geography made two signal contributions. First, it provided a set of compelling and persuasive criticisms of Sauerian cultural geography beginning with James Duncan's (1980) superorganic paper. The upshot was that by the 1990s, the Sauerian stamp no longer marked cultural geography. The change was not always smooth—Duncan (1994) speaks of "civil war"—and once the heat of debate cooled, it was possible to see a number of continuities and overlaps between old and new cultural geographies. The important point, though, was that Sauerian cultural geography, and its antipathy to urbanism both in word and in deed, was no longer obstructive. One could practice cultural geography in the city without "the old man" wagging a disapproving finger. Second, the new cultural geography insisted that culture was found everywhere from the most mundane to the most spectacular, and was not a miscellaneous category for ill-fitting, awkward and otherwise lost elements of social life. As Cosgrove and Jackson (1987, p. 99) wrote in their manifesto: "Culture is not a residual category, the surface variation left unaccounted for by more powerful economic analyses; it is the very medium through which change is experienced, contested and constituted." And where that change is most often experienced, contested, and constituted is in the city.

## Urban Cultural Geography

The consequence was that urban geography during the 1990s became increasingly urban cultural geography. A casual inspection of the journal *Urban Geography*, illustrates the change over the decade. For the first time in the 1990s, it included special issues (e.g., Vol. 17, No. 6), progress reports (e.g., Mitchell, 1999), and editorials (e.g., Wheeler, 1998) in favor of cultural urban geography. In addition, in 2000, Bridge and Watson published their edited *A Companion to the City*, a volume in size and intent to rival *Readings in Urban Geography*, with 52 separate essays, and divided into five sections. The difference is that only one *Companion* section is on the economy, but three sections cover culture: "Imagining Cities," "Cities of Division and Difference," and "Public Cultures and Everyday Spaces." The cultural turn clearly has turned urban.

Understanding culture is not easy, however. Raymond Williams (1976, p. 74) famously said that culture is "one of the two or three most complex words in the English language." And once joined with the urban, the combination is dense and tangled. Perhaps the way forward is less the single road of grand theoretical statement, but paths that are more piecemeal, less defined and limited, and which join bits of empirical study with pieces of different cultural theories. I would argue that it is precisely this kind of "promiscuous mingling," to use Dick Walker's (1997, p. 173) term, that characterizes much of substantive writings within recent urban cultural geography, and which accounts for its vibrancy and success.

For reasons of brevity, I cannot provide a comprehensive review of that literature. Instead, I will mention five current substantive urban research areas in which culture now figures large. In each case, there is an attempt to recognize, and to varying degrees to integrate culture with the economy. I say varying degrees because different researchers bring to these topics their own past intellectual baggage, and in which some cases has been economistic. In such cases, the past can still weigh heavy. That said, in each of the five examples, it is never pure economism (or pure culture), but always an attempt to have both urban culture and urban economy: it is both/and and not either/or.

The first literature is on public space, and which shades into discussions of consumption. Found initially in Harvey's (1989) *Condition of Postmodernity*, and clearly different from *Limits* in its recognition of culture, the argument is that the urban public space has become increasingly commodified, hybridizing dollar signs of profit with cultural signs of meaning.

Certainly, one can still detect strains of economism in the argument. But it is not the all-powerful capital logic of Harvey's earlier works. The subtitle of *The Condition of Postmodernity* is *An Inquiry into the Origins of Cultural Change*, and evident in his concern to identify changing sources of cultural meaning in everyday lives as people move about and use the city including its public spaces. In a similar vein, is Zukin's (1995) work on *The Cultures of Cities*. Softening her earlier Marxism, Zukin made a concerted attempt in the book to deal with a suite of cultural issues including urban symbolism, identity, and the meaning of urban public space. "Culture is, arguably, what cities 'do' best," Zukin (1995, p. 264) observed. But it is not an innocent culture. It is marked by the hieroglyph of the commodity. She wrote, "since the 1970s ... collective space—public space—has been represented as a consumer good" (Zukin, 1995, p. 260). In this view, public urban spaces like Baltimore's Inner Harbor or Sony Plaza in Manhattan or the French Quarter in New Orleans are utterly entangled in both culture and economy.

The second is on urban cultural industries, and associated especially with Scott's (1996, 1997) work. Originally coming from a political economy perspective, Scott extended his earlier regulationist-inspired idea of post-Fordist industrial agglomerations, and resulting synergies within particular urban places, to cultural industries, and their location within the world's most powerful metropolitan centers such as Los Angeles, New York, London, and Paris. His argument is that cultural products, that is, goods and services "infused ... with broadly aesthetic or semiotic attributes" (Scott, 1997, p. 323) are "one of the critical pulses of the economy and cultural conditions of twentieth-century capitalism" (Scott, 1996, p. 306). In this view, "culture and economy are highly symbiotic" (Scott, 1997, p. 325), meaning that they are mutually constitutive: the economy increasingly is defined by the generation of cultural products, and cultural products emerge because they are defined as economic commodities. As a result, the dividing line between culture and economy is blurred, with the necessity of dealing with both.

The third is on gentrification, and interesting theoretically because at least initially there were two competing interpretations: one associated with Smith's (1984) political economic approach, and another associated with Ley (1980, 1987) that focused not on the economy but on cultural meaning and the everyday. Smith's "rent gap thesis" was the economistic interpretation, with gentrifying neighborhoods the precipitate of the "seesaw" of capital. In contrast, Ley emphasized gentrification as part of a wider lifestyle and consumption choice stemming from the emergence of a postindustrial new middle class, and who took their pro-urban cultural values in part from the critical youth movements of the 1960s. Over the

1990s, however, there has been give on both side, as each author sought to accommodate both culture and economy. Smith now says there "an intimate connection between economic change and sociocultural expression in the explanation of gentrification" (Smith, 1996; Smith and Defilippis, 1999, p. 639), and Ley (1996b, chapter 4) is much more explicit about the role of the economy, couched in terms of the central city labor markets. Clearly, there are still differences between the two, but the important point is that both have moved to include *both* culture and economy.

The fourth focuses on economic services and innovation and associated with Amin and Thrift (2002; Amin and Graham, 1997; Thrift, 2000b). Drawing upon an eclectic range of theoretical sources institutional economics, performativty, and actor network—they argue that the urban economy increasingly operates as a discursive construction blending economy and culture. As Thrift (1997, p. 136) noted, "capitalism seems to be undergoing its own cultural turn as increasingly business is about the creation, fostering, and distribution of knowledge." Where this happens primarily is in the city. Its peculiar set of institutional assets enable cultural performance to be inextricably bound to economic performance. As a result, "the easy separation between the 'social,' 'political,' 'cultural,' and 'economic' becomes more and more problematic" (Amin and Graham, 1997, p. 419) as the city itself becomes "variegated and multiplex" (Amin and Graham, 1987, p. 418).

The final literature is on international urban migration and transnationalism. In urban geography it is associated especially with work on the Chinese diaspora and cities of the Pacific Rim (Mitchell, 1993; Olds, 2000; Ley, 2003). Katharyne Mitchell (1999, p. 671) summarized this work in her review essay for *Urban Geography*, "What's Culture Got to Do with It?," saying that "the main body of this research incorporates both economic and cultural factors in explaining the types and changes occurring in the urban environment, the patterns of resistance to those changes, and the general outcome of struggles based on varying intersections of class, ethnicity and gender." To answer Mitchell's question: culture has everything to do with it. From rhizomatic Chinese business networks, to socially embedded patterns of entrepreneurship and migration, to symbolic meanings of house form and urban landscape, culture is everywhere, intermingling, and inseparable from the economic.

## Conclusion

Mae West once said, "I used to be Snow White but I drifted." So too has urban geography over the 1990s, and me as well. Urban geography used to study the pure and unsullied—wholesome patterns of mid-western central places, decent Toronto bid-rent curves, upright Chicago population density gradients, unadulterated Baltimore class-conflict, and a proper Philadelphia rent gap. But no more. Things now seem less straightforward, messier, mixed up, and contaminated. Reaching for pure economic entities as an urban explanation seems less satisfactory. There are various reasons for such a change, but one, as I argued here, is because of the increasing recognition of the cultural, and an attempt to theories its relation especially to the economy. Urban geography has drifted.

I don't think this should be bemoaned, morally impugned, or haughtily ignored. Why it took me so long to figure out the point of Lukermann's seminar, and in the end had to be told, was precisely because I was operating with a Snow-White purity. My Road-to-Damascus experience in Lukermann's office was the start of my own drifting, the recognition that purity should be soiled, and the realization that Snow White is too good to be true. I hope in its "cultural turn," urban geography like Mae West becomes too true to be good. As Martin Luther said, if one is going to sin one should sin boldly.

## References

Adams, J. S., 2001, The quantitative revolution in urban geography. *Urban Geography*, Vol. 22, 530–539.

Amin, A. and Graham, S., 1997, The ordinary city. *Transactions of the Institute of British Geographers*, Vol. 22, 411–429.

Amin, A. and Thrift, N. J., 2002, *Cities: Reimagining the Urban*. Cambridge, UK: Polity.

Barnes, T. J., 1996, *Logics of Dislocation: Meanings, Metaphors and Models of Economic Space*. New York, NY: Guilford.

Barnes, T. J., 2003, "Never mind the economy. Here's culture." In K. Anderson, M. Domosh, S. Pile, and N. J. Thrift, editors, *Handbook of Cultural Geography*. London, UK: Sage, 89–97.

Bradley, H. and Fenton, S., 1999, Reconciling culture and economy: Ways forward in the analyses of ethnicity and gender. In L. Ray and A. Sayer, editors, *Culture and Economy After the Cultural Turn*. London, UK: Sage, 112–134.

Bridge, G. and Watson, S., 2000, editors, *A Companion to Urban Geography*. Oxford, UK: Blackwell.

Castells, M., 1977 [1972], *The Urban Question: A Marxist Approach*. Translated by Alan Sheridan. London, UK: Edward Arnold.

Clark, W. A. V., 2001, Pacific views of urban geography in the 1960s. *Urban Geography*, Vol. 22, 540–548.

Cosgrove, D. E., 1993, Commentary on "The reinvention of cultural geography" by Price and Lewis. *Annals of the Association of American Geographers*, Vol. 83, 515–517.

Cosgrove, D. E., 2000, Cultural geography. In R. J. Johnston, D. Gregory, G. Pratt, and M. J. Watts, editors, *The Dictionary of Human Geography*. Oxford, UK: Blackwell, 135–138.

Cosgrove, D. E. and Jackson, P., 1987, New directions in cultural geography. *Area*, Vol. 19, 95–101.

Cressey, P. G., 1932, *The Taxi-Dance Hall: A Sociological Study in Commercialized Recreation and City Life*. New York, NY: Greenwood.

Dear, M. and Flusty, S., 1998, Postmodern urbanism. *Annals of the Association of American Geographers*, Vol. 88, 50–72.

Dickinson, R. E., 1947, *City, Region and Regionalism*. London, UK: K. Paul, Trench, Trubner & Co.

Duncan, J. S., 1980, The superorganic in American cultural geography. *Annals of the Association of American Geographers*, Vol. 70, 181–198.

Duncan, J. S., 1994, After the civil war: Reconstructing cultural geography as heterotopia. In K. Foote, P. J. Hugill, K. Mathewson, and J. M. Smith, editors, *Re-reading Cultural Geography*. Austin, TX: University of Texas Press, 401–408.

Eagleton, T., 2000, *The Idea of Culture*. Oxford, UK: Blackwell.

Elster, J., 1985, *Making Sense of Marx*. Chicago, IL: Chicago University Press.

Engels, F., 1987 [1844], *The Condition of the Working Class in England*. Harmondsworth, UK: Penguin.

Gibson-Graham, J. K., 1996, *The End of Capitalism (As We Knew It): A Feminist Critique of Political Economy*. Oxford, UK: Blackwell.

Gibson-Graham, J. K., 2000, Poststructural interventions. In E. Sheppard and T. J. Barnes, editors, *A Companion to Economic Geography*. Oxford, UK: Blackwell, 95–110.

Hall, S., 1986, The problem of ideology: Marxism with out guarantees. *Journal of Communication Inquiry*, Vol. 10, 28–44.

Harris, C. D., 1990, Urban geography in the United States: My experience of the formative years. *Urban Geography*, Vol. 11, 403–417.

Harvey, D., 1973, *Social Justice and the City*. London, UK: Edward Arnold.

Harvey, D., 1982, *The Limits to Capital*. Chicago, IL: University of Chicago Press.

Harvey, D., 1989, *The Condition of Postmodernity: An Inquiry into the Origins of Cultural Change*. Oxford UK: Blackwell.

Hoggart, R., 1957, *The Uses of Literacy*. Oxford, UK: Oxford University Press.

Johnston, R. J., 2001, Robert E. Dickinson and the growth of urban geography: An evaluation. *Urban Geography*, Vol. 22, 702–736.

Katz, S., 1986, Towards a sociological definition of rent: Notes on David Harvey's Limits to Capital. *Antipode*, Vol. 18, 64–78.

Law, J., 2000, Economics as interference. Center for Science Studies and the Department of Sociology, Lancaster University, at www.comp.lancaster.ac.uk/sociology/soc034jl.html

Ley, D., 1980, Liberal ideology and the postindustrial city. *Annals of the Association of American Geographers*, Vol. 70, 238–258.

Ley, D., 1987, Styles of the times: Liberal and neoconservative landscapes in inner Vancouver, 1968–1986. *Journal of Historical Geography*, Vol. 13, 40–56.

Ley, D., 1996a, Urban geography and cultural studies. *Urban Geography*, Vol. 17, 475–477.

Ley, D., 1996b, *The New Middle Class and the Remaking of the Central City*. Oxford, UK: Oxford University Press.

Ley, D., 2003, Seeking homo economicus: The strange story of Canada's Business Immigration Program. *Annals of the Association of American Geographers*, Vol. 93, 426–441.

Marcus, S., 1974, *Engels, Manchester, and the Working Class*. New York, NY: Random House.

Martin, G., 1987, Foreword. In M. Kenzer, editor, *Carl O. Sauer: A Tribute*. Corvalis, OR: University of Oregon Press, ix-xvi.

Marx, K., 1904, *A Contribution to the Critique of Political Economy*. Translated by N. I. Stone. Chicago, IL: C. H. Kerr.

Marx, K. and Engels, F., 1967, *The Communist Manifesto* (originally published in 1848). Harmondsworth, UK: Penguin.

Mathewson, K., 1987, Sauer south by southwest: Antimodernism and the austral impulse. In M. Kenzer, editor, *Carl O. Sauer: A Tribute*. University of Oregon Press: Corvalis, OR, 90–111,

Mayer, H. M., 1954, Urban geography. In P. E. James and C. F. Jones, editors, *American Geography: Inventory and Prospect*. Syracuse, NY: Syracuse University Press.

Mayer, H. M. and Kohn, C., editors, 1959, *Readings in Urban Geography*. Chicago, IL: Chicago University Press.

Mayer, H. M., 1990, A half-century of urban geography in the United States. *Urban Geography*, Vol. 11, 418–421.

Mitchell, K., 1993, Multiculturalism or the united colors of capitalism. *Antipode*, Vol. 25, 263–294.

Mitchell, K., 1999, "What's culture got to do with it?" *Urban Geography*, Vol. 20, 667–677.

Mumford, L., 1938, *The Culture of the City*. New York, NY: Harcourt and Brace.

Olds, K, 2000, *Globalization and Urban Change: Capital, Culture, and Pacific Rim Mega Projects*. Oxford, NY: Oxford University Press.

Park, R. E., Burgess, E. W., McKenzie, R. D., 1925, *The City*. Chicago, IL: University of Chicago Press.

Park, R. E., 1952, *Human Communities: The City and Human Ecology. The Collected Writings of Robert E. Park. Vol. 2.* E. C. Hughes, editor. The Free Press: Glencoe, IL.

Philo, C., 2000, More words, more worlds: Reflections on the "cultural turn" and human geography. In I. Cook, D. Crouch, S. Naylor, and J. R. Ryan, editors, *Cultural Turns/Geographical Turns: Perspectives on Cultural Geography.* Harlow, UK: Prentice Hall, 26–53.

Sauer, C. O., 1963 [1925], The morphology of landscape. Reprinted in J. Leighly, editor, *Land and Life: Selections from the Writings of Carl Ortwin Sauer.* Berkeley, CA, and Los Angeles, CA: University of California Press, 315–350.

Scott, A. J., 1996, The craft, products, and cultural products industries of Los Angeles: Comparative dynamics and policy dilemmas in a multisectoral image-producing complex. *Annals of the Association of American Geographers,* Vol. 86, 306–323.

Scott, A. J., 1997, The cultural economy of cities. *International Journal of Urban and Regional Research,* Vol. 21, 323–339.

Smith, N., 1984, *Uneven Development.* Oxford, UK: Blackwell.

Smith, N., 1996, *The New Urban Frontier: Gentrification and the Revanchist City.* London, UK: Routledge.

Smith, N. and Defilippis, J., 1999, The reassertion of economics: 1990s gentrification in the Lower East Side. *International Journal of Urban and Regional Research,* Vol. 23, 638–653.

Speth, W. W., 1987, Historicism: The disciplinary world view of Carl O. Sauer. In M. Kenzer, editor, *Carl O. Sauer: A Tribute.* Corvalis, OR: University of Oregon Press, 11–39.

Taaffe, E. J., 1990, Some thoughts on the development of urban geography in the United Stated during the 1950s and 1960s. *Urban Geography,* Vol. 11, 422–431.

Thomas, W. I. and Znaniecki, F., 1918, *The Polish Peasant in America.* Chicago, IL: University of Chicago Press.

Thrasher, F., 1936, *The Gang: A Study of 1,313 Gangs in Chicago.* Chicago, IL: University of Chicago Press.

Thrift, N. J., 1997, The rise of soft capitalism. *Cultural Values,* Vol. 1, 29–57.

Thrift, N. J., 2000a, Pandora's box? Cultural geographies of economics. In G. L. Clark, M. Feldman, and M. S. Gertler, editors, *The Oxford Handbook of Economic Geography.* Oxford, UK: Oxford University Press, 689–701.

Thrift, N. J., 2000b, Performing cultures in the new economy. *Annals of the Association of American Geographers,* Vol. 91, 674–692.

Wagner, P. and Mikesell, M., editors, 1962, *Readings in Cultural Geography.* Chicago, IL: University of Chicago Press.

Walker, R., 1997, Unseen and disbelieved: A political economist among cultural geographers. In P. Groth and T. Bressi, editors, *Understanding Ordinary Landscapes.* New Haven, CT: Yale University Press, 163–174.

Wheeler, J. O., 1998, Urban cultural geography: Country cousin comes to the city. *Urban Geography,* Vol. 19, 585–590.

Wheeler, J. O., 2001a, Introduction to special issue: Urban geography in the 1960s. *Urban Geography,* Vol. 22, 511–513.

Wheeler, J. O., 2001b, Assessing the role of spatial analysis in urban geography in the 1960s. *Urban Geography,* Vol. 22, 549–558.

Williams, R., 1976, *Keywords: A Vocabulary of Society and Nature.* London, UK: Fontana.

Williams, R., 1977, *Marxism and Literature.* Oxford, UK: Oxford University Press.

Yeates, M., 2001, Yesterday as tomorrow's song: The contribution of the 1960s "Chicago School" to urban geography. *Urban Geography,* Vol. 22, 514–529.

Zukin, S., 1995, *The Cultures of Cities.* Cambridge, MA: Blackwell.

# CHAPTER 24

# The Los Angeles School of Urbanism
## *An Intellectual History*[1]

MICHAEL DEAR[2]

## ABSTRACT

Los Angeles, or more precisely, the Southern California region, has many claims on our attention, but until recently it has been regarded as

[1]An earlier version of the chapter was presented at the Annual Meeting of the Association of American Geographers in New Orleans, 2003; I thank the conference participants for their critiques. This essay draws on ideas developed in my book, *The Postmodern Urban Condition* (2000), and also the edited collection by M. Dear titled *From Chicago to LA; Making Sense of Urban Theory* (2001). I am especially indebted to Steven Flusty, who has helped immeasurably in the development of these ideas (see Dear and Flusty, 1998, 2001; Flusty, 2003). Thanks also to Tony Orum, who arranged an exchange on the idea of an LA School, which was published in the inaugural issue of the new journal, *City and Community* (Abbott, 2002; Clark, 2002; Dear, 2002; Molotch, 2002; Sampson, 2002; Sassen, 2002). I have also benefitted greatly from conversations on this topic with Bob Beauregard and Bob Lake.
[2]Correspondence concerning this chapter should be addressed to Michael Dear, Department of Geography, University of Southern California, Los Angeles, CA 90033; telephone: 213-740-0743; fax: 213-740-0056; e-mail: mdear@usc.edu

an exception to the rules governing American urban development. Since the mid-1980s, a remarkable outpouring of scholarship has given birth to a "Los Angeles School" of urbanism. This essay outlines the intellectual history of the LA School, explains the distinctiveness of its break with previous traditions (especially those of the Chicago School), and advocates the need for a comparative urban analysis that utilizes Los Angeles not as a new urban "paradigm," but as one of many exemplars of contemporary urban process.

The Los Angeles School of urbanism emerged as a coherent challenge to established urban theory during the mid-1980s. Needless to say, there had been much work on past and present urbanisms in LA and the broader Southern California region before that date, but never before had that work been transformed into larger claims about the prototypicality of the LA experience. In fact, quite the reverse was true: LA was almost universally regarded as an exception to the rules governing urban growth and change. The proposition that LA was somehow emblematic of urban process on a broad national (even international) stage was truly revolutionary, in intellectual terms.

In this essay, I shall examine the intellectual history of the LA School. I am less interested in the substantive theoretical domains of the School; in any event, it is too early to properly assess that contribution. Nor will I be concerned with my personal place in the discourses on LA. Instead, I am concerned with how an intellectual movement was formed, entered the public realm, and what response it generated. In a sense, this is primarily an essay on the social construction of knowledge. In the introductory section, I outline the chronology of the LA School. Next, I describe the intellectual faultlines separating the LA School from its predecessors, most especially the Chicago School, in order to demonstrate just how radical the break offered by the LA School is. Finally, I assess some of the responses to the ways in which the LA School has challenged our mental and material understanding of the city.

## "Schools" in Academic Discourse

Academic discourse seems to favor the pretense that intellectual progress occurs in a reasonably ordered way, with one paradigm replacing its outmoded predecessor as a consequence of accumulated anomalies that prove the predecessor's obsolescence. This mind-set encourages the belief that the search for knowledge is characterized above all by the existence of a single dominant framework, within which "normal" science is practiced. This is a characterization that I reject. Academic discourse, at least outside the realms of a strictly defined "scientific method," tends to proceed as a consequence of a variety of impulses, most notably the influence of charismatic disciplinary leaders, fashion, plus a healthy dose of anarchy. As a consequence, I prefer to embrace the notion of a "school" of thought, which emphasizes the plurality of discourses occurring within and between disciplines.

In the case of Los Angeles, it may surprise some that a region notorious for an apparent contempt for its own history should, in fact, possess a rich heritage of intellectual, cultural, and artistic heritage. For many decades, these traditions have spawned a variety of "LA Schools," involving (for instance) art, music, poetry, literature, and of course, filmmaking. Closest to my concern is the contemporary "LA School of architecture," which has enjoyed a rigorous documentation due to the efforts of that most intrepid chronicler, Charles Jencks. There are many LA Schools of architecture, both past and present. These include Richard Neutra, Rudolph Schindler, Gregory Ain et al. in the 1930s, as well as members of the LA Forum on Architecture and Urban Design. According to Jencks, the current LA School of architecture includes such luminaries as Frank Gehry and Charles Moore, and was founded amid acrimony in 1981:

> The L.A. School was, and remains, a group of individualized mavericks, more at home together in an exhibition than in each other's homes. There is also a particular self-image involved with this Non-School which exacerbates the situation. All of its members see themselves as outsiders, on the margins challenging the establishment with an informal and demanding architecture; one that must be carefully read (Jencks, 1993, p. 34).

Jencks concurs with architectural critic Leon Whiteson that LA's cultural environment is one that places the margin at its core: "The ultimate irony is that in the L.A. architectural culture, where heterogeneity is valued over conformity, and creativity over propriety, the periphery is often the center" (Jencks, 1993, p. 34). Jencks's interpretation is of particular interest here because of its explicit characterization of a "school" as a group of marginalized individuals incapable of surrendering to a broader collective agenda. This is hardly the distinguishing feature I had in mind when I began this inquiry into an LA school of urbanism. My search was originally for some notion of an identifiable cohort knowingly engaged on in a collaborative enterprise. Jencks' vision radically undermines this expectation as, in retrospect, has my personal experience of the LA school of urbanism.

A large part of the difficulty involved in identifying a "school" is etymological. The *Shorter Oxford English Dictionary* (Trumble et al., 1999, p. 2714) provides 14 principal categories, including a "group of gamblers or of people drinking together," and a "gang of thieves or beggars working together" (both 19th-century coinages). Also from the mid-19th century is something closer to the spirit of this discussion: "a group of people who share some principle, method, style, etc. ...a particular doctrine or practice as followed by such a body of people." The dictionary gives goes on to give as an example, the "Marxist school of political thought."

In a broad examination of a "second" Chicago School, Jennifer Pratt used the term "school" in reference to:

> A collection of individuals working in the same environment who at the time and through their own retrospective construction of their identity and the impartations of intellectual historians are defined as representing a distinct approach to a scholarly endeavor (Pratt, quoted in Fine, 1995, p. 2).

Such a description suggests four elements of a working definition of the term "school." The adherents of a school should be:

(1) engaged on a common project (however defined);
(2) geographically proximate (however delimited);
(3) self-consciously collaborative (to whatever extent); and
(4) externally recognized (at whatever threshold).

The parentheses associated with each of the four characteristics underscore the contingent nature of each trait. Conditions 1–3 may be regarded as the minimum, or least restrictive components of this definition. Second-order criteria for defining a school could include the following:

(5) that there exists broad agreement on the program of research;
(6) that adherents voluntarily self-identify with the school and/or its research program; and
(7) that there exists organizational foci for the school's endeavors (such as a learned journal, meetings, or book series).

Most of these traits should be relatively easy to recognize, even though no candidate for the "school" appellation is likely to satisfy all these criteria.

Verifying the existence of a school must always remain unfinished business, not least because we, who would identify such a phenomenon, are ourselves stuck in those particular circumstances of time and place to which our bodies have been consigned. But of greater practical concern is the fourth identifying characteristic, i.e., the external recognition deemed necessary to warrant the title of School. Outside recognition traditionally arrives only after most (if not all) school instigators are dead, simply because there are so many incentives to deny the existence of a school. Accolades from outsiders are routinely refused because of professional rivalries, or routinely attacked as crass careerism. Outsiders also appeal to alternative standards of evidence in rejecting a challenge, most commonly seen in appeals to the "hardness" of existing paradigms, (as in "hard science"). Yet another variant of denial is the unthinking, perverse pleasure taken by many in puncturing a novice's enthusiasm with claims like: "There's nothing new in that. It's all been said and done before." With such curt put-downs, existing orders and authority remain undisturbed, and old hegemonies once again settle about us like an iron cage.

The refusal to even contemplate the existence of a distinctive (intellectually focused, place-based) school of thought is both intellectually and politically reprehensible. It stifles the development of a critical gaze, both in epistemological and material terms; and it inhibits the growth of intellectual and political alliances. In short, the unexamined dismissal of a school's claims is a denial of new ways of seeing and acting. Thus, members of the LA School can be forgiven if they did not wait for outsiders' recognition or permission before declaring the School's existence.

## The LA School Emerges

Most births are inherently messy, and the arrival of an LA School of urbanism is no exception. The genetic imprint of the School lies in some unrecoverable past, though we can identify its traces in the work of inveterate city-improver Charles Mulford Robinson. In his 1907 plan to render LA as *The City Beautiful*, Robinson conceded that: "The problem offered by Los Angeles is a little out of the ordinary" (p. 4). A peculiarly Angeleno urban vision was more convincingly established in 1946, with the publication of Carey McWilliams' *Southern California: An Island on the Land* (1973). This work remains the premier codification of the narratives of Angeleno (sur)reality, and served to establish LA's status as "the great exception." It has since colored both popular and scholarly perceptions of the city. McWilliams emphasized LA's uniqueness, asserting that the region reversed almost any proposition about the settlement of western America. He described Southern California as an engineered utopia attracting pioneers from faraway places like Mexico, China, Germany, Poland, France, Great Britain. Among the most exotic immigrants, however, were families from the American Midwestern states who, baked beneath "a sun that can beat all sense from your brains" (McWilliams, 1973, p. 8), were crushed beneath the heel of an open shop industrial system, and generated a hot-house of segregated communities. In McWilliams' account, local communities were rife with bizarre philosophies, carnivalesque politics, and a confused cultural melange of immigrant influences imperfectly adapted to local conditions. The whole enterprise was pervaded by apocalyptic undercurrents suitable to a fictive paradise situated within a hostile yet simultaneously fragile desert environment.

McWilliams' exceptionalism was confirmed and consolidated by Robert Fogelson's *The Fragmented Metropolis*, which in 1967, the year of its publication, was the only account of the region's urban history between 1850 and 1930. Fogelson summarized the exceptionalist credo: "The essence of Los Angeles was revealed more clearly in its deviations from [rather] than its similarities to the great American metropolis of the nineteenth and early twentieth centuries" (p. 134). But perhaps the canonical moment in the prehistory of the LA School came with the publication of Reyner Banham's *Los Angeles: The Architecture of Four Ecologies* (1971). Responding to the notion that Southern California was devoid of cultural or artistic merit, Banham was the first to assert that Los Angeles should not be "rejected as inscrutable and hurled as unknown into the ocean" (p. 23). Rather, he argued, the city should be taken seriously and read on its own terms instead of those used to make sense of other American cities. But while LA was an object worthy of serious study, according to Banham its structure remained exceptional: "Full command of Angeleno dynamics qualifies one only to read Los Angeles. ... [The] splendors and miseries of Los Angeles, the graces and grotesqueries, appear to me as unrepeatable as they are unprecedented" (p. 24). More that any other single volume to that date, Banham's celebration of LA landscapes served to legitimize the study of Los Angeles, and to temporarily neutralize (in some small extent) the propensity of East Coast media and scholars to chart with mock amazement the eccentricities of their West Coast counterparts with mock amazement.

During the 1980s a group of loosely associated scholars, professionals, and advocates based in Southern California became convinced that what was happening in the region was somehow symptomatic of a broader sociogeographic transformation taking place within the United States as a whole. Their common, but then unarticulated project was based on certain shared theoretical assumptions, and on the view that LA was emblematic of a more general urban dynamic. One of the earliest expressions of an emergent "LA School" came with the appearance of a 1986 special issue of the journal *Society and Space*, which was entirely devoted to understanding Los Angeles. In their prefatory remarks to that issue, Allen Scott and

Edward Soja (1986) referred to LA as the "capital of the 20th century," deliberately invoking Walter Benjamin's designation of Paris as capital of the 19th. They predicted that the volume of scholarly work on Los Angeles would quickly overtake that on Chicago, the dominant model of the American industrial metropolis.

In this same journal issue Ed Soja's celebrated tour of Los Angeles was published (1986; 1989). In that essay, Soja most effectively achieved the conversion of LA from the exception to the rule—the prototype of late 20th-century postmodern geographies:

> What better place can there be to illustrate and synthesize the dynamics of capitalist spatialization? In so many ways, Los Angeles is the place where "it all comes together"...one might call the sprawling urban region ...a prototopos, a paradigmatic place; or ...a mesocosm, an ordered world in which the micro and the macro, the idiographic and the nomothetic, the concrete and the abstract, can be seen simultaneously in an articulated and interactive combination (p. 191).

Soja went on to assert that L.A. "insistently presents itself as one of the most informative palimpsests and paradigms of 20th-century urban development and popular consciousness," comparable to Borges' Aleph: "the only place on earth where all places are seen from every angle, each standing clear, without any confusion or blending" (p. 248).

As ever, Charles Jencks quickly picked up on this trend quite quickly, taking care to distinguish its practitioners from the LA School of architecture:

> The L.A. School of geographers and planners had quite a separate and independent formulation in the 1980s, which stemmed from the analysis of the city as a new postmodern urban type. Its themes vary from L.A. as the post-Fordist, postmodern city of many city of many fragments is search of a unity, to the nightmare city of social inequities (p. 132).

This very same group of geographers and planners (accompanied by a few dissidents from other disciplines) gathered at Lake Arrowhead in the San Bernardino Mountains on October 11–12, 1987, to discuss the wisdom of engaging in an LA School. The participants included, if memory serves, Dana Cuff, Mike Davis, Michael Dear, Margaret FitzSimmons, Rebecca Morales, Allen Scott, Ed Soja, Michael Storper, and Jennifer Wolch. Davis later provided a wry description of the putative School:

I am incautious enough to describe the "Los Angeles School." In a categorical sense, the twenty or so researchers I include within this signatory are a new wave of Marxist geographers or, as one of my friends put it, "political economists with their space suits on" although a few of us are also errant urban sociologists, or, in my case, a fallen labor historian. The "School," of course, is based in Los Angeles, at UCLA and USC, but is includes members in Riverside, San Bernardino, Santa Barbara, and even Frankfurt, West Germany (1989, p. 9).

The meeting, I can attest, was insightful as it was inconclusive, as exhilarating as hilarious. Davis described one evening as a:

...somewhat dispiriting retreat ...spent wrestling with ambiguity: "Are we the LA School as the Chicago School was the Chicago School, or as the Frankfurt School was the Frankfurt School?" Will the reconstruction of urban political economy allow us to better understand the concrete reality of LA, or is it the other way around? Fortunately, after a night of heavy drinking, we agreed to postpone a decision on this question. ...So in our own way we are as "laid back" and decentralized as the city we are trying to explain (pp. 9–10).

Despite these ambiguities and tensions (with their curious echoes of the experiences in the LA School of architecture recorded by Jencks), Davis is clear about the school's common theme:

One of the nebulous unities in our different research—indeed the very ether that the LA School mistakes for oxygen—is the idea of "restructuring." We all agree that we are studying "restructuring" and that it occurs at all kinds of discrete levels, from the restructuring of residential neighborhoods to the restructuring of global markets or whole regimes of accumulation (p. 10).

Davis also recorded some of the substantive contributions made by the school's early perpetrators:

To date [1989], the LA School has contributed original results in four areas. First, particularly in the work of Edward Soja and Harvey Molotch, it has given "placeness," as a social construction, a new salience in explaining the political economy of cities.

Secondly, via the case studies by Michael Storper, Suzanne Christopherson, and Allen Scott, it has deepened our understanding of the economies of high-tech agglomeration, producing some provocative recent theses about the rise of a new regime of "flexible accumulation." Thirdly, through both the writing and activism of Margaret FitzSimmons and Robert Gottlieb, it has contributed a new vision of the environmental movement, with emphasis on the urban quality of life. And, fourthly, through the collaboration of Michael Dear and Jennifer Wolch, it is giving us a more realistic understanding of the homeless and indigent, and their connection to the decline of unskilled inner city labor markets (p. 10).

Davis was, to the best of my knowledge, the first to mention a specific LA School of urbanism, and he repeated the claim in his popular 1990 contemporary history of Los Angeles, *City of Quartz* (1990). But truth be told, following those strange days of quasi-unity at Lake Arrowhead, the LA School had already begun to fracture, even as the floodgates opened and tentative claims for a prototypical LA began to flow.

Journalist Joel Garreau understood more clearly than most urban scholars where the country was heading. The opening sentences in his 1991 book, *Edge City*, proclaimed: "Every American city that is growing, is growing in the fashion of Los Angeles" (p. 3). By 1993, the trickle of Southern California studies had grown to a continuous flow. In his careful, path-breaking study of high technology in Southern California, Allen Scott noted:

Throughout the era of Fordist mass production, [LA] was seen as an exception, as an anomalous complex of regional and urban activity in comparison with what were then considered to be the paradigmatic cases of successful industrial development … [Yet] with the steady ascent of flexible production organization, Southern California is often taken to be something like a new paradigm of local economic development, and its institutional bases, its evolutionary trajectory, and its internal locational dynamics … providing important general insights and clues (1993, p. 33).

Charles Jencks added his own spin on the social forces underlying LA's architecture when he argued that:

Los Angeles, like all cities, is unique, but in one way it may typify the world city of the future: there are only minorities. No single ethnic group, nor way of life, nor industrial sector dominates the scene. Pluralism has gone further here than in any other city in the world and for this reason it may well characterize the global megalopolis of the future (p. 7).

The foundations of a putative school were completed in Marco Cenzatti's 1993 examination of the thing called an LA School of urbanism. Responding to Davis, he underscored the fact that the School's practitioners combined precepts of both the Chicago and Frankfurt Schools:

Thus Los Angeles comes ...into the picture not just as a blueprint or a finished paradigm of the new dynamics, but as a laboratory which is itself an integral component of the production of new modes of analysis of the urban (p. 8).

During the 1990s, the rate of scholarly investigations into LA accelerated, just as Scott and Soja predicted it would. For instance, in their 1993 study of homelessness in Los Angeles, Wolch and Dear situated their analysis within the broader matrix of LA's urbanism. However, the pivotal year in the maturation of the LA School may prove to be 1996, which saw the publication of three edited volumes on the region: *Rethinking Los Angeles* (Dear et al., 1996); *The City: Los Angeles and Urban Theory at the End of the Twentieth Century* (Scott and Soja, 1996); and *Ethnic Los Angeles* (Waldinger and Bozorgmehr, 1996). The 40 or more essays in these volumes represent a quantum leap in the collective understanding of the region and the implications of their new insights for national and international urbanisms. By 1996, there were also a growing number of university-based centers that legitimized scholarly and public-policy analyses of the region, among them USC's Southern California Studies Center, UCLA's Lewis Center for Regional Policy Studies, and Loyola Marymount University's Center for the Study of Los Angeles. Other institutions consolidated parallel interests in regional governmental and nongovernmental agencies, including the Getty Research Institute, the Public Policy Institute of California, and RAND.

But what were the substantive visions being offered by the Angelistas? How much did they differ from conventional wisdom?

## From Chicago to LA

The basic primer of the Chicago School was *The City*. Originally published in 1925, the book retains a tremendous vitality far beyond its interest as a historical document. I regard the book as emblematic of a modernist analytical paradigm that remained popular for most of the 20th century. Its assumptions included:

- A modernist view of the city as a unified whole, i.e., a coherent regional system in which the center organizes its hinterland;
- An individual-centered understanding of the urban condition; urban process in *The City* is typically grounded in the individual subjectivities of urbanites, their personal choices ultimately explaining the overall urban condition, including spatial structure, crime, poverty, and racism; and
- A linear evolutionist paradigm, in which processes lead from tradition to modernity, from primitive to advanced, from community to society, and so on.

There may be other important assumptions of the Chicago School, as represented in *The City*, that are not listed here. Finding them and identifying what is right or wrong about them is one of the tasks at hand, rather than excoriating the book's contributors for not accurately foreseeing some distant future.

The most enduring of the Chicago School models was the *zonal* or *concentric ring theory*, an account of the evolution of differentiated urban social areas by E. W. Burgess (1925). Based on assumptions that included a uniform land surface, universal access to a single-centered city, free competition for space, and the notion that development would take place outward from a central core, Burgess concluded that the city would tend to form a series of concentric zones. The main ecological metaphors invoked to describe this dynamic were invasion, succession, and segregation, by which populations gradually filtered outward from the center as their status and level of assimilation progressed. The model was predicated on continuing high levels of immigration to the city.

At the core of Burgess' schema was the Central Business District (CBD), which was surrounded by a transitional zone, where older private houses were being converted to offices and light industry, or subdivided to form smaller dwelling units. This was the principal area to which new immigrants were attracted; and it included areas of vice and unstable or mobile social groups. The transitional zone was succeed by a zone of working-men's homes, which included some of the city's oldest residential buildings inhabited by stable social groups. Beyond this, newer and larger dwellings were to be found, occupied by the middle classes. Finally, the commuters' zone was separate from the continuous built-up area of the city, where much of the zone's population was employed. Burgess' model was a broad generalization, and not intended to be taken too literally. He anticipated, for instance, that his schema would apply only in the absence of "opposing factors" such as local topography (in the case of Chicago, Lake Michigan).

He also anticipated considerable internal variation within the different zones.

Other urbanists subsequently noted the tendency for cities to grow in star-shaped rather than concentric form, along highways that radiate from a center with contrasting land uses in the interstices. This observation gave rise to a *sector theory* of urban structure, an idea advanced in the late 1930s by Homer Hoyt (1933, 1939), who observed that once variations arose in land uses near the city center, they tended to persist as the city expanded. Distinctive sectors thus grew out from the CBD, often organized along major highways. Hoyt emphasized that "nonrational" factors could alter urban form, as when skillful promotion influenced the direction of speculative development. He also understood that older buildings could still reflect a concentric ring structure, and that sectors may not be internally homogeneous at one point in time.

The complexities of real-world urbanism were further taken up in the *multiple nuclei* theory of C. D. Harris and E. Ullman (1945). They proposed that cities have a cellular structure in which land uses develop around multiple growth-nuclei within the metropolis as a consequence of accessibility-induced variations in the land-rent surface and agglomeration (dis)economics. Harris and Ullman also allowed that real-world urban structure is determined by broader social and economic forces, the influence of history, and international influences. But whatever the precise reasons for their origin, once nuclei have been established, general growth forces reinforce their preexisting patterns.

Much of the urban research agenda of the 20th century has been predicated on the precepts of the concentric zone, sector, and multiple nuclei theories of urban structure. Their influences can be seen directly in factorial ecologies of intra-urban structure, land-rent models, studies of urban economies and diseconomies of scale, and designs for ideal cities and neighborhoods. The specific and persistent popularity of the Chicago concentric ring model is harder to explain, however, given the proliferation of evidence in support of alternative theories. The most likely reasons for its endurance are related to its beguiling simplicity and the enormous volume of publications produced by adherents of the Chicago School (e.g., Abbott, 1999; Fine, 1995).

In the final chapter of *The City*, Louis Wirth (1925) provided a magisterial review of the field of urban sociology, titled (with deceptive simplicity, and astonishing self-effacement) "A Bibliography of the Urban Community." But what Wirth did in this chapter, in a remarkably prescient way, was to summarize the fundamental premises of the Chicago School, and to isolate two fundamental features of the urban condition that were to rise to prominence at the beginning of the 21st century. Specifically, Wirth established that the city lies at the center of, and provides the organizational logic for, a complex regional hinterland based on trade:

> Far from being an arbitrary clustering of people and buildings, the city is the nucleus of a wider zone of activity from which it draws its resources and over which it exerts its influence. The city and its hinterland represent two phases of the same mechanism which may be analyzed from various points of view (p. 182).

He also noted that the development of satellite cities is characteristic of the latest phases of city growth, and that the location of such satellites can exert a determining influence upon the direction of growth:

> One of the latest phases of city growth is the development of satellite cities. These are generally industrial units growing up outside of the boundaries of the administrative city, which, however, are dependent upon the city proper for their existence. Often they become incorporated into the city proper after the city has inundated them, and thus lose their identity. The location of such satellites may exert a determining influence upon the direction of the city's growth. These satellites become culturally a part of the city long before they are actually incorporated into it (p. 185).

Wirth further observed that modern communications have transformed the world into a single mechanism, where the global and the local intersect decisively and continuously:

> With the advent of modern methods of communication the whole world has been transformed into a single mechanism of which a country or a city is merely an integral part. The specialization of function, which has been a concomitant of city growth, has created a state of interdependence of world-wide proportions. Fluctuations in the price of wheat on the Chicago Grain Exchange reverberate to the remotest part of the globe, and a new invention anywhere will soon have to be reckoned with at points far from it origin. The city has become a highly sensitive unit in this complex mechanism, and in turn acts as a transmitter of such stimulation as it receives to a local area. This is as true of economic and political as it is of social and intellectual life (p. 186).

And there, in a sense, you have it. From a few, relatively humble first steps, we gaze out over the abyss—the yawning gap of an intellectual fault line separating Chicago from Los Angeles. In a few short paragraphs, Wirth anticipated the pivotal moments that characterize Chicago-style urbanism—those primitives that eventually will separate it from an LA-style urbanism. He effectively foreshadowed *avant la lettre* the shift from what I term a "modern" to a "postmodern" city, and, in so doing, the necessity of the transition from the Chicago to the LA School. For it is no longer the center that organizes the urban hinterlands, but the hinterlands that determine what remains of the center. The imperatives of fragmentation have become the principal dynamic in contemporary cities; the 21st century's emerging world cities (including LA) are ground-zero loci in a communications-driven globalizing political economy.

The shift toward an LA School may be regarded as a move away from modernist perspectives on the city (à la Chicago School) to a postmodern view of urban process. We are all by now aware that the tenets of modernist thought have been undermined, discredited; in their place, a multiplicity of new ways of knowing have been substituted. Analogously, in postmodern cities, the logics of previous urbanisms have evaporated; and, absent a single new imperative, multiple (ir)rationalities clamor to fill the vacuum. The LA School is distinguishable from the Chicago precepts (as noted above) by the following counterpropositions:

- Traditional concepts of urban form imagine the city organized around a central core; in a revised theory, the urban peripheries are organizing what remains of the center.
- A global, corporate-dominated connectivity is balancing, even offsetting, individual-centered agency in urban processes.
- A linear evolutionist urban paradigm has been usurped by a nonlinear, chaotic process that includes pathological forms such as common-interest developments (CIDs), and life-threatening environmental degradation (e.g., global warming).

In empirical terms, the urban dynamics driving these tendencies are by now well known. They include: *World City*: the emergence of a relatively few centers of command and control in a globalizing economy; *Dual City*: an increasing social polarization, i.e., the increasing gap between rich and poor, between nations, between the powerful and the powerless, between different ethnic, racial, and religious groupings, and between genders; *Hybrid City*: the ubiquity of fragmentation both in material and cognitive life, including the collapse of conventional communities, and the rise of new cultural categories and spaces, including especially cultural hybrids;

and *Cybercity*: the challenges of the information age, especially the seemingly ubiquitous capacity of connectivity to supplant the constraints of place.

"Keno capitalism" is the synoptic term that Steven Flusty and I have adopted to describe the spatial manifestations that are consequent upon the (postmodern) urban condition implied by these assumptions. Urbanization is occurring on a quasi-random field of opportunities in which each space is (in principle) equally available through its connection with the information superhighway (Dear and Flusty, 1998). Capital touches down as if by chance on a parcel of land, ignoring the opportunities on intervening lots, thus sparking the development process. The relationship between development of one parcel and nondevelopment of another is a disjointed, seemingly unrelated affair. While not truly a random process, it is evident that the traditional, center-driven agglomeration economies that have guided urban development in the past no longer generally apply. Conventional city form, Chicago-style, is sacrificed in favor of a noncontiguous collage of parcelized, consumption-oriented landscapes devoid of conventional centers yet wired into electronic propinquity and nominally unified by the mythologies of the (dis)information superhighway. In such landscapes, "city centers" become almost an externality of fragmented urbanism; they are frequently grafted onto the landscape as a (much later) afterthought by developers and politicians concerned with identity and tradition. Conventions of "suburbanization" are also redundant in an urban process that bears no relationship to a core-related decentralization.

I am insisting on the term "postmodern" as a vehicle for examining LA urbanism for a number of reasons, even though many protagonists in the debates surrounding the LA School have explicitly distanced themselves from the precepts of postmodernism. I have long understood postmodernism as a concept that embraces three principal referents:

- A series of *distinctive cultural* and *stylistic* practices that are in and of themselves intrinsically interesting;
- The totality of such practices, viewed as a *cultural ensemble characteristic* of the contemporary *epoch of capitalism* (often referred to as postmodernity) ; and
- A set *philosophical* and *methodological discourses antagonistic to the precepts of Enlightenment thought*, most particularly the hegemony of any single intellectual persuasion.

Implicit in each of these approaches is the notion of a "radical break," i.e., a discontinuity between past and present political, sociocultural and economic trends. My working hypothesis is that there is sufficient evidence to support the notion that we are witnessing a radical break in each of these three categories. This is the fundamental promise of the revolution prefigured by the LA School; this is why it is so revolutionary in its recapitulation of urban theory.

The localization (sometimes literally the concretization) of these diverse dynamics is creating the emerging time-space fabric of a postmodern society. This is not to suggest that existing (modernist) rationalities have been obliterated from the urban landscape or from our mind-sets; on the contrary, they persist as palimpsests of earlier logics, and continue to influence the emerging spaces of postmodernity. For instance, they are presently serving to consolidate the power of existing place-based centers of communication technologies, even as such technologies are supposed to liberate development from the constraints of place. However, newer urban places, such as LA, are being created by different intentionalities, just as older places such as Chicago are being overlain by the altered intentionalities of postmodernity. Nor am I suggesting that earlier theoretical logics have been (or should be) entirely usurped. For instance, in his revision of the Chicago School, Andrew Abbott (1999, p. 204) claimed that the "variables paradigm" of quantitative sociology has been exhausted, and that the "cornerstone of the Chicago vision was location"—points of departure that I regard as totally consistent with the time-space obsessions of the LA School of postmodern urbanism. Another example of overlap between modern and postmodern in current urban sociology is Michael Peter Smith's evocation of a transnational urbanism (Smith, 2001).

## Comparative Urbanism

Since its inception, the writings on the LA School have generated a significant criticism, as should be expected. I will not attempt to survey these critiques, since this would require a separate essay in order to do justice to the volume of work. Suffice it to say that the complaints have sprung from many sources, including those who persist in the belief that Los Angeles adds nothing that is not already known to contemporary urban theory, those who are opposed to postmodernism, those who object to the perceived dystopianism of the LA School practitioners, and those who simply dislike Los Angeles. Despite their attacks, the literature on Los Angeles has flourished. What began as inquiry into economic restructuring has blossomed into a wide-ranging critique of urban history (Hise, 1997), environmentalism (Pincetl, 1999; Wolch and Emel, 1998), race (Bobo et al., 2000; Pulido, 2000; Roseman et al., 1996), cultural diversity (James, 2003; Kenny, 2001), and internationalism (Cartier, 2002; Flusty, 2003; Heikkila and Pizarro, 2002). Moreover, the theoretical challenge posed by the LA

School has generated a far-ranging debate on general urban theory (Harris, 1997; Keil, 1998; Beauregard, 2003; Brenner, 2003; Dear, 2003), which has spilled over into the general media (Miller, 2000).

The opening created by the LA School has also been exploited by others anxious to challenge the traditions of Chicago and LA. Recently, a "Miami School" (Nijman, 1996, 1997; Portes and Stepnick, 1993); a "Las Vegas School" (Gottdeiner et al., 1999); and even an "Orange County School" have arisen. For example, Gottdeiner and Kephart (1991, p. 51) claimed that in Orange County:

> We have focused on what we consider to be a new form of settlement space the fully urbanized, multinucleated, and independent county. ...As a new form of settlement space, they are the first such occurrence in five thousand years of urban history.

While those who are familiar with Orange County might regard this assertion as a somewhat exaggerated, if not entirely melodramatic, gesture such counterclaims are in fact an important piece of the comparative urban discourse that I believe to be the single most important research item in contemporary urban theory.

Let me conclude by elaborating on the challenges posed by a polyvocal urban discourse. In these postmodern times, the gesture to an LA School might appear to be a deeply contradictory intellectual strategy. And yet, despite its inherent plurality, the notion of a "school" has semantic overtones of codification and hegemony; it has structure and authority. Modernists and postmodernists alike might shudder at the irony implied by these associations. And yet, ultimately, I am comfortable in proclaiming the existence of an LA School of urbanism or two reasons. First, the LA School *exists as a body of literature*. It exhibits an evolution through time, beginning with analysis of Los Angeles as an aberrant curiosity distinct from other forms of urbanism. The tone of that history has shifted gradually to the point that the city is now commonly represented as indicative of a new form of urbanism, supplanting the older forms against which Los Angeles was once judged deviant. Second, the LA School *exists as a discursive strategy* demarcating a space both for the exploration of new realities and for resistance to old hegemonies. It is proving to be far more successful than its detractors at explaining the form and function of the urban.

Still, I acknowledge the danger that an LA School could be perceived as yet another panoptic fortress from whence a new totalizing urban model is manufactured and marketed, running roughshod over divergent ways of seeing like the hegemonies it supplanted. The danger of creating a new "master" narrative stands at every step of this project: in defining the very boundaries of an LA School itself; in establishing a unitary model of Los Angeles; and in imposing a template of Los Angeles upon the rest of the world. Let me consider these threats in turn.

The fragmented and globally oriented nature of the Los Angeles School will counter any potential for a new hegemony. Those who worry about the hegemonic intent of an LA School may rest assured that its adherents are in fact pathologically antileadership. Nor will everyone who writes on LA readily identify as a member of the LA School; some adamantly reject such a notion (e.g., Ethington and Meeker, 2001). The programmatic intent of the LA School remains fractured, incoherent, and idiosyncratic even to its constituent scholars, who most often perceive themselves as occupying a place on the periphery rather than at the center. The LA School promotes inclusiveness by inviting as members all those who take Los Angeles as a worthy object of study on a contemporary urbanism. Such a School evades dogma by including divergent empirical and theoretical approaches rooted in philosophies both modern and postmodern, from Marxist to Libertarian. Admittedly, such a school will be a fragmentary and loosely connected entity, always on the verge of disintegration—but, then again, so is Los Angeles itself.

A unified, consensual description of Los Angeles is equally unlikely, since it would necessitate excluding a plethora of valuable readings on the region. For instance, numerous discursive battles have been fought since the events of April 1992 to decide what term best describes them. Those who read the events as a spontaneous reaction to the acquittal of Rodney King employ the term *riot*. For those who read the events within the context of economic and social polarization, the term *uprising* is preferred. And those who see in them a more conscious political intentionality apply the term *rebellion*. For its part, civic authority skirts these issues by relying upon the supposedly depoliticized term, *civil unrest*. But others concerned with the perspective of the Korean participants deploy the Korean tradition of naming an occurrence by its principal date, and so make use of the term, *Sa-I-Gu*. The loosely-constituted polyvocality of the Los Angeles School permits us to replace the question "Which name is definitive?" with "Which names should we use, at what stage in the unfolding events, at which places in the region, and from whose perspective?" Such an approach may well entail a loss of clarity and certitude, but in exchange it offers a richness of description and interpretation that would otherwise be forfeited in the name of achieving a homogeneous narrative.

Finally, the temptation to adopt LA as a world city template is avoidable because the urban landscapes of Los Angeles are not necessarily original to LA: the luxury compound atop a matrix of impoverished misery or self-contained communities of fortified homes can also be found in places like Manila and São Paulo. The LA School justifies a presentation of LA not as *the* model of contemporary urbanism, nor as the privileged locale from whence a cabal of solipsistic theorists issues proclamations about the way things are, but as one of a number space-time geographical prisms through which current processes of urban (re)formation may be advantageously viewed. Thus, the School categorically does not represent an emerging vision of contemporary urbanism via a single, hegemonic "paradigm;" instead it is but one component in an emerging new comparative urban studies working out of Los Angeles but inviting the participation of (and placing equal importance upon) the on-going experiences and voices of Tijuana, São Paulo, Hong Kong, and the like. One consequence of a postmodern perspective is the insistence that all theoretical voices should be heard. And to put it bluntly, we (as urbanists) need all the help we can get, if we are to understand contemporary cities.

## References

Abbott, A., 1999, *Department and Discipline: Chicago Sociology at One Hundred*. Chicago, IL: University of Chicago Press.

Abbott, A., 2002, Los Angeles and the Chicago School: A comment on Michael Dear. *City and Community*, Vol. 1, 33–38.

Banham, R., 1971, *Los Angeles: The Architecture of Four Ecologies*. Harmondsworth, England: Penguin.

Beauregard, R., 2003, City of superlatives. *City and Community*, Vol. 2, No. 3, 183–199.

Bobo, L. D., Oliver, M. L., Johnson, J. H., and Valenzuela, A., editors, 2000, *Prismatic Metropolis: Inequality in Los Angeles*. New York, NY: Russell Sage.

Brenner, N., 2003, Stereotypes, archetypes, and prototypes: Three uses of superlatives in contemporary urban studies. *City and Community*, Vol. 2, 205–216.

Burgess, E. W., 1925, The growth of the city. In R. E. Park, E. W. Burgess, and R. McKenzie, editors, *The City: Suggestions of Investigation of Human Behavior in the Urban Environment*. Chicago, IL: University of Chicago Press, pp. 47–62.

Cartier, C., 2002, Transnational urbanism in the reform-era Chinese city: Landscapes from Shenzhen. *Urban Studies*, Vol. 39, 1513–1532.

Cenzatti, M., 1993, *Los Angeles and the L. A. School: Postmodernism and Urban Studies*. Los Angeles, CA: Los Angeles Forum for Architecture and Urban Design.

Clark, T., 2002, Codifying LA chaos. *City and Community*, Vol. 1, 51–57.

Davis, M., 1989, Homeowners and homeboys: Urban restructuring in L.A. *Enclitic*, Summer, 9–16.

Davis, M., 1990, *City of Quartz: Excavating the Future in Los Angeles*. New York, NY: Verso.

Dear, M., 2000, *The Postmodern Urban Condition*. Oxford, UK: Blackwell.

Dear, M., editor, 2001, *From Chicago to LA: Making Sense of Urban Theory*. Thousand Oaks, CA: Sage.

Dear, M., 2002, Los Angeles and the Chicago School: Invitation to a debate. *City and Community*, Vol. 1, 5–32.

Dear, M., 2003, Superlative urbanisms: The necessity for rhetoric in social theory. *City and Community*, Vol. 2, 201–204.

Dear, M. and Flusty, S., 1998, Postmodern urbanism. *Annals of the Association of American Geographers*, Vol. 88, 50–72.

Dear, M. and Flusty, S., editors, 2001, *The Spaces of Postmodernity: A Reader in Human Geography.* Oxford, UK: Blackwell.

Dear, M. J., Schockman, H. Eric, and Hise, G., editors, 1996, *Rethinking Los Angeles.* Thousand Oaks, CA: Sage.

Ethington, P. and Meeker, M., 2001, Saber y conocer: The metropolis of urban inquiry. In M. Dear, editor, *From Chicago to LA: Making Sense of Urban Theory.* Thousand Oaks, CA: Sage, 403–420.

Fine, G. A., editor, 1995, *A Second Chicago School?: The Development of a Postwar American Sociology.* Chicago, IL: University of Chicago Press.

Folgelson, R. M., 1993 [1967], *The Fragmented Metropolis: Los Angeles 1850–1970.* Berkeley, CA: University of California Press.

Flusty, S., 2003, *De-Coca-Colonization: Making the World From the Inside Out.* New York, NY: Routledge.

Garreau, J., 1991, *Edge City: Life on the New Frontier.* New York, NY: Anchor.

Gottdeiner, M., Collins, C. C., and Dickens, D. R., 1999, *Las Vegas: The Social Production of an All-American City.* Oxford, UK: Blackwell.

Gottdeiner, M. and Klephart, G., 1991, The multinucleated metropolitan region. In R. Kling, O. Spencer, and M. Poster, editors, *Postsuburban California: The Transformation of Orange County since World War II.* Berkeley, CA: University of California Press, 31–54.

Harris, C. D. and Ullman, E. L., 1945, The nature of cities. *Annals of the American Academy of Political and Social Science,* Vol. 242, pp. 7–17.

Harris, C. D., 1997, The nature of cities, *Urban Geography,* Vol. 18, 15–35.

Heikkila, E. J. and Pizarro, R., editors, 2002, *Southern California and the World.* Westport, CT: Praeger.

Hise, G., 1997, *Magnetic Los Angeles: Planning the Twentieth-Century Metropolis.* Baltimore, MD: The Johns Hopkins University Press.

Hoyt, H., 1933, *One Hundred Years of Land Values in Chicago.* Chicago, IL: University of Chicago Press.

Hoyt, H., 1939, *The Structure and Growth of Residential Neighborhoods in American Cities.* Washington, DC: United States Federal Housing Administration.

James, D., editor, 2003, *The Sons and Daughters of Los: Culture and Community in L.A.* Philadelphia, PA: Temple University Press.

Jencks, C., 1993, *Heteropolis: Los Angeles, the Riots, and the Strange Beauty of Hetero-architecture.* New York, NY: St. Martin's.

Keil, R., 1998, *Los Angeles.* New York, NY: Wiley.

Kenny, M. R., 2001, *Mapping Gay L.A.* Philadelphia, PA: Temple University Press.

McWilliams, C., 1973 [1946], *Southern California: An Island on the Land.* Santa Barbara, CA: Peregrine Smith.

Miller, D. W., 2000, The new urban studies. *Chronicle of Higher Education,* Vol. 50, August 18, A15–A16.

Molotch, H., 2002, School's out: A response to Michael Dear. *City and Community,* Vol. 1, 39–43.

Nijman, J., 1996, Breaking the rules: Miami in the urban hierarchy. *Urban Geography,* Vol. 17, 5–22.

Nijman, J., 1997, Globalization to a Latin beat: The Miami Growth Machine. *Annals of the American Academy of Political and Social Science,* Vol. 551, 163–176.

Pincetl, S., 1999, *Transforming California: A Political History of Land Use and Development.* Baltimore, MD: The Johns Hopkins University Press.

Portes, A. and Stepick, A., 1993, *City on the Edge: The Transformation of Miami.* Berkeley, CA: University of California Press.

Pulido, L., 2000, Rethinking environmental racism: White privilege and urban development in Southern California. *Annals of the Association of American Geographers,* Vol. 90, 12–40.

Robinson, C. M., 1907, *The City Beautiful.* Report to the Mayor, City Council and Members of the Municipal Art Commission. Los Angeles, CA: City of Los Angeles.

Roseman, C., Laux, H., and Thieme, G., editors, 1996, *EthniCity.* Lanham, MD: Rowman and Littlefield.

Sampson, R. J., 2002, Studying modern Chicago. *City and Community,* Vol. 1, 45–48.

Sassen, S., 2002, Scales and spaces. *City and Community,* Vol. 1, 48–50.

Scott, A..J. 1993, *Technopolis: High-Technology Development and Regional Development in Southern California*. Berkeley, CA: University of California Press.

Scott, A. J., and Soja, E. W., 1986, Los Angeles: Capital of the late 20th century. *Society and Space*, Vol. 4, 249–254.

Scott, A. J. and Soja, E. W., editors, 1996, *The City: Los Angeles and Urban Theory at the End of the Twentieth Century*. Berkeley, CA: University of California Press.

Smith, M. P., 2001, *Transnational Urbanism: Locating Globalization*. Oxford, UK: Blackwell.

Soja, E. W., 1986, Taking Los Angeles apart: Some fragments of a critical human geography. *Environment and Planning D: Society and Space*, Vol. 4, 255–272.

Soja, E. W., 1989, *Postmodern Geographies: The Reassertion of Space in Critical Social Theory*. New York, NY: Verso.

Trumble, W. R., Brown, L., Stevenson, A., and Siefring, J., editors, 1999, *Shorter Oxford English Dictionary*. Oxford, UK: Oxford University Press.

Waldinger, R. and Bozorgmehr, M., editors, 1996, *Ethnic Los Angeles*. New York, NY: Russell Sage.

Wirth, L., 1925, A bibliography of the urban community. In R. E. Park, E. W. Burgess, and R. McKenzie, editors, *The City: Suggestions of Investigation of Human Behavior in the Urban Environment*. Chicago, IL: University of Chicago Press, pp. 161–228.

Wolch, J. and Dear, M., 1993, *Malign Neglect: Homelessness in an American City*. San Francisco, CA: Jossey Bass.

Wolch, J. and Emel, J., editors, 1998, *Animal Geographies*. London, UK: Verso.

CHAPTER 25

# Unbounding Critical Geographic Research on Cities
## *The 1990s and Beyond*[1]

HELGA LEITNER AND ERIC SHEPPARD[2]

ABSTRACT

Critical urban geography came to dominate knowledge production in urban geography during the 1990s. This extremely fruitful research program nevertheless faces certain bounds to knowledge production that hinder its ongoing vitality. With respect to conceptual approaches, there has been considerable unbounding from an earlier focus on class and commodity production, recognizing the importance of other axes of

[1]We thank Ryan Holifield for background research, and him and Bob Lake for useful comments on an earlier version of this chapter.
[2]Correspondence concerning this chapter should be addressed to Helga Leitner, Department of Geography, University of Minnesota, Minneapolis, MN 55455; telephone: 612-625-9010; fax: 612-624-1044; e-mail: Helga.Leitner-1@umn.edu

social differentiation for urban processes and urban lives. Transcending a bounded conception of cities as objects of inquiry has been slower. With respect to philosophical foundations, strenuously contested philosophical boundaries persist between critical and other urban geography and within critical urban geography. With respect to participation in knowledge production, disciplinary boundaries marginalize knowledge situated in non-White communities and the global south, and critical urban geographers face criticism for failing to broaden knowledge production beyond the Ivory Tower. Notwithstanding a good track record of policy-oriented research, critical urban geographers need to decenter knowledge production, by including disadvantaged communities as full partners in this process. Helen Longino provides critical urban geographers a model for unbounded, nonrelativist and nonmonist knowledge production. [Key words: Cities, critical geography, knowledge production, activist research.]

It is some 30 years since David Harvey published *Social Justice and the City*, thereby inaugurating what has since become the critical urban geographic research program, the subject of this paper (Harvey, 1973). The 1990s was a period during which the research program flourished, as its more prominent proponents came to occupy positions of power and influence within English-speaking geography, but it was also a period of sweeping internal change as post-structuralism and feminism challenged political economy, reflecting the broader cultural turn in human geography in these countries.[3] We do not seek to provide here a comprehensive review of critical urban geography during the 1990s, and still less of urban geography more generally. The literature is by now far too large to receive this kind of treatment. In addition, any assessment of necessity reflects the situated perspective of its authors. Thus our treatment is avowedly and consciously partial. We focus on critical urban geographic research published in English, with particular attention to those aspects that have intersected with our own work. We will focus more on the future than the past, seeking to initiate a conversation about the future of both critical urban geographic research and of urban geography more generally.

We examine here the production of knowledge within the critical urban geographic research program. Like any active research program, critical

---

[3]We use the term poststructuralism to capture a variety of research that also includes significant strands of feminism, postmodernism, and postcolonialism, as all are informed by a similar poststructural philosophical sensibility.

urban geography has experienced an evolution of themes, questions, and conceptual and philosophical approaches, some becoming fashionable while others fade or never get taken up. A research program only remains lively, however, if existing presuppositions and blind spots are reflexively identified, challenged and transcended: a process that we call unbounding (noting, but not sharing David Sibley's (2001) pessimism about the difficulty of unbounding). We highlight three aspects of knowledge production in critical urban geography: conceptual approaches, forming a starting point for studying cities; philosophical and disciplinary foundations, deployed to justify particular conceptual approaches; and participants, referring to who is involved in knowledge production. We will discuss the variable efforts devoted to unbounding in these three areas, and the work remaining to be done. While each section of the paper focuses on one of these, the three are of course closely interrelated.

In terms of conceptual approaches, some unbounding is already underway. The internal changes within the critical urban research program that came to fruition during the 1990s sought to release the program from preexisting presumptions that urban analysis must begin with class and commodity production, and was in many ways successful. We trace this shift briefly in the first section of the paper. Yet we also argue that the critical urban research agenda retains a spatially bounded conception of cities. There has been a tendency to treat cities as isolable objects of inquiry: More attention is given to their internal functioning, and the diversity of social relations and livelihoods within cities, than to connectivities extending beyond the city. Cities have also been treated largely as separated from nature, and rural areas.

By contrast, as we discuss in section two, philosophical and disciplinary boundaries have been energetically defended and contested. Boundaries persist around geography, between critical and other schools of urban geography and within critical urban geography. Much energy is spent not only on policing such boundaries, but also on extending them to incorporate other approaches in order to construct hegemonic perspectives against which competing approaches can be measured and found wanting. Notwithstanding the dangers of unreflective eclecticism, urban geography too often takes the form of a destructive competition among competing world views. Such debates keep the field lively, but can also undermine constructive progress. In seeking to unbound ourselves from excesses of criticism and self-promotion, we argue that it is important to recognize that all knowledge production is situated, shaped both by researchers' social location (e.g., their gender, race, class), and by their location in time (e.g., when socialization into geography occurred, changes along the career life-path) and space (located predominantly in Anglo-American institutions). Drawing on Helen Longino (2002), we offer a vision for engaging differently situated intellectual perspectives, including those outside critical urban geography, in a rigorous but nonrelativist intellectual debate. This offers the possibility of replacing power struggles and rivalries with a strongly democratic exchange of ideas, and forced consensus with life-long learning. We see this as a model for helping critical theorists to face up to our biggest challenge: critically and reflexively reevaluating our own preconceptions.

Third, bounds persist in terms of who participates in knowledge production. Some limits are imposed by geography's inability to unbound itself from a White and Anglo-American bias in its practitioners (section two). Knowledge production also needs to be taken beyond the Ivory Tower, however (section three). While critical urban geography has engaged in policy-relevant research and writing, we suggest that this is not enough. Critical urban geographers must also work in and with disadvantaged urban communities, helping them gain full voice in debates about urban futures. This requires critical urban geographers to decenter their knowledge production outside the academy, accepting members of disadvantaged communities as full partners in this process. This can facilitate a process through which the voices of these communities are brought to bear on policy-making, unmediated by categories and discourses imposed by researchers, politicians and the media. Longino's scheme is also relevant here, although extending it to nonacademic contexts poses particular difficulties.

## Unbounding Class, Commodity Production and Cities

Critical urban geographic research, as it emerged out of radical geography in the 1980s, was to a large degree conceptualized within a Marxian political economy problematic. In this view, capitalism produced its own distinctive spatial forms, one of which was the capitalist city. Critical urban geographers thus sought to unravel how the capitalist mode of production shaped the dynamics and spatial structure of the capitalist city. In the first instance, therefore, the capitalist city was seen as a locus of production: a place where capital and labor gathered to engage in commodity production, and where, therefore, capital-labor relations were particularly intense and spatially concentrated (Harvey, 1985; Sheppard and Leitner, 1989). In short, the capitalist city was treated as a laboratory for making sense of the capitalist space economy. One central set of questions addressed how capitalist commodity production shaped both the spatial structure of the city and spatial segregation and inequality along class lines, and how the ensuing spatial arrangements themselves enhance and shape capitalist social relations (Scott, 1980). For example, research examined how the urban built environment contributes to and embodies the capitalist dynamic, as well as the role of different tiers of the state in this process (Dear and Scott, 1981; Harvey, 1985; Smith and Williams, 1987). Conceptualizing the state in Marxian terms meant that its role was seen as seeking to negotiate the competing interests of capital and labor, while unable to transcend the contradictions of capitalism. A second set of questions emerged around cities as places of collective consumption, including struggles over the distribution of collective goods and the implications of an unequal distribution of public resources for such processes as suburbanization (Castells, 1977; Walker, 1981).

Beginning in the late 1980s, and throughout the 1990s, critical urban geographic research progressively moved away from commodity production and class. It also moved away from an emphasis on the spatial as an outcome of the social/economic (e.g., the capitalist city as a consequence of capitalism), to recognize that the relationship between the two is a mutually constitutive one in which the spatial also shapes the social (Soja, 1989; Massey, 1994). Indeed, through their theorization of the problematic of space, critical urban geographers have been on the forefront in developing critical urban theory not just for the sake of explanation and understanding, but also for the purpose of constructing both a conceptual apparatus to guide the production of urban space, and strategies for achieving more socially just cities (Harvey, 1989; Harvey, 1992; Smith, 1994; Amin and Thrift, 1995).

Beginning with feminist debates about the relationship between gender and class in the city, critical urban researchers sought to move beyond class to emphasize the importance of other axes of social differentiation for understanding urban processes and urban lives (McDowell, 1999). More generally, the introduction of a difference perspective, recognizing the diverse bases of social affiliation (gender, ethnicity, race, class, sexual orientation, nationality) and multiple oppression, has not only afforded a deeper understanding of the diversity of urban experiences and differentials of power, but also brought greater attention to everyday practices, meanings and identities. This move is part of what has been referred to as the cultural turn, in which notions of difference and identity, performance, representation, meanings and symbolism are seen as at least as important as political economic processes in understanding cities and urban lives (Bridge and Watson, 2000; Dear, 2000; see also Barnes, this issue). While itself a very diverse body of scholarship, the cultural turn has produced new questions and understandings of difference and the material foundations of cities and the urban experience, extending insights produced by political economy approaches (Fincher and Jacobs, 1998; Fincher et al.,

2002; Watson, 2002). To date, however, race has received less attention than other aspects of difference (but see Anderson, 1991; Delaney, 1998; Wilson, 2000; Peake and Kobayashi, 2002).

Critical urban geographers in the 1990s also developed more sophisticated understandings of the relationships between the economy, the state, and civil society, whereby urban restructuring and politics are not seen as simply reducible to capitalist economic imperatives. Drawing on and extending conceptions of capitalist regulation, urban regimes and growth machines, scholars have highlighted the crucial role of social and institutional factors in understanding urban governance and restructuring. Most recently, critical geographers have examined the profound ramifications of the rise of neoliberalism as a dominant ideology of statecraft for shaping urban change (Brenner and Theodore, 2002a; Peck and Tickell, 2002). Attention is also now being paid to social movements and grassroots initiatives contesting dominant urban development paradigms and their implications for city life and livelihoods, thereby attempting to redefine the boundaries and power relations between the state, market and civil society (Merrifield, 2002; Mitchell, 2003).

One characteristic emerging from this body of research is that it extends the subject matter of urban geography beyond the bounded territory of the city (or metropolitan area). Rather than treating the city as an isolable laboratory for analysis, this research emphasizes the ways in which cities, urban life and politics are embedded in supra-local processes, particularly those of globalization. At the center of this research is a focus on geographic scale, and relations between scales (e.g., local-global relations), showing that urban politics cannot be separated from analysis of other scales, or from processes of rescaling, e.g., the implications of glocalization for cities and urban politics (Brenner, 1997; Swyngedouw, 1997; Leitner, 2003).

Scale and the embeddedness of cities in larger-scale processes also figure prominently in research on the world city and global cities. Initially, this research, like that on the capitalist city, focused on global cities as an outcome of globalization, specifically studying how the economic aspects of globalization (international flows of capital, commodities and labor) shape global cities (Sassen, 1991). More recently increased attention has been paid to cultural and discursive aspects of globalization and the global-local dynamic, and their impact on social and cultural relations within global cities, as well as how local events and politics in global cities influence the trajectory of globalization (Thrift, 1994; Knox and Taylor, 1995). The focus on global cities, however, has been at the cost of a large-scale neglect of the millions of other cities, except for some work on "second tier" cities (Markusen et al., 1999; Sassen, 2002). This neglect implies that it is sufficient to study global cities in order to understand the reciprocal relationship between global and urban processes, but other cities also shape globalization. Consider, for example, how antiglobalization protests in Seattle and Genoa have influenced international debates on globalization, or how the transnational ties of new immigrants in small towns across the USA and Europe are shaping globalization from below.

This emphasis on how cities are embedded in larger-scale processes has not been matched by an equivalent attention to the many connectivities that stretch across city and metropolitan boundaries, positioning cities within broader networks that shape intra- and inter-urban change (Amin and Thrift, 2002). It is only recently, for example, that positionality in global networks has been taken seriously as shaping the emergence of global cities, challenging previous attempts to define the global status of cities in terms of such internal characteristics as the number of TNC headquarters (Beaverstock et al., 2000). Indeed critical urban political economy research was dominated during the 1990s by the idea that the prosperity of cities is largely shaped by their internal characteristics (industrial districts, relational assets, social capital) without fully taking into account the importance of trans-local, inter-urban connections (Sheppard, 2002). Amin and Thrift (2002) have identified similar gaps in poststructural critical urban geography.

Understanding cities as spatially open and connected also requires that critical urban geographers put further effort into unbounding such firmly entrenched dualisms as urban and rural, and cities and nature. Treating cities as if they were isolated from nature and the countryside leads to a neglect of their mutual constitution, and its implications for cities and urban life. Critical human geographers have studied nature-society relations in political ecology, but largely in rural and Third World geographic contexts. It is only recently that urban geographers have pointed out that cities not only have effects on nature, but exist within and are permeated by nature (Swyngedouw, 1996; Pile, 1999; Lynn and Sheppard, 2003). The same applies to the urban rural divide. There is an emerging literature on animals in the city (Wolch, 2002), urban gardening (Kurtz, 2001), and urban-rural commodity chains connecting urban consumers back to rural economies (Cook and Crang, 1996; Whatmore and Thorne, 1997; Hartwick, 1998), In addition, research on environmental justice in cities is highlighting social struggles over the right for a clean urban environment for all urban residents (Pulido, 2000). While this literature extends the previous emphasis in urban geography on the city as a purely economic, political, social and cultural space, it still only scratches the surface of a fertile research agenda on cities and nature, and city-related human-environment interactions. With the exception of environmental justice research, the proliferation of conceptual statements has yet to be matched by a rich body of empirical research.

## Challenging Disciplinary and Philosophical Boundaries

In this section, we examine the bounds that have emerged around and within critical urban geographic academic knowledge production. Building on the principle that knowledge production is situated, we discuss three scales of boundary-making and boundary battles: around geography as a discipline, between critical and other urban geographic research, and between philosophical schools within critical urban geography. While contestation is a vital ingredient in knowledge production, the entrenched boundary battles that often ensue both reduce the richness of intellectual exchange and marginalize certain voices and research questions. We offer Helen Longino's normative vision as a way forward, to transgress entrenched boundaries and facilitate a more productive intellectual exchange.

It is now widely accepted by philosophers of science that all knowledge production is situated, meaning that every producer of knowledge looks at the world through a particular set of ontological, epistemological and methodological spectacles that s/he generally treats as rose-tinted (Hacking, 1983; Haraway, 1991; Harding, 1991; Longino, 2002). It follows that the diversity of approaches to knowledge production necessarily reflects the diversity of experiences of those able to gain recognition as legitimate participants in this process. As noted above, making more space for women in geography catalyzed feminist urban geography, and thereby critical urban geography. While men could and did study women in the city, it was female geographers' situated knowledge, grounded in their own gendered lives and upbringing, that catalyzed a new set of questions and methods that have vastly enriched our understanding of cities and urban lives. The same can be said about challenging geography's homophobic past and our now much richer understanding of sexuality and urban space (cf. Pile, 1996; Valentine, 2001). Yet other groups have not been brought into Anglo-American geography's tent, impoverishing our understanding of other issues.

For example, with a total of 44 African American faculty in U.S. geography departments (Joe Darden, professor, geography and urban affairs, Michigan State University, pers. comm., January 24, 2003), it is not surprising that research on race in the city remains an area where critical geographic research lags behind other critical urban research, as noted above. Despite the centrality of race to daily urban life and to the socio-spatial ordering of cities, the paucity of research on these questions reflects the paucity of non-White geographers (Pulido, 2002; a comparison to Susan Christopherson's [1989] plea at the dawn of feminist urban geography is instructive). Similarly, cities of the global south have received much less sustained attention, particularly those located outside English-speaking regions of the global south, than have European, North American or Australian cities. While leading Anglo-American critical urban geographers occasionally call attention to such biases (Jackson, 1988; McGee, 1991; Robinson, 2003), English-speaking urban geography includes a dearth of the scholars from the global south who can articulate a distinctive research agenda grounded in their situated perspectives on urban livelihoods. There has been a recent surge of work on Asian cities, for example, correlating with the increased number of geographers of Asian origin working within Anglo-American institutions, as well as the presence of Anglo-American geographers in academic institutions in the former colonies of Singapore and Hong Kong (Yeung, 1998; Logan, 2002). Notwithstanding its concern for bringing attention to the disempowered, critical urban geography shares exclusions that are endemic to the discipline as a whole. Yet critical geographers should be particularly motivated to help overcome them.

Within geography, critical urban geography has gained influence in large part by effectively constructing and critiquing its Other: mainstream urban geography. "Critical" is potentially open to a variety of meanings, like the adjective "scientific" frequently deployed by its Other, but critical geography's proponents have harnessed the term to represent a particular perspective on society and space, thereby discursively situating themselves within the human and social sciences . Genealogically, "critical geography" became increasingly popular as a replacement for radical geography in the 1980s. Critical theory was coined in the 1930s by the Frankfurt School of social theorists (Horkheimer, Adorno, Marcuse, Habermas, Offe) to describe their variant of radical social science (Held, 1980). They saw themselves as departing substantially from Marx, while maintaining his concern for social and political economic structures and human emancipation. In comparison to French social theorists, the Frankfurt School has received limited attention in critical urban geographic research, with the notable exceptions of Klaus Offe's research on the state and Walter Benjamin's writings on urban life. Critical geography has retained, however, the School's concern for emancipation and empowerment of the disadvantaged. Thus, when Moss, Berg and Desbiens introduce the critical geography e-journal ACME, they define critical and radical geography "as for example, anarchist, anti-racist, environmentalist, feminist, Marxist, postcolonial, poststructuralist, queer, situationist, and socialist. By critical thinking and radical analysis we mean that the work is part of the praxis of social and political change aimed at challenging, dismantling, and transforming prevalent relations, systems, and structures of capitalist exploitation, oppression, imperialism, neo-liberalism, national aggression, and environmental destruction." (Moss et al., 2001, p. 3). As radical social theory took new turns in Europe and North America, "critical" became an umbrella term to represent this post-Marxist spectrum, although Noel Castree (2000) protested that the shift from "radical" to "critical" marks an undesirable institutionalization: a taming and even cooptation of radical geography into the mainstream.

Critical urban geographic research is presaged on a vigilant examination and critique of the logic and assumptions underlying preexisting mainstream theoretical accounts of cities, narratives of urban processes and urban life, and the urban policies reflecting these. At present, for example, critical urban geographic research is critiquing neoliberal theories, narratives and policies directed at cities, seeking to expose their logical, empirical and practical shortcomings (Brenner and Theodore, 2002b). During the 1990s, however, critical urban geographic research moved beyond critique to place the bulk of its effort into explicating alternative theories, narratives and prescriptions. Our own research on urban entrepreneurialism, for example, has sought to identify shortcomings in the economic theories of Michael Porter and Paul Krugman that provide a justification for urban entrepreneurialism and inter-urban competition; to point out that really existing inter-urban competition in a geographically differentiated world leads to very different outcomes (characterized by persistent geographical and social inequalities within and between cities) than those suggested by proponents of urban entrepreneurialism; and to suggest alternative strategies such as intra- and inter-urban collaboration by city governments and civil society (Leitner, 1990; Leitner and Sheppard, 1999; Leitner and Sheppard, 2002). In short, as it has flourished, critical urban geographic research has become more activist and positive in outlook, not just demonstrating the undesirable effects of ongoing economic and political change, but also actively contesting the thinking behind these and developing alternatives. Critical urban geographic research has sought to accent emergent or persistent opportunities for alternative practices, identify spaces of hope, and articulate a normative vision for public engagement and strong democracy (Amin and Thrift, 1995; Amin et al., 2000; Amin and Thrift, 2002).

There is little doubt that the critical urban geographic research program has become influential within urban geography during the 1990s, as the other papers in this special issue document. This influence extends beyond geography to research on cities more generally. Both in the U.S. and the U.K., recent assessments share the view of Aitken et al. (2003) that "geographers are now well positioned to say something important about the urban issues that are shaping the new millennium. This sea-level change occurred in the 1990s and now places many aspects of geographic research at the forefront of urban analysis." (See also Lees, 2002). Yet success is always dangerous, particularly for an approach founded on the principle of vigilantly critiquing the status quo. As critical urban geography becomes the status quo within urban geography, how does it critique itself?

One approach is to downplay and neglect urban geographic research that does not self-identify with the critical geography program. As noted above, this has long been a successful strategy. Early articulations of the political economy approach were founded on a rejection of spatial science, including vigorous attacks on its most visible and unrelenting proponent, Brian Berry (cf. Harvey, 1972). The common use of "postpositivist" to define what critical urban geographic research has in common indicates how important this othering remains. Such boundaries are policed on both sides. The clearest recent example of this is Brian Berry's "last man standing" attacks on critical urban geographic research, which at various points he has dubbed as anti-American, anticapitalist, armchair, contrarian, relativist and as pseudo-science (Berry, 2002a; 2002b; 2002c). Belief in the epistemological and methodological superiority of one's chosen approach, nostalgia for a time when others agreed, sour grapes at declining influence, and concern about the barbarians at the gates typify such responses.

It may seem easy for critical urban geographers to laugh off such rhetoric, particularly now that they are situated in a position of intellectual strength within contemporary urban geography. Yet, while such ignorance may be bliss, it is not productive. As the critical theorist Theodore Adorno once put it: "Genuine refutation must penetrate the power of the opponent and meet him on the ground of his strength; the case is not won by attacking him somewhere else and defeating him where he is not" (1982, p. 5). Grounding the identity of critical urban researchers on rejection of its spatial scientific "other" can be self-limiting. For example, the quantitative and empirical methods typically associated with spatial science have been excluded from the critical geographic research program, on questionable philosophical and political grounds (Plummer and Sheppard, 2001; Sheppard, 2001). Thus othering the mainstream can undermine the vitality of critical urban geographic knowledge production.

Vitality can also be undermined by factionalism within critical urban geography. A professional hazard of academia is that differently situated perspectives coalesce into cliques, separated by debates that often generate more heat than light. Such cliques are by no means stable, and the different approaches that emerge often struggle for power and influence, instead of seeking to learn from one another. Recruits are sought either by conversions from other cliques or by recruiting new scholars. Occasionally prominent urbanists have undergone whole-scale conversions. Sometimes they are convinced of the shortcomings of their previous position; at other times, unfortunately, they may seek to keep up with the latest fashions (Sheppard, 1995). Young scholars are typically inculcated into a particular clique as a result of when, how and by whom they are socialized into the profession, particularly when that socialization process discourages them from taking other approaches seriously. Cliques also undergo internal transformation.

Critical urban geographic research has been characterized by such debates between different emergent schools, notwithstanding a shared emancipatory vision inherited from critical theory. The broadest divide has separated political economy, Marxian and realist urban geographers from postmodern, poststructural and cultural urban geographers. The shift from political economy to poststructuralism and cultural theory (section 1) reflected the shifting context of intellectual debate (the rise of feminist, poststructural and postcolonial theory during the 1980s and 1990s in English-speaking academia); changes in who is allowed to participate (affirmative action facilitating the rise of feminist urban geography); and inter-generational rivalry (new generations seeking to make their name by attacking the status quo, as in the rise of quantitative urban geography (cf. Berry, 1993). Cultural theorists currently have the discursive upper hand, from which complaints recur about the unwillingness of the political economic group to take on board fully the poststructural turn: "while they get into the complex spirit of the urban, the tendency to generalize from prevalent phenomena or driving processes remains strong" (Amin and Thrift, 2002, p. 8). Charges of economism abound. Poststructuralists see political economists as rooted in an outdated, and often raced and gendered, politics of class and of struggles for control over the means of production (Gibson-Graham, 1996; Jones, 1999). In a rear-guard action, political economists accuse their poststructural and cultural theoretic colleagues for their insistence (or grand claim) that no grand claims or overarching narratives are possible; for their prioritization of discourse and representation over material processes; and for substituting culture for the economy as the fulcrum for analysis (Sayer, 1997). They accuse poststructuralists of paying too much attention to difference and identity, replacing a politics of claims-making based on redistribution with a politics of recognition, thereby undermining the progressive intent of critical urban geography (Harvey, 2000). Under poststructuralism, it is argued, justice becomes "just us" (Merrifield, 1997).

Attempts to break down this divide within critical urban geography seem hard to initiate and have been slow in coming. One way is to seek to persuade the opposition. Harvey, for example, insists that his dialectical approach is capable of taking on board questions of difference and identity, whereas Thrift (2002) argues equally strenuously that emancipatory politics remains central to his approach. Others, however, are beginning to articulate the complementary strengths of the two approaches, and the need to build on these in order to move critical urban geography forward (Mitchell, 1999; Fincher et al., 2002; Lees, 2002; Watson, 2002). Significantly, these more positive overtures come from women, and we worry whether such more complex interventions will receive the same attention as the big boys' boundary battles.

Other disputes exist within the critical urban geography research program. While feminist research has been a highly productive area within critical urban geography, its proponents remain skeptical that their work has much impact on their male colleagues. Questions of private space, patriarchy and house-work are often not at the center of critical urban theorizing (Deutsche, 1991; McDowell, 1999). In addition, the successful articulation of a Los Angeles School of postmodern urban geography, positioning itself in opposition to both the Chicago School of the 1960s and 1970s and the political economic research of the 1970s and 1980s, has received external receptions ranging from suspicion to hostility (Lake, 1999; Dear, 2001; Gottdiener, 2002).

Both internal and external boundaries surrounding intellectual communities produce a rhetoric dominated by dismissal and alienation. In such circumstances situated knowledge too often takes on a personal dimension. Sooner or later almost all scholars stop rethinking their approach to knowledge production, settling into the paradigm that we convince ourselves is best, and from which all future change appears degenerative. Although this runs counter to the ethos of critical geography, it remains a very real danger.

We wish to argue that it is necessary to develop a critical sensibility that can prevent critical urban geographic research from becoming entangled in such self-defeating dynamics. Helen Longino (2002) provided such a vision, seeking to move beyond a dualism separating philosophers of science and science studies. She noted that science studies has provided compelling accounts of how scientific knowledge is socially produced, but at the expense of being unable and often unwilling to make normative judgments about knowledge production. By contrast, philosophers of science seek normative statements to separate science and knowledge from opinion and belief, but limit themselves to individualist, cognitive/rational principles of knowledge production (such as those of coherence or correspondence in logical empiricism) that neglect the obviously social nature of scientific practice. Longino argued that a "plurality of adequate and epistemically acceptable explanations or theories can be generated by a variety of different factors in any situation of inquiry" (Longino, 2002, p. 184). She dubbed these "local epistemologies." Each of these is a situated understanding of the subject at hand, grounded in a set of methodological and substantive assumptions with respect to which the account is persuasive. She argued that this plurality of explanations need not be reducible to a single, monistic truth about the world. Indeed she suspected that such monistic accounts, of the kind that now dominate mainstream science, often become hegemonic by pushing aside competing explanations, or by excluding differently situated actors from scientific debate.

In order to improve our understanding of the world, she envisioned a social approach to science quite different from current practices. This is centered on a vision of a forum for scientific debate with four principles: venues, uptake, public standards and tempered equality (Appendix). The principle of tempered equality is particularly important, echoing feminist attempts to radically diversify the community of scientists and thereby deconstruct and decenter the knowledge claims emanating from this largely male, White and first world community (Harding, 1991). Longino's vision seems reminiscent of Habermas' ideal speech community, but has an important difference. Whereas Habermas argued that communicative action within such a community can eventually result in consensus (in this case, a monist understanding of the world), Longino argued that such strong consensus is unnecessary. In her view, monist accounts can only be finally accepted as qualifying as knowledge/science after they have been opened up to criticism from the full range of situated stakeholders under the conditions described in the Appendix, and an ongoing debate between different situated understandings that never results in consensus is equally valid.

This implies that the goal of critical scholarship is not an agreed truth about the world, but can be a restless target: a ceaseless debate between different local epistemologies that nevertheless provides more reliable and justifiable knowledge about the world than any artificial resolution into a single hegemonic viewpoint. This means that science cannot be reduced to a single foolproof epistemology, but also that science cannot be separated from political debate and action: debates between situated perspectives are inevitably also political. Data are always theory-laden and theories are always politics- and opinion-laden, articulated on the basis of differing assumptions and presumptions about how the world works, and how it could work better. A strongly democratic forum of the kind envisioned by Longino provides the possibility for recognizing such links, and engaging in debates with both the rigor of science and the openness of a politics of difference (Young, 1990).

## Unbounding the Academy

Critical urban geographic researchers have come under steady criticism in recent years for failing to make good on their commitment to social/urban change. They have been criticized for failing to engage in mainstream policy analysis or enter debates on urban policy; for use of a theoretical language that is inaccessible to nonspecialists, even within the discipline; and for reproducing the Ivory Tower mentality they profess to reject. These criticisms each have an element of truth to them, and yet the issues are much more complicated than these claims suggest (Imrie, 2003). Intervention on behalf of the disempowered is central to critical social science, and the critical urban geography research program has in fact engaged in high-profile policy discussions and socially relevant research with this aim in mind. Empowering the disadvantaged requires more than this, however. It is necessary to undertake the difficult task of decentering knowledge production out of the academy, making urban residents full participants in this process.

Critical social science has always been motivated, of course, by an abiding interest in social change to the benefit of the disempowered. Indeed, applications of critical urban geographic research to prominent policy issues are not difficult to find. One recent example is the book *Cities for the Many Not the Few* (Amin et al., 2000), a critical response to the British Government's Rogers Report on urban design in the United Kingdom. Persuasively written in an accessible style, brief, and modestly priced, this critical assessment of a national urban policy White Paper brings cutting-edge critical urban geography to bear on the question of making cities better for all their residents. A second example is *Sprawl Hits the Wall* (Southern California Studies Center, 2001), analyzing sprawl in Los Angeles for the Brookings Institution. Indeed, well-funded empirical critical urban geographic research, paying explicit attention to issues of social relevance, has been increasing throughout the English-speaking academy. This reflects a combination of selective incentives favoring socially relevant research, in state and private foundation funding; and increased pressure on faculty to seek external funding for their research, and on universities

364 • Helga Leitner and Eric Sheppard

to rely on nonprofit and private-sector funds, and to provide immediate benefits to society. Critical social scientists have thereby demonstrated that they are as skilled and adept as anyone at responding to the shifting incentives and pressures of an increasingly corporatized academia.

At the same time, however, it is still the prevailing view among critical social scientists that the academy remains the preeminent place of knowledge production for the good of society. They have the same confidence in the legitimacy and rigor of their philosophical foundations as every other philosophical school in the human and social sciences, appointing themselves as providers of expert knowledge for "the real world" outside the academy. Such attitudes reflect their own situatedness as successful academics, often located in well-supported elite institutions of higher learning, whose steadily accelerating and increasingly over-burdened life-worlds are focused on and validated through teaching and research, leaving little energy even for critical analysis of their own institutions (Castree, 2000).

Feminist urban geographers have taken the lead in challenging this presumption that researchers are privileged sources of understanding relative to their informants. This argument reflects the emphasis in feminist philosophy on situated knowledge, itself born of the experience that women's differentiated worldviews have enriched our collective understanding of the world. In our view, critical urban geographic research should take this insight on board. This means not only talking to those outside the academy—firms, city governments, think-act tanks, workers, housewives, children—but welcoming them as full participants in knowledge production about cities, urban livelihoods and policy-making that affects them. In short, we envision extending the culture of intellectual debate proposed by Longino beyond the academy. In this vision, academics would be one of many groups of participants in debates about cities and urban change, and policy-making to effectuate such change. Academics have a particular situated expertise to bring to such a process of deliberative democracy, but urban residents' experiential knowledge, and their visions of the good city, are equally important.

Critical urban geographers should seek to ensure that all urban residents, and not only the political and economic elite, have a chance to have their voices heard—particularly the disadvantaged. This would include participatory action research, in which critical scholars collaborate with urban residents to define and investigate research questions that incorporate residents' concerns, views and knowledge. It should also entail collaboration with think-act tanks: policy institutions now emerging to shift public discourse away from its hegemonic focus on competition, winners-take-all and individual responsibility and toward cooperation, caring and mutual responsibility.[4]

However, it is hard to overstate the difficulty of decentering knowledge production out of the academy and speaking across worlds. On the one hand, bringing academics and urban residents together poses the practical difficulty of bridging the often conflicting priorities and time demands of very different lifeworlds. On the other hand are communications difficulties. We focus here primarily on academic perspectives on these issues, as this is where our knowledge is situated.

Academic institutions place little value in activist-oriented attempts to reach out beyond the academy. Socially relevant research is by and large equated with funded policy research and/or unpaid consulting work with policy-makers and policy shapers. Activist research is typically seen as subjective and of little social value, other than as a form of academic charity addressing problems in the community within which a university is located, for example. In addition, steadily rising demands on academics' time and expectations of increased productivity together mean that academics have little energy to spare for their home lives, let alone for activist research.

At the same time, as we learned in our own research on environmental justice in the Twin Cities, urban residents question whether such collaboration is in their interest and worth their time investment. In our experience, academics' belief that we have others' best interests in mind is often not shared by community audiences, who are skeptical of academics' intentions and politics. When it is possible to overcome such skepticism, other difficulties may arise. In our case, for example, neighborhood residents sometimes assumed that the university researchers are the experts, and delegated to us the responsibility of undertaking the research project, once they had helped set its broad parameters. This departed from our ideal of a community-university partnership (Kurtz et al., 2001). Faced by such challenges, the best-intended attempts at community-university research may often deteriorate into short-lived experiments from which academics and urban residents alike gain little long-term satisfaction.

Furthermore, difficulties of communicating with the very different communities making up a city, who should be involved in decentered knowledge production, seem at times overwhelming. Besides the effort of keeping in touch with all kinds of people, it requires becoming comfortable

---

[4]It is high time that progressive social scientists learned from conservatives' success in using think tanks to shift the focus of public discourse, as the key to social change, a leaf taken from the book of poststructuralism.

with diverse languages, dialects and lifestyles, as well as an ability to write for different audiences. Nagar (2002) emphasized the importance of producing scholarly analysis that can be accessed, used and critiqued by audiences in multiple geographical, social and institutional locations. Speaking and writing across worlds is enormously challenging, however. It is not only about language, but also about the situated theoretical approaches that inevitably shape our speaking and writing. For academics, "the framing of problems—itself an inherently theoretical act—conditions their ability to talk with different people and in different settings" (Staeheli and Nagar, 2002, p. 170). This generates a certain jargon, necessary to express ideas precisely and gain peers' respect, which is often inaccessible to other academics, let alone outside the academy. In addition, irrespective of the care taken to write in a broadly accessible manner, any text is liable to be interpreted quite differently in different time and place specific contexts. It also may have the unintended effect of disempowering those whose interests the author seeks to support (Raju, 2002).

Self-reflexivity is crucial in approaching these diverse challenges. It is vital to reflect on how our positionality as researchers (and within the community of researchers), and our multiple and shifting identities and agendas, shape community-university collaboration. We must be constantly aware of unequal power relations, on the lookout for unintended undesirable effects of our actions, and willing to completely rethink our approach when necessary. Otherwise, sooner or later the academic researcher begins to isolate herself from, and elevate herself above, those others who are supposed to have an equal voice.

## Conclusion

We have argued that unbounding critical urban geographic research is necessary in three areas (conceptual approaches, philosophical foundations and participants), in order for critical urban geographic knowledge production both to reproduce the vitality that has recently characterized it, and to make good on its intention to improve urban livelihoods for the many, and not just the few (Amin et al., 2000). Interestingly, conceptual approaches have experienced substantial unbounding, whereas philosophical debate still is too often characterized by counterproductive entrenched battles for the hearts and minds of urban geographers, over the superiority of particular philosophical and ontological positions. It is ironic that such negative competition characterizes the philosophical debate among critical geographers, given the intellectual and political common ground and the oft-discussed goal of avoiding dualisms and othering. The unbounding of conceptual approaches provides grounds to believe, however, that more productive exchanges are possible, within and beyond critical urban geography, if we can commit ourselves to realizing the culture of communication envisioned by Helen Longino.

The effort put into using critical urban geography to improve the lifestyles of the disadvantaged pales, however, by comparison to that devoted to defending philosophical positions. Critical urban geographers' policy interventions and socially relevant empirical work are extremely important. Critical social scientists are only too aware that neither the state nor the market is motivated primarily to serve the needs of all urban residents, nor can these needs be met indirectly through the operation of Adam Smith's hidden hand or political pluralism. The current dominance of neoliberal discourse in urban policy-making makes it likely, however, that policy prescriptions of critical urban geographers will be either ignored, or tweaked to serve very different ends than were intended. Thus it is particularly important now that critical urban geographers go beyond such efforts, and extend participation in knowledge production beyond the academy to include grassroots communities—for example through participatory action research. Longino's vision is again relevant, although extension outside the academic arena, together with the stark differences between the lifeworlds and situated knowledge emerging from academic institutions and impoverished communities, compound the difficulties of implementing her scheme, and should not be overlooked.

It is hard enough to persuade academics to step outside their particular presumptions and intellectual communities, notwithstanding an academic canon that supposedly values open-mindedness. Engaging nonacademics in such debates throws up even more difficulties, as discussed above. Communication across worlds is further tested by militant particularism, and by social groups (e.g., religious fundamentalists) with world-views that preclude alternative opinions. Consequently, a residue of incommensurability can undermine agreement even on Longino's four publicly recognized standards (Appendix).[5] Nevertheless, attention to Longino's vision can form a basis for rethinking cultures of communication and knowledge production in critical urban geography, and beyond.

---

[5]We are indebted to Ryan Holifield for this comment.

# References

Adorno, T. W., 1982, *Against Epistemology*. Oxford, UK: Blackwell.
Aitken, S., Staeheli, L., and Mitchell, D., 2003, Urban Geography. In G. Gaile and C. Wilmott, editors, *Geography in America at the Dawn of the 21st Century*. Oxford, UK: Oxford University Press.
Amin, A., Massey, D., and Thrift, N., 2000, *Cities for the Many Not the Few*. London, UK: Polity.
Amin, A. and Thrift, N., 1995, Institutional issues for the European regions: From markets and plans to socioeconomics and powers of association. *Economy and Society*, Vol. 24, 41–66.
Amin, A. and Thrift, N., 2002, *Cities: Reimagining the Urban*. Cambridge, UK: Polity.
Anderson, K., 1991, *Vancouver's Chinatown: Racial Discourse in Canada, 1875–1980*. Montreal, Canada: McGill-Queen's University Press.
Beaverstock, J., Smith, R., and Taylor, P. J., 2000, World-city network: A new metageography? *Annals of the Association of American Geographers*, Vol. 90, 123–34.
Berry, B. J. L., 1993, Geography's quantitative revolution: Initial conditions, 1954–1960: A personal memoir. *Urban Geography*. Vol. 14, 343–441.
Berry, B. J. L., 2002a, Big tents or firm foundations? *Urban Geography*, Vol. 23, 501–502.
Berry, B. J. L., 2002b, My Cheshire cat's smile. *Urban Geography*, Vol. 23, 1–2.
Berry, B. J. L., 2002c, Paradigm lost. *Urban Geography*, Vol. 23, 441–445.
Brenner, N., 1997, State territorial restructuring and the production of spatial scale: Urban and regional planning in the FRG 1960-1990. *Political Geography*, Vol. 16, 273–306.
Brenner, N. and Theodore, N., 2002a, Cities and the geographies of "actually existing neoliberalism." In N. Brenner and N. Theodore, editors, *Spaces of Neoliberalism: Urban restructuring in North America and Western Europe*. Oxford, UK: Blackwell, 2–33.
Brenner, N. and Theodore, N., editors, 2002b, *Spaces of Neolioberalism: Urban Restructuring in North America and Western Europe*. Antipode book series. Oxford, UK: Blackwell.
Bridge, G. and Watson, S., editors, 2000, *A Companion to the City*. Oxford, UK: Blackwell.
Castells, M., 1977, *The Urban Question*. London, UK: E. Arnold.
Castree, N., 2000, Professionalisation, activism, and the university: Whither "critical geography"? *Environment and Planning A*, Vol. 32, 955–970.
Christopherson, S., 1989, On being outside "the project." *Antipode*. Vol. 21, 83–89.
Cook, I. and Crang, P., 1996, The world on a plate—Culinary culture, displacement, and geographical knowledges. *Journal of Material Culture*, Vol. 1, 131–153.
Dear, M., 2000, *The Postmodern Urban Condition*. Oxford, UK: Blackwell.
Dear, M., 2001, The politics of geography: Hate mail, rabid referees, and culture wars. *Political Geography*. Vol. 20, 1–12.
Dear, M. and Scott, A. J., editors, 1981, *Urbanization and Urban Planning in Market Societies*. London, UK: Methuen.
Delaney, D., 1998, *Race, Place and the Law: 1836-1948*. Austin, TX: University of Texas Press.
Deutsche, R., 1991, Boys Town. *Environment and Planning D: Society and Space*, Vol. 9, 5–30.
Fincher, R. and Jacobs, J. M., editors, 1998, *Cities of Difference*. New York, NY: Guilford.
Fincher, R., Jacobs, J. M., and Anderson, K., 2002, Rescripting cities with difference. In J. Eade and C. Mele, editors, *Understanding the City: Contemporary and Future Perspectives*. Oxford, UK: Blackwell, 27–48.
Gibson-Graham, J. K., 1996, *The End of Capitalism (as we know it)*. Oxford, UK: Blackwell.
Gottdiener, M., 2002, Urban analysis as merchandising: The "LA School" and the understanding of metropolitan development. In J. Eade and C. Mele, editors, *Understanding the City: Contemporary and Future Perspectives*. Oxford, UK: Blackwell, 159–180.
Hacking, I., 1983, *Representing and Intervening: Introductory Topics in the Philosophy of Natural Science*. Cambridge, UK: Cambridge University Press.
Haraway, D., 1991, *Simians, Cyborgs, and Women: The Reinvention of Nature*. New York, NY: Routledge.
Harding, S., 1991, *Whose Science? Whose Knowledge?* Ithaca, NY: Cornell University Press.
Hartwick, E. R., 1998, Geographies of consumption: A commodity-chain analysis. *Environment and Planning D: Society and Space*, Vol. 16, 423–437.
Harvey, D., 1972, Revolutionary and counter revolutionary theory in geography and the problem of ghetto formation. *Antipode*, Vol. 6, No. 2, 1–13.

Harvey, D., 1973, *Social Justice and the City*. London, UK: Edward Arnold.

Harvey, D., 1985, *The Urbanization of Capital*. Oxford, UK: Basil Blackwell.

Harvey, D., 1989, *The Condition of Postmodernity*. Oxford, UK: Basil Blackwell.

Harvey, D., 1992, Social justice, postmodernism and the city. *International Journal of Urban and Regional Research*. Vol. 16, 558–601.

Harvey, D., 2000, *Spaces of Hope*. Berkeley, CA: University of California Press.

Held, D., 1980, *Introduction to Critical Theory: Horkheimer to Habermas*. Berkeley, CA: University of California Press.

Imrie, R., 2003, *Urban Geography, Relevance, and Resistance to "Policy Turns."* Paper presented at the Annual Meeting of the Association of American Geographers, New Orleans, LA.

Jackson, P., 1988, *Race and Racism*. London, UK: Unwin Hyman.

Jones, A., 1999, Dialectics and difference: Against Harvey's dialectical "post-Marxism." *Progress in Human Geography*. Vol. 23, 529–556.

Knox, P. and Taylor, P. J., editors, 1995, *World Cities in a World-System*. Cambridge, UK: Cambridge University Press.

Kurtz, H., 2001, Differentiating multiple meanings of garden and community. *Urban Geography*, Vol. 22, 656–670.

Kurtz, H., Leitner, J., Sheppard, E., and McMaster, R. B., 2001, Neighborhood environmental inventories on the Internet: Creating a new kind of community resource for Phillips Neighborhood. *CURA Reporter*, Vol. 31, 20–26.

Lake, R. W., editor, 1999. Postmodern Urbanism? *Urban Geography*, Vol. 20, 393–416.

Lees, L., 2002, Rematerializing geography: the "new" urban geography. *Progress in Human Geography*, Vol. 26, 101–112.

Leitner, H., 1990, Cities in pursuit of economic growth. *Political Geography Quarterly*, Vol. 9, 146–170.

Leitner, H., in press, Geographic scales and networks of spatial connectivity: Transnational inter-urban networks and the rescaling of political governance in Europe. In E. Sheppard and R. McMaster, editors, *Scale and Geographic Inquiry*. Oxford, UK: Blackwell.

Leitner, H. and Sheppard, E., 1999, Transcending interurban competition: Conceptual issues, and policy alternatives in the European Union. In A. Jonas and D. Wilson, editors, *The Growth Machine: Critical Perspectives Twenty Years Later*. Albany, NY: State University of New York Press, 227–246.

Leitner, H. and Sheppard, E., 2002, The city is dead, long live the network: Harnessing networks for a neoliberal era. *Antipode*. Vol. 31, 495–518.

Logan, J., editor, 2002, *The New Chinese City*. Oxford, UK: Blackwell.

Longino, H., 2002, *The Fate of Knowledge*. Princeton, NJ: Princeton University Press.

Lynn, W. and Sheppard, E., in press, Cities. In S. Pile, S. Harrison, and N. Thrift, editors, *Patterned Ground*. London, UK: Reaktion.

Markusen, A., Lee, Y.-S., and DiGiovanna, S., editors, 1999, *Second Tier Cities: Rapid Growth beyond the Metropolis*. Minneapolis, MN: University of Minnesota Press.

Massey, D., 1994, *Space, Place and Gender*. Minneapolis, MN: University of Minnesota Press.

McDowell, L., 1999, *Gender, Identity, Place: Understanding Feminist Geographies*. Minneapolis, MN: University of Minnesota Press.

McGee, T. R., 1991, Presidential Address: Eurocentrism in geography—The case of Asian urbanization. *The Canadian Geographer*, Vol. 35, 332–344.

Merrifield, A., 1997, *Social Justice and Communities of Difference: A Snapshot from Liverpool*. In A. Merrifield and E. Swyngedouw, editors, The Urbanization of Injustice, New York, NY: New York University Press, 200–222.

Merrifield, A., 2002, *Dialectical Urbanism: Social Struggles in the Capitalist City*. New York, NY: Monthly Review Press.

Mitchell, D., 2003, *The Right to the City: Social Justice and the Fight for Public Space*. New York, NY: Guilford.

Mitchell, K., 1999, What's culture got to do with it? *Urban Geography*, Vol. 20, 667–677.

Moss, P., Berg, L., and Desbiens, C., 2001, The political economy of publishing in geography. *ACME: An International E-Journal For Critical Geographies*. Vol. 1, 1–7.

Nagar, R., 2002, Footloose researchers, "traveling" theories, and the politics of transnational praxis. *Gender, Place and Culture*. Vol. 9, 179–86.

Peake, L. and Kobayashi, A., 2002, Policies and practices for an antiracist geography at the millenium. *The Professional Geographer.* Vol. 54, 50–61.

Peck, J. and Tickell, A., 2002, Neoliberalizing space. In N. Brenner and N. Theodore, editors, *Spaces of Neoliberalism: Urban restructuring in North America and Western Europe.* Oxford, UK: Blackwell, 34–57.

Pile, S., 1996, *The Body and the City.* London, UK: Routledge.

Pile, S., 1999, What is a city? In D. Massey, J. Allen, and S. Pile, editors, *City Worlds.* London, UK: The Open University.

Plummer, P. and Sheppard, E., 2001, Must Emancipatory Economic Geography be Qualitative? A Response to Amin and Thrift. *Antipode.* Vol. 30, 758–763.

Pulido, L., 2000, Rethinking environmental racism: White privilege and urban development in Southern California. *Annals of the Association of American Geographers,* Vol. 90, 12–40.

Pulido, L., 2002, Reflections on a white discipline. *The Professional Geographer,* Vol. 54, 42–49.

Raju, S., 2002, We are different, but can we talk? *Gender, Place and Culture.* Vol. 9, 173–177.

Robinson, J., 2003, *Cities between Modernity and Development.* Paper presented at the Annual Meeting of the Association of American Geographers, New Orleans, LA.

Sassen, S., 1991, *The Global City: New York, London, Tokyo.* Princeton, NJ: Princeton University Press.

Sassen, S., editor, 2002, *Global Networks, Linked Cities.* London, UK: Routledge.

Sayer, A., 1997, The dialectic of culture and economy. In R. Lee and J. Wills, editors, *Geographies of Economies.* London, UK: Arnold, 16–26.

Scott, A. J., 1980, *The Urban Land Nexus and the State.* London, UK: Pion.

Sheppard, E., 1995, Dissenting from spatial analysis. *Urban Geography,* Vol. 16, 283–303.

Sheppard, E., 2001, Quantitative geography: Representations, practices, and possibilities. *Environment and Planning D: Society and Space.* Vol. 19, 535–554.

Sheppard, E., 2002, The spaces and times of globalization: Place, scale, networks, and positionality. *Economic Geography.* Vol. 78, 307–330.

Sheppard, E. and Leitner, H., 1989, The city as a locus of production: The changing geography of commodity production within the capitalist metropolis. In R. Peet and N. Thrift, editors, *The New Models in Geography.* London, UK: George Allen & Unwin, 55–83.

Sibley, D., 2001, The binary city. *Urban Studies.* Vol. 38, 239–250.

Smith, D. M., 1994, *Geography and Social Justice.* Oxford, UK: Basil Blackwell.

Smith, N. and Williams, P., 1987, *Gentrification of the City.* London, UK: Allen & Unwin.

Soja, E., 1989, *Postmodern Geographies: The Reassertion of Space in Critical Social Theory.* London, UK: Verso.

Southern California Studies Center, 2001, *Sprawl Hits the Wall.* Los Angeles, CA: University of Southern California, and Brookings Institution.

Staeheli, L. and Nagar, R., 2002, Feminists talking across worlds. *Gender, Place and Culture.* Vol. 9, 167–172.

Swyngedouw, E., 1996, The city as hybrid: On nature, society and cyborg urbanization. *Capitalism, Nature, Society.* Vol. 7, 65–80.

Swyngedouw, E., 1997, Neither global nor local: "Glocalization" and the politics of scale. In K. Cox, editor, *Spaces of Globalization: Reasserting the Lower of the Local.* New York, NY: Guilford, 137–166.

Thrift, N., 1994, On the social and cultural determinants of international financial centers. In S. Corbridge, R. Martin and N. Thrift, editors, *Money Power and Space.* Oxford, UK: Blackwell, 327–355.

Thrift, N., 2002, Summoning Life. Unpublished paper, School of Geographical Sciences, University of Bristol, UK.

Valentine, J., editor, 2001, *Social Geographies: Space and Society.* Harlow, UK: Prentice Hall.

Walker, R., 1981, A theory of suburbanization. In M. Dear and A. J. Scott, editors, *Urbanization and Urban Planning in Market Societies.* London, UK: Methuen, 383–430.

Watson, S., 2002, The public city. In J. Eade and C. Mele, editors, *Understanding the City: Contemporary and Future Perspectives.* Oxford, UK: Blackwell, 49–65.

Whatmore, S. and Thorne, L., 1997, Nourishing networks: Alternative geographies of food. In D. Goodman and M. Watts, editors, *Globalizing Food: Agrarian Questions and Global Restructuring.* London, UK: Routledge, 287–304.

Wilson, B., 2000, *America's Johannesburg: Industrialization and Racial Transformation in Birmingham*. Lanham, MD: Rowman and Littlefield.

Wolch, J., 2002, Anima urbis. *Progress in Human Geography*, Vol. 26, 721–742.

Yeung, Y.-M., editor, 1998, *Urban Development in Asia: Retrospect and Prospect. Research Monograph #38*. Hong Kong: Hong Kong Institute of Asia-Pacific Studies, Chinese University of Hong Kong.

Young, I. M., 1990, *Justice and the Politics of Difference*. Princeton, NJ: Princeton University Press.

# APPENDIX
## Longino's Conditions for Knowledge Production

| | |
|---|---|
| Venues | Publicly recognized forums for criticisms of evidence, methods, assumptions, and reasoning; criticism to be given the same weight as original research. |
| Uptake | Criticism must be taken seriously, and theories adjusted in the face of adequate criticism. |
| Public standards | There must be publicly recognized standards for evaluating knowledge claims and the relevance of a criticism to a particular knowledge claim, to which criticisms must refer in order to obtain a hearing. |
| Tempered equality | Communities must be characterized by equality of intellectual authority. The social position or power of a community should not determine which perspectives are taken seriously. Participation is tempered by the side-condition that full recognition of participants requires that their acts conform to the responsibilities and standards discussed above. |

*Source*: Authors, based on Longino (2002, pp. 128–131).

# Index: Paradigms

# Index: Personalities

T - #0137 - 071024 - C0 - 229/152/22 - PB - 9780415951913 - Gloss Lamination